The Culinary Herpetologist

ERNEST A. LINER

THE CULINARY HERPETOLOGIST

Compiled by
Ernest A. Liner
Houma, Louisiana

Bibliomania!
Salt Lake City, Utah

Photo credits:
Front cover: Breck Bartholomew
Back cover: Preparing lizards to eat in Thailand © Michael Freeman/CORBIS

Layout & Editing by Ruthe Smith
Design by Breck Bartholomew

Published 2005 by: Bibliomania!
P.O. Box 58355
Salt Lake City, UT 84158
USA

www.Herplit.com

ISBN: 1-932871-05-5 Hardcover; 1-932871-06-3 Paperback

Printed in Hong Kong

The paper used in this publication meets the minimum requirements of the American National Standard for Information Sciences—Permanence of Paper for Printed Library Materials. ANSI Z39.48-1992

This book is dedicated to two longtime friends and colleagues
who shared many pleasant campfires with me in the field:
Dr. Allan H. Chaney and the late Dr. Richard M. Johnson.

publisher's note

As a person who strongly supports conservation efforts to protect amphibians and reptiles as well as other animals, you might wonder why on earth I would publish a cookbook of amphibian & reptile recipes. The main reasons are because Ernie Liner is the author and I don't view this as a cookbook *per se*.

I first met Ernie in 1996 at the meeting of the Society for the Study of Amphibians and Reptiles (SSAR), a professional meeting for scientists who study these animals. Ernie had brought a few books from his library to sell, and as a book collector and dealer I was interested in what he had. From our first meeting I knew Ernie was unique in his kindness and generosity.

In 1998 the University of Colorado at Boulder awarded Ernie a Honorary Doctor of Science for his contributions to herpetology. Ernie's contributions include numerous scientific publications on the amphibians and reptiles of México and his home state of Louisiana as well as other papers. He has also been a long supporter of the scientific societies in both the USA and México. Ernie has meticulously compiled several useful bibliographies for herpetologists. This cookbook is yet another meticulous compilation of Ernie's which stems from both his love of cooking and his love of herpetology. Although this cookbook may not seem like an academic pursuit, it really is one.

The introduction of the first edition of this book, *A Herpetological Cookbook: How to cook Amphibians and Reptiles*, begins with "Dear Colleague," and goes on to explain the book was published for the 1978 joint meeting of the SSAR, Herpetologists' League, and the American Society of Ichthyologists and Herpetologists — hardly a group likely to start throwing their study animals into their soup. Disguised as a cookbook this is a contribution to culinary folklore and anthropology. In this light, the significance of such a collection of recipes became important to preserve and make available. Thus I agreed to publish it for its value as documentation of one aspect of the human–amphibian/reptile relationship.

I have never eaten an amphibian or reptile and I never intend to eat one. I realize some people will wish to use this book as a cookbook — several of the recipes look quite interesting. My only hope is they decide to substitute a more common reptile* for the meat — a chicken. After all, IT ALL TASTES LIKE CHICKEN anyway.

Breck Bartholomew

*Evolutionarily birds are feathered reptiles. In fact, crocodilians are more closely related to birds than they are to snakes.

TABLE OF CONTENTS

INTRODUCTION

In 1978 I published a little booklet titled *A Herpetological Cookbook: How to Cook Amphibians and Reptiles*, which was based on recipes that I had casually accumulated over the years. This publication was "released" at the joint meeting of the Society for the Study of Amphibians and Reptiles (SSAR) and The Herpetologists' League (HL) in Tempe, Arizona, and 1000 copies were printed. Not long after that it went out of print. An additional printing of 500 copies was made and in a short time these were exhausted. To reprint it again a third time would have cost more than they were being sold for. I had also actively started searching for more recipes because many colleagues suggested that I follow-up on the initial publication.

Since 1978 I have searched through hundreds of cookbooks, both general and game, sports magazines, newspaper cookbooks, newspapers, pamphlets, scientific journals and books, conservation magazines, government publications, and any other source I thought might have something. Many people gave their personal recipes or sent me recipes, sometimes anonymously, from various sources, sometimes known, others unknown. I have made no effort to search out European, Asian, Australian, or South American cookbooks except those that I "stumbled" upon or had sent to me. Many, many recipes undoubtedly are still to be found from these sources. Chinese and Japanese sources could be very profitable since these animals are common in their diets. Europeans, particularly in France, consume an enormous number of frogs. I understand that they are now large consumers of alligator and crocodile meat. South and Central Americans consume many iguanas in their diet but recipes are hard to obtain. Turtles are eaten worldwide, especially sea turtles, and some have been eaten to extinction such as some subspecies of giant tortoises. Note that presently all sea turtles and giant tortoises are protected as well as many other exploited species, and others will be soon. The proliferation of alligator and crocodile farms has made the meat from these species now readily available, bringing in more money than the hides in some areas.

Only the use of amphibians and reptiles as food is of concern here. Many are exploited for other reasons. Indigenous people of many regions use the animals somewhat as staples in their diets with no records of their methods of preparation, some very primitive, with ingredients unknown to "civilized" peoples. Even in this country some salamander species such as the Hellbender (*Cryptobranchus*) and Mudpuppy (*Necturus*) are eaten but I can find no written recipes, possibly because people are embarrassed to admit they eat them.

The varied cuisine available in stores today suggests that many new methods to prepare amphibians and reptiles would be worthwhile. I have been tempted to start "making up" new recipes using many of these animals but I have refrained from doing so. The recipes included here are those that have already been tried and in many cases appear in many different publications, sometimes identical or with some slight variation. Many that I found are not included here as I felt that the slight differences were not significant. Although many of these recipes included are "basically" similar, the differences warrant their inclusion. I have tried to make the presentation of these recipes fit a simple standard pattern but I was not always successful. Some originals were very complicated and difficult to follow. Hopefully I have been able to eliminate this. Many of the basic amphibian or reptile recipes have auxiliary recipes with them or are referred to a glossary. Many standard well-known commercial forms were not included. The auxiliary recipes sometimes entailed having several to make one. Some are significantly different from the standard, so were included. Some were included as they were recommended to be served with a particular dish. Each recipe was treated as if it was the only recipe available and as if the person preparing the dish has no real basic knowledge of cooking, thus the asterisk (*) refers the reader to the glossary for a definition, gives a short recipe, or tells where to find it in the text. A suitable name for each recipe was attempted as many of the recipes had the same name. This was not always possible to do. As a result some recipes have a number placed after the basic name to make them different and this method is used often in cookbooks.

An attempt was made to try to categorize the various recipes as to class but I was not able to totally do so. Some recipes could fall into several categories. There are 13 mock turtle soup recipes included for those who would like turtle soup but do not have the turtle. These are included in the turtle recipe accounts instead of the auxiliary. Many of the recipes can be prepared either with turtle or alligator, but then the cooking times would have to be altered. In fact, some restaurants have added alligator soup to their menus in place of turtle due to turtle scarcity but use their same turtle soup recipes.

A small section included is "Recipes of Indigenous People." These are recipes acquired on how some primitive peoples prepare these animals.

In breaking down the recipes in this book I found that 952 were accumulated with 103 auxiliary recipes. The auxiliary recipe counts are deceiving as many are used a number of times in different recipes but are cross referenced.

RECIPE BREAKDOWN

Basic	Recipes	Auxiliary Recipes
Salamanders	26	0
Frogs and Toads	193	52
Lizards	9	1
Snakes	34	5
Turtles	281	20
Crocodilians	379	25
Recipes of Indigenous People (Mixed)	30	0
Totals	**952**	**103**

In this country, frogs are the most easily obtainable amphibian for food being readily available in seafood markets, supermarkets, and specialty stores, usually frozen. Salamanders are not available commercially for food. In the past turtle was easily obtainable in seafood markets but now, due to its scarcity, alligator, which was once largely unobtainable, has taken its place. Sale of alligator meat was illegal until fairly recently. Lizards have never been a commercial food item, but snakes to a small degree, especially rattlesnake, were available from a few specialty outlets. Some recipes are included in the book for commercial rattlesnake preparations put out by Ross Allen's Reptile Institute, in Silver Springs, Florida, but it is no longer being produced.

Survival books point out that many amphibians and reptiles are edible but give no specific recipes or preparation methods and one must assume that primitive methods are utilized. Several "Road Kill" cookbooks available with bonafide recipes and written with tongue-in-cheek are on the market. Other publications in this same vein have appeared. At the 2nd World Congress of Herpetology in Adelaide, Australia in 1993-94 a "menu" was prepared consisting of soups as Filesnake Slough Consomme and Frog Spawn Soup; entrees of Hemipenis Cocktail with Parotid Sauce, Froglegs au Adelaide; maincourses of Grilled Fillet of Sleepy Lizard, Hatchling Turtle Casserole, and *Sphaerodactylus* Omelette followed by a dessert of Spermatophore Custard--all mouth-watering recipes written with tongue-in-cheek.

TERMINOLOGY

I have attempted to use a standard terminology for the ingredients. I use scallions* instead of green onions or shallots by which they are also known. In following these recipes, the term "to taste" is used often for various seasonings, especially peppers (black, white, red) and salt.

Considering people's diets and possible health restrictions is left up to the cook. Adding herbs and spices when serving is easier than trying to take them out later. What is too hot, salty, or whatever for one may be wrong for another so this gives the cook some freedom to experiment. Also many items can be substituted for others. I have consistently used "plain flour" except when another is specifically called for because if you use "self-rising flour" to make a roux* you will make glue instead. Many cooks understand this but beginners may not. I have checked many of the can weights from my own larder or stores and most are of standard size. However, different manufacturers in different parts of the country may use slightly different volumes. Many brand names are specified, but if another brand or a generic is available it may be used.

Butter is used throughout the book but in most cases margarine is acceptable, except in certain recipes where the real thing is necessary. When just onion is noted use yellow onions except when specifically red or white is called for. Tabasco Sauce is used in most cases when a hot sauce is called for but many of the different red pepper sauces available can be used. Keep in mind some are hotter than others. Cayenne pepper is used when a ground red pepper is called for. I prefer it to the flaked dried pepper available; the latter usually has seeds and takes away from the smoothness of many dishes. In some it doesn't matter.

As for the herbs, parsley is used frequently throughout the book. Use fresh herbs when available. When using dried herbs in place of fresh, use half the amount called for in the recipe. When chicken, beef bouillon, or broth is specified, you may use cubes or powder diluted to manufacturer's directions or you can make your own.

The term diced, cubed, and minced is used often for some of the meats. "Cubed" means cut in about 1-inch squares, diced in even smaller pieces, and minced is chopped fine. In vegetables diced means cut in about ½-inch squares, chopped is even smaller than diced, and minced means chopped fine.

Whenever sherry, wine, or liquor is used, the varieties specifically used for cooking are not the best. Cooking wines and liquors typically have added salt and are not intended for drinking. If you must use cooking wines, you may have to adjust the salt in the recipe. Typically, do not cook with wine and liquors that you would not want to drink. When just oil is specified in a recipe, a good vegetable oil of your choice can be used unless a specific oil is called for. Bell pepper is used throughout for sweet pepper or green pepper.

In the use of tsp or tbsp, the standard measuring teaspoon or tablespoon is used instead of household teaspoons and tablespoons as these vary in size. In many cases the cooking time is given in ranges instead of specific times, others not. This is somewhat flexible in case you are using gas or electric and to the individual stoves. Oven temperatures also vary and you cannot always depend on the thermostat. Know your stove. Cooking times may also vary because of the amount of liquids given off by various vegetables or meats. These are not consistent.

ABBREVIATIONS

As is standard with cookbooks, certain abbreviations are used. This one is no exception. Abbreviations are only used in the list of ingredients section except for teaspoon (tsp) and tablespoon (tbsp) which are also used in the directions section.

Abbreviations used are given below and are fairly standard.

Bu = Bunch(s)
C = Cup = 8 ozs
Doz = Dozen
Ft = Foot
Gal(s) = Gallon(s)
Lb(s) = Pound(s)
Lge = Large
Med = Medium
Oz(s) = Ounce(s)
Pkg(s) = Package(s)
Pr = Pair
Pt(s) = Pint(s)
Qt(s) = Quart(s)
Sm = Small
Tbsp = Tablespoon
Tsp = Teaspoon

CONSERVATION

In compiling this book my intention is not that these animals be decimated in the wild. We have a worldwide decline in amphibian populations. Alligators and crocodilian populations were rapidly declining because of their trade in skins and some were on the protected lists. With this protection and the proliferation of alligator and crocodile farms, many have made a remarkable comeback and now the trade in meat may be even greater than for their skins. Sea turtles are being successfully raised in the Cayman Islands. Commercial farming of the Bullfrog (*Rana catesbeiana*) has been tried in this country in the past but has failed. It has been introduced into areas outside of its normal range and wild populations now exist. Frozen frog legs usually come from Asia and may be a native species from that part of the world. Iguana farming is now underway in Central and South America with success. With the large number of amphibians and reptiles now being bred in this country and elsewhere for the pet trade by herpetoculturists some of this may spill over into the food industry. Some turtles have been introduced (escaped) into other areas and have become food sources or pests. The Cane Toad (*Bufo marinus*) comes to mind as one amphibian introduced into Australia (as pointed out in another section of this book) which has become a pest and is now being eaten.

In the 1800's the giant Asian Salamander (*Andrias*) was introduced into California for food but was unsuccessful. An accidental introduction of the African Clawed Frog (*Xenopus laevis*) in the American West and now in Mexico has played havoc with our local species, many of which were being used as a food source.

With some of these thoughts in mind many of these recipes are of historical value only and become strictly academic. So be it.

ACKNOWLEDGEMENTS

I tried to avoid citing the source of each recipe, even when known, but in many cases it is unknown. In many garden club, church, club, fair, etc. books, the same recipe appears to have been submitted by different people. My thanks are extended to the many cooks of the past and present who remain anonymous for their contributions. My special thanks to Robert L. Bezy who goaded me to put out the first feeble effort and for continuing to goad me for this one, to Harold A. Dundee for editorial help on the introduction, and to Breck Bartholomew for putting the manuscript in order.

Also my thanks and gratitude to a number of people for sending recipes, encouragement, and other courtesies. They are: Kraig Adler, Oscar Alvarez, Kathy Arcement, Beatrice Aucoin, Breck Bartholomew, Aaron Bauer, Robert L. Bezy, Ernest Birn, the late Jeffrey H. Black, Joe Black, Richard M. Blaney, Ronald A. Brandon, Bayard H. Brattstrom, Pat Breaux, the late Leo D. Brongersma, Agapito Castro, Allan H. Chaney, Joseph T. Collins, Herbert C. Dessauer, James R. Dixon, C. Kenneth Dodd, Jr., Adonis Domingue, Harold A. Dundee, Cecile Dupre, the late Mary Anne Fayette, Thomas H. Fritts, Carl Gans, J. Whitfield Gibbons, Thomas H. Givens, Vaughn L. Glasgow, Bill Gleeson, Sam Goolsby, the late Robert E. Gordon, Lawrence M. Hardy, Raymond B. Huey, Victor H. Hutchison, Fui Lian Inger, Robert Inger, Paxton Johnson, the late Richard M. Johnson, Harry Kami, Sakinh Kim, Trip Lamb, Jeffrey E. Lovich, the late Clarence J. McCoy, Jr., the late Madge Minton, the late Sherman A. Minton, Joseph C. Mitchell, Fernando E. Palomeque, Joseph H. K. Pechmann, Jaime Pineda, Jane Rabalais, Douglas A. Rossman, Gerald Roy, Frank Savoie, Jimmy Serpas, the late Ivan Siekmann, the late Joyce M. Siekmann, Hobart M. Smith, Eric Thiss, Ann Valure, Paulo E. Vanzolini, Richard Wahlgren, William Wilder, Dolores Williams, John W. Wright, and George R. Zug.

SALAMANDERS

Preparation—In the preparation of salamanders they must be skinned, which is not an easy task. To skin them you have to nail or hook it to an upright and with a sharp knife and a pair of pliers peel the skin off as you would an eel or catfish. This only works on the larger forms. Most salamanders in the United States are too small. Then you gut it and cut off the head. Wash thoroughly. *Amphiuma* can also be prepared by a method I call coring. Cut the *Amphiuma* into cross-sections about 1–1½ inches long. Remove the intestines and wash thoroughly and then partially freeze. Then with a sharp pointed knife work it around the meat and the skin to remove the skin. Skin has to be removed from salamanders because many have toxic or unpleasant secretions.

AMPHIUMA A LA POULETTE

- 2 lbs *Amphiuma**, dressed
- 1 sprig sweet basil
- dash of cayenne pepper
- 1 8 oz can mushrooms, drained
- 1 onion, minced
- lemons, sliced, seeded
- 1 tbsp butter
- 2 sprigs parsley
- 1 bay leaf
- croutons
- 1 glass white wine
- juice of 2 lemons
- 2 tbsp plain flour
- 3 egg yolks
- 1 c hot vinegar
- 1 sprig thyme
- salt and pepper to taste

Clean and skin the *Amphiuma* and put in a pot of boiling water and add the vinegar. After boiling 15 minutes, take out and cut into 3 inch pieces. Put butter rubbed smoothly with flour into a frying pan and add the *Amphiuma* and the can of mushrooms and fry. When this begins to brown pour over the wine and add the parsley, thyme, bay leaf, sweet basil, and onion. As the grease rises skim it off. Add a dash of cayenne and salt and pepper to taste. When quite done, take the egg yolks and mix thoroughly with the lemon juice. Take off fire and add the egg mixture and be careful not to set back on the fire once these are added or the eggs will curdle. Place on a dish and garnish with croutons and slices of lemon. Serve hot.

FRIED AMPHIUMA

1 *Amphiuma**, dressed
bread or cracker crumbs
1-2 cloves garlic
1 egg
olive oil
1 bay leaf

Clean *Amphiuma* in 1 inch lengths and wipe dry. Dip pieces in beaten egg and then roll them in the crumbs. Place the bay leaf and garlic in the oil and heat until hot. Fry *Amphiuma* in hot oil, 300°F until brown. Remove from oil and drain. Serve hot.

CONGO EEL WITH CUCUMBERS

1 2–3 lb congo eel*, dressed
salted water
2 tbsp plain flour
1 tsp chervil, chopped
4-5 sprigs parsley
1 tsp lemon juice
1 tbsp parsley, chopped
½ c sour cream
1 bay leaf, crumbled
2 tbsp butter
1 tbsp fennel, chopped
cucumber salad

Cut congo eel (another name for *Amphiuma*) into 2 inch pieces. Place in pot with parsley, bay leaf, lemon juice, and salted water to cover. Simmer gently until congo eel is tender, 15-20 minutes. Remove to platter. Boil down stock to 1 cup. Melt butter, stir in flour, then slowly add the stock through a sieve. Cook until slightly thickened. Add the chopped parsley, fennel, chervil, and then the sour cream. Cook over the lowest heat a few minutes longer. Serve the sauce over the congo eel. Serve cucumber salad with it. Serves 4.

MATELOTE OF CONGO EEL

1½ lbs congo eel*, dressed, cut in 2 inch lengths
2 med carrots, sliced
2 tbsp plain flour
3 tbsp parsley, chopped
1 bay leaf
½ tsp thyme
1½ c white wine
1 clove garlic, minced
½ c light cream
12 sm onions
1½ c chicken broth
2 tbsp butter
salt and pepper to taste

In a large pot combine the onions, carrots, wine, chicken broth, bay leaf, thyme and garlic. Bring to a boil, then reduce heat, cover and simmer 10 minutes or until the onions are tender. Add the congo eel and simmer gently 15-20 minutes until the congo eel flakes when prodded. Remove congo eel and vegetables from liquid, discarding bay leaf, and reserve the liquid. Arrange congo eel and vegetables in a deep serving dish, cover and keep warm while you make the sauce. Melt the butter in a pot over medium-low heat. Stir in the flour and cook until bubbly. Gradually stir in the cream and stock and continue cooking about 10-15 minutes, stirring constantly until thickened. Salt and pepper to taste and pour over congo eel and vegetables and sprinkle with parsley. Serves 4.

DEEP FRIED BRANDIED AMPHIUMA

3 lbs *Amphiuma**
2 tsp salt
½ c milk
dill or parsley, chopped
juice of 1 lemon
½ tsp pepper
1 c plain flour
3 tbsp cognac
1 egg, beaten
oil

Clean and skin *Amphiuma* and cut into 4 inch pieces. Combine the lemon juice, cognac, 1 tsp salt, and the pepper, and marinate the meat in this mixture for an hour or so, turning often. Combine the egg and milk, sprinkle in flour, add remaining salt, and mix into a smooth batter. Dip the *Amphiuma* pieces into the batter and fry 2-3 at a time in deep oil until golden brown. Serve garnished with the dill or parsley. Serves 4.

FRIED AMPHIUMA IN BEER BATTER

*Amphiuma**
lemon wedges
beer batter*
creole tartar sauce*
oil

Dress *Amphiuma* and cut into 2 inch pieces. Dip in Beer Batter and coat well. Fry in hot oil turning often until golden. Serve with lemon wedges and creole tartar sauce.

AMPHIUMA IN HERB SAUCE

1 lb *Amphiuma**
1 pt fish stock*
¼ c parsley, chopped
a little mint, chopped
½ c olive oil
1 c dry white wine
¼ c spinach, chopped
salt and pepper to taste
1 clove garlic
2 tbsp lemon juice
¼ c chives or scallions, chopped

Saute *Amphiuma* (cleaned, skinned, and cut into 1 inch pieces) in olive oil and the garlic for 5 minutes. Pour in the fish stock, cook 5 minutes longer and drain. Add wine, lemon juice, parsley, spinach, chives or scallions, mint, and season to taste with salt and freshly ground pepper. Mix well and cook for 5 minutes. Chill and serve. Serves 6.

ACHOQUES A LA VERACRUZANA

2 lbs Achoques*, dressed
4 tbsp olive oil
¼ tsp cinnamon
3 tbsp capers
1 tsp salt
1 onion, chopped
¼ tsp cloves
1 tbsp lemon juice
1 clove garlic, crushed
5 tomatoes
⅓ c jalapeno pepper strips
⅓ c stuffed olives, halved

Rub Achoques (larval form of *Ambystoma dumerilii*) with salt and garlic. Heat oil and add the onion and saute for 5 minutes until soft and golden. Add the tomatoes (peeled, seeded, and chopped), cinnamon and cloves and simmer gently for 5 minutes. Place the Achoques in a buttered oven dish and cover with the jalapeno peppers, capers, lemon juice and olives. Cover with the tomato mixture. Bake in preheated 350°F oven for 30–40 minutes or until Achoques are tender. Serve with frijoles and crisp tortillas.

ACHOQUES RELLENO

12 Achoques*
½ c chicken broth
freshly ground black pepper
12 cooked sliced mussels
6 tbsp butter
¼ c light cream
juice of ½ lemon
2 tbsp parsley, minced
3 tbsp plain flour
¼ tsp salt
1 c mushrooms, sliced

Heat half the butter in a saucepan. Add the flour and cook, stirring, 1 minute. Add the broth and cream gradually, stirring constantly, until the sauce is thick and smooth. Season with salt, pepper and the lemon juice. Heat the remainder of butter in a skillet and saute the mushrooms 2–3 minutes over high heat. Remove from heat and combine the mushrooms with the mussels and half cup of the sauce to make the stuffing. Sprinkle dressed Achoques with salt and pepper and spread each with a portion of the stuffing in the body cavity and place in a buttered shallow baking dish and top with the remaining sauce. Bake in a 375°F oven 15-20 minutes. Garnish with parsley and serve with rice.

FRIED ACHOQUES

Achoques*
egg
corn meal
oil
plain flour
lemon wedges

Clean Achoques by gutting and skinning. Make a batter of equal amounts of corn meal and flour and add egg. Dip Achoques into batter and coat well and fry in hot oil until golden brown. Serve with lemon wedges.

ACHOQUE RELLENO CON CAMARON

Achoques*
⅓ c plain flour
mayonnaise
parsley
1 bay leaf
½ c butter
2 c milk
salt and pepper to taste
pimentos
1 med onion, sliced
1 tsp oregano
1 med onion, minced
¼ lb cooked shrimp
hard boiled eggs
1 clove garlic

Clean Achoques by gutting and skinning. Make a cavity for stuffing. Simmer Achoques in water to cover with the sliced onion, garlic, bay leaf, oregano, salt and pepper until cooked. Drain, cool and set aside. Make stuffing by melting butter and adding the minced onion and blending in the flour. Then add the milk and cook until smooth and thick. Coarsely chop the shrimp and add, then cool. Place Achoques on serving platter and then spread the filling in and over it. Chill in refrigerator for 2 hours. Before serving cover with mayonnaise, garnish with strips of pimento, slices of egg and parsley. Serve.

Note: The following 15 recipes for the Axolotl and Achoques are included here in Spanish just the way they were sent to me. Many of the ingredients are of local origin and may not be available in all stores, if at all. Leaving the recipes in their native language instead of translating them adds to their historical value.

AXOLOTE EN JITOMATE

Se abre el Axolote en canal y se le sacan las viceras y se laba, se pone a hervir en jitomato molido y condimentado con ajo, cebolla y perejil a que se haga una salsa espesa y se pone a cocer ahí mismo el Axolote hasta que este cocido, pero cuidando que no se desbarate.

AXOLOTE EN DULCE

Se pone a hervir el Axolote ya limpio hasta que este cocido y se le despegue el esqueleto, se le afrega azúcar en bastante cantidad o alguna miel.

PILETE DE AXOLOTE

Se muelen chiles cuaresmeños con cebolla y poco ajo hasta que la salsa quede muy espesa. El Axolote se enjuaga y se pone en salmuera con poco limón unos cinco mínutos. Se pone un pedazo de hoja de plátano más o menos grande, encima una hoja de Acuyo un hoja Santa (Piper sanctum), se coloca el Axolote que se saca de la Salmuera y se baña con la salsa (muy espesa). Se le pone otra hoja santa encima y luego se le ponen rebanadas de plátano macho verde y se envuelve con el resto de la hoja de plátano a manera de tamal, y se amarran y se cuecen a vapor approximadamente una hora.

AJOLOTES EN FLOR DE CALABAZA

6 a 8 Ajolotes de regular tomaño	Tomates verdes
Habas verdes tiernas	3 manojos de flor calabaza
Chiles verdes serranos	Ajo, cebolla, y sal al gusto

Se ponen a cacer los Ajolotes con sal y cebolla, por separado se cuecen los tomates verdes y los chiles verdes. En otro recipiente se cuecen las habas peladas, ya teniendo todo cocido se muelen los tomates con un ajo, un poco de cebolla, y los chiles. En una cazuela se pone aceite se agrega el tomate molido. se agragan la habas ye cocidas y los Ajolotes y se le incorporan las flores limpias y picadas. Se le pone un poco de agua sal (o concentrado de pollo) al gusto, se tapa hasta cue la flor de calabaza está bien cocida. Se sirve en seguida.

AJOLOTES EN ADOBO

6 a 8 Ajolotes de regular tamaño	4 chiles anchos
2 chiles pasilla	vinagre
Cebolla	Ajo y sal al gusto

Se pelan los Ajolotes. se les corta la cabeza, se lavan muy bien, y se ponen a cocer con ajo y cebolla. Por separado se ponen a remojar los chiles en agua caliente durante una media hora, ya que están suaves, se muelen en el molcajete con un pedacito de ajo y otro de cebolla; después de molido secuela. En una Cacerola se pone aceite (o manteca). Cuando el aceite esté caliente se la agrega el chile, se sazona con concentrado de pollo o sal, y un poco de vinagre. Se le agregan los Ajolotes y un poco de agua. Se tapa y se deja hevir hasta que el chile no esté ni espeso ni aguado. Se sirven my calientes, encima se le ponen unas rodajas de cebolla blanca como adorno.

QUESADILLAS DE AJOLOTE

Masa	Chile ancho
buena crema	Flores de Calabaza
Chiles cuaresmenos	hepazete
Ajolotes occidos y deshebrados	

Relleno para las quesadillas. En una caserola se pone aceite, se da una pasadita a un poco de cebolla en banadas o picada cuando esta acitronada se le ponen los chiles cuares-meños en rajas muy delgadas después la carne de ajolote deshebrada, la flor de calabaza picada lavada (sin cáliz), después un poco de hepazote y muy po/ca agua. Se tapa y se deja hasta que la flor esté cocida. Procurando que esto quede seco.

Preparación de la masa. El chile ancho se pone a remojar en agua caliente, cuando esté suave se muele y se mezcla con la masa.

Después se hacen las tortillas. Se les pone el relleno y se doblan. Si se hacen en esta forma deben dorarse en aceite. Se hacen las tortillas se ponen en el comal se voltean y se les pone el relleno pegandolas de las orillas. Una vez cocidas, se abren y se les pone un poco de crema yse sirven calientes. A continuación se presentan las formas más comunes de preparer Ajolotes en la localidad de Xochilmilco, D. F.

TORTA (CAPEADOS)

Los Ajolotes orejones perfectamente limpios se secan y se capean-con huevo a manera de ceiles rellenos. A continuación es muestra le manera frecuente de preparer Achoque en Pátzcuaro, Michoacán.

MIXMOLE

ajo
cilantro (finamente picado)
espinacas
epazote
cebolla
chiles verdes
nopales hervidos
Ajolotes (todo esto al gusto)

Los tomates y los chiles ya hervidos se muelen con ajo y cebolla, se sasone vaciando después espinacea y nopales picados junto con los Ajolotes; dejandose hervir 30 minutos.

TLAPIQUES O TAMALES

hojas de maiz
chiles verdes en rejas o venas
cebolla
epazote
nopales picados y hervidos
de chile Ajolotes (uno por cada hoja de sal maiz)

El Ajolote ya limpio se envuelve con hojas de maiz junto con todos los ingredientes y se cuecan en el comal. Otra variente es quisarlos con manteca y papas. En seguida se ilustre las formas más usuales de preparer alimentos con Ajolotes en la región de Zumpango, Edo de México.

TLATONILE (MOLÉ)

tomate o jitomate (al gusto)
manteca o aceite
cebolla
Ajolotes orajones sin visceres, perfectamente limpios y sin piel
chiles verde (al gusto)
cilantro
ajo sal (al gusto)

Tomates y chiles cocidos se muelen con ajo, cebolla y cilantro o si prefiere se puede agregar este después finementa picado. Se pone a calenter el aceite y se sasone le salsa y posteriormente se agregan los Ajolotes, dejándose cocinar -- por 30 minutos.

TLAPIQUES (TOMALES)

hojas de maiz
chile picado al gusto
sal al gusto
cilantro
jitomate picado
Ajolotes orejones sin visceras y limpios

En una hoja de maiz colocamos al Ajolote un-tado de sal, agregando además el jitomate, chile y cilantro se unvuelven y se colocan en el comal.

ACHOQUE EN CALDO

Verduras como papas	zanahorias
cilantro	cebolla
ajo	jitomate
chile	sal (todo esto al gusto)
Achoques	

Quita la piel y visceras después lávelos. Prosiga a picar las verduras, moliendo solo jitomate, chile, cebolla y ajo juntos. En une olla agregue la cantidad suficiente de agua para poner a cocer las verduras que tardan más en cocerse, ya que estén un poco cocidas aproximadamente 15 minutos agregue los - achoques junto con las otras verduras y deje hervir el caldo por 30 minutos. Sirvase caliente. Otra variante en la elaboración del caldo es untar limón al achoque y usar tomate en lugar de jitomate. También se pueden eliminar papas y zanahorias. Otra forma de preparer el caldo es picando cebolla, ajo, chile y jitomate o tomate; se acitronan y se agregan los achoques ya limpios junto con agua y sal al gusto; se dejan hervir. Ahora se presentan las formas más comunes de elaborer platillos - con ajolote en la region de Lerma, Edo. de México.

EN CHILE VERDE

jitomate	chile verde en rajas
epazote	Ajolotes

Se le quita la piel poniéndolos en la ceniza del comel, se desvisceran, se lavan y se securren; se sazone el - jitomate picado o molido junto con las rejas y el epazote, se agregan los Ajolotes y se deja hervir por espacio de 30 minutos.

TAMAL

hojas de maiz	sal
cebolla	epazote
venas de chile	Ajolotes

Coloque al Ajolote untado de sal, cebolla, epazote y venas en una hoja de maiz, envuélvalo y ásslo en un comal.

TORTA

Harine	huevo
sal	aceite o manteca
Ajolotes	

Quita piel y visceres, lávelos y escúrralos mazclelos con herina y huevo y frialos. También los consumen solo fritos, asados con sal ó fritos en chile quajillo con epazote.

FROGS AND TOADS

Preparation—Once you have frogs of edible size, you will need to prepare them for the pot. This is easy. Decapitate the frogs, peel the skin backwards, and discard all intestines. Wash thoroughly. The whole frog can be eaten but if just the legs are wanted for a particular dish separate as a pair or singly at the pelvis. Use the bodies in another dish. Toads are prepared basically the same way, being careful of the large glands on the sides of the head. These contain unpleasant and sometimes toxic secretions. Decapitate behind these glands and peel the skin back carefully. Be very sure to wash thoroughly several times after skinning and gutting in clean water each time as the skin in some forms also has unpleasant secretions. In both the frogs and toads cut off the feet as they have no food value, just bones. Some species of frogs (*Rana*) also have unpleasant secretions in the skin.

FROG SOUP

2 doz lge frogs
1 carrot
½ c olive oil
1 8 oz can mushrooms, drained
2 cloves garlic
1 celery stem
dash of pepper
croutons
6 sprigs parsley
6 sprigs sweet basil
1 tomato, peeled and chopped
water

Cut off the frog legs and lay them aside. Grind and mash together well the garlic, parsley, carrot, celery and basil. Place the mixture on a slow fire with the olive oil. Add pepper and when sautéed, add the frog bodies. Turn the frogs often enough to prevent them from sticking to the pot. When the frogs have absorbed a good deal of the oil, add the tomato. Allow this to cook for a while and then add enough water to cover the contents. Boil until the frogs are well done, in fact overdone. Strain the soup through a sieve so that only the frog legs are left. Now place this strained soup in a small pot together with the frog legs and allow it to boil for a while. Remove the bones from the legs and add the drained mushrooms to the soup together with the legs. Heat and serve with croutons. Serves 6.

SOUPE DE GRENOUILLES CRESSONIERE

18 frog legs
1 shallot, minced
2 c chicken stock
6 egg yolks
salt and pepper to taste
2 tbsp butter
2 c dry white wine
1 c whipping cream
2 bu watercress, pureed
¼ c fresh chives, chopped

Melt the butter in a large saucepan over medium heat. Add shallots and sauté until softened, about 3-5 minutes. Add frog legs and sauté 5 minutes. Stir in wine, stock, ½ c cream and simmer for 30 minutes. Strain mixture through a sieve lined with cheesecloth and return the liquid to the saucepan. Discard the bones, dice the meat, and set aside. Beat yolks with the rest of the cream in a medium bowl. Blend in the watercress and stir into the stock mixture. Place over low heat and stir constantly until the mixture thickens. Season to taste and arrange the meat in the bottom of a tureen and pour the soup over. Sprinkle with chopped chives and serve. Serves 6.

CREAM OF TOMATO SOUP WITH FROG LEGS

6 pr frog legs, boned, diced
1 med stalk celery, diced
2 tbsp water
2 lbs tomatoes, chopped
1 thyme branch
pinch of sugar
1 tsp shallot, minced
salt and pepper to taste
1 med onion, diced
1 sm carrot, diced
5 tbsp butter
2 lge basil leaves
1 lge garlic clove, crushed
2½ c chicken stock
½ c whipping cream
fresh chives, snipped

Combine onion, celery, carrot, water and 1 tbsp butter in a heavy large saucepan over medium-low heat and cook until water evaporates, stirring occasionally, for about 10 minutes. Stir in the tomatoes, basil, thyme, garlic and sugar and cook 5 minutes. Add the chicken stock and cook for 10 minutes, stirring occasionally. Melt 3 tbsp butter in a heavy large skillet over medium high heat and add the shallots and stir until softened, about 3 minutes. Add finely diced frog leg meat and sear quickly on all sides. Remove the thyme from the soup. Puree the soup in a blender and strain into a saucepan pressing on vegetables with the back of a spoon. Whisk in the cream (reduced to ¼ cup by boiling) and the remaining butter. Season with salt and pepper. Ladle soup into bowls. Garnish with frog meat and chives. Serves 6.

FROG VELOUTÉ SOUP A LA SICILIENNE

48 med frog legs
3 egg yolks
4 c consommé
chervil leaves
4 tbsp white roux*
8 tbsp pistachio butter (see recipe below)
1 tbsp onion, chopped
8 artichoke hearts
butter
½ c cream

Prepare velouté using white roux and 4 cups consommé. Blanch and dice the artichoke hearts, then cook by simmering lightly in butter in a covered pan. Add the frog legs and onions to the velouté and simmer gently for 10 minutes. Drain the frog legs, bone, pound them in a mortar, then rub all ingredients through a sieve and add to the velouté. Dilute with additional consommé if needed. Heat and thicken with a liaison of the egg yolks and cream. At the last moment incorporate the pistachio butter. Garnish with the diced artichoke hearts and chervil.

PISTACHIO BUTTER

6 ozs pistachios
few drops of water
9 ozs butter

Pound freshly blanched pistachios into a paste while moistening with a few drops of water. Add the butter and press through a sieve.

FROG LEGS HORS D'OEUVRES

12 pr sm frog legs
bread crumbs
juice of 1 lemon
3 eggs, beaten
salt and pepper to taste
cornstarch
4 ozs butter
milk
cayenne pepper
oil

Separate the legs into single pieces like small chicken drumsticks. Dip into milk, then cornstarch, then eggs, and then seasoned bread crumbs. Fry in oil at 365°F until golden brown. Keep them hot. Melt the butter in a skillet and cook until light brown and then add the lemon juice. Pour over legs and place into a chafing dish. Serve hot. Serves 3-4.

FLORIDA SPINACH AND FROG LEG SALAD

8 frog legs
1 c carrots, julienne strips
2 cloves garlic, minced
1 sprig thyme
1 c dry white wine
salad dressing
2 lbs leaf spinach, fresh
2 tbsp butter
1 bay leaf
½ sm onion, minced
salt & pepper to taste

Wash and drain thoroughly the spinach leaves and remove the stems and pat dry. Tear into pieces and chill until they are crisp. Sauté the onions, garlic and frog legs in the butter. Add seasoning and the white wine. Cover and simmer for 15-20 minutes or until tender. When cool remove the meat from the bones. To serve toss the spinach, carrots and frog legs with the dressing and serve on pretty glass plates. Serves 4.

SALAD DRESSING

½ c wine vinegar
1 tbsp dijon mustard
1 egg yolk
salt & pepper to taste
¼ c white wine
1½ tsp salad oil

Combine the vinegar, salt, pepper, mustard and white wine and mix well with 2 forks. Add the egg yolk and oil and beat dressing until it thickens.

DEEP FRIED FROG LEGS

24 prs med frog legs
salt and pepper to taste
¼ c heavy cream
oil
2 eggs, well beaten
nutmeg, grated
1 c bread crumbs

Put the eggs into a bowl and mix in the salt, pepper, nutmeg and cream and stir well. Dip the legs in this mixture and then in the bread crumbs. Fry them in very hot oil so that they are nice and crisp, about 4-5 minutes. Serves 4-5.

ITALIAN STYLE FRIED FROG LEGS

47 pr med frog legs
4 eggs, lightly beaten
salt and white pepper to taste
milk
plain flour
8 tbsp butter
cold water
white wine marinade

Soak the legs in equal amounts of cold water and milk for at least 1 hour. Drain and pat dry. Place the legs in the white wine marinade in a large dish and marinate for 1 hour, stirring occasionally. Remove them and pat them dry again. Coat them with flour and dip them in the eggs. Melt the butter in two large skillets and when it is hot sauté the frogs until golden all over, about 5 minutes. Reduce the heat to low, cover, and continue cooking for about 15 minutes. Season the legs with salt and pepper and serve them at once. Serves 12.

WHITE WINE MARINADE

2 c dry white wine
salt
1 onion, very thinly sliced
white pepper, freshly ground

Combine all ingredients and mix well.

ANCAS DE RANA A LA VALENCIANA

1½ lbs med frog legs
4 peppercorns
1 tbsp olive oil
½ tsp dried thyme, crumbled
black pepper
2 eggs, lightly beaten
oil (½ olive, ½ vegetable)
½ c dry white wine
1 clove garlic, crushed
1 tbsp parsley, chopped
juice of 1 lemon
plain flour
bread crumbs
salt

Combine the wine, peppercorns, garlic, olive oil, parsley, thyme, lemon juice and salt in a small bowl and mix well. Place the frog legs in a single layer in a shallow casserole and pour the mixture over the top. Cover and refrigerate for several hours. Drain and pat dry the frog legs and sprinkle well with salt and pepper and dust with flour. Dip in eggs and then bread crumbs. Add the oil mixture ½ inch deep in a skillet and heat to about 375°F. Cook the frog legs in batches until golden. Remove with a slotted spoon and drain on paper towels. Serve immediately. (Can be prepared ahead and kept warm in a 200°F oven up to 30 minutes.) Makes 10-12 appetizers.

FROG OMELET

12 pr frog legs
2 tbsp butter
1 clove garlic, chopped
1 tbsp salt
½ c mushrooms, chopped
water
2 tbsp olive oil
½ c onion, chopped
½ c chablis
1 tbsp black pepper
1 doz eggs, lightly beaten

Place half the oil and half the butter in a skillet with the onions and sauté until tender. Add the garlic and sauté for 2 minutes. Add the chablis and simmer for 20 minutes. Add frog legs, salt and pepper and enough water to cover. Cover and simmer until the legs are tender and then remove and bone legs. Reserve. Cook down the liquid in the skillet until saucy. In a large omelet pan add the remaining oil and butter and the mushrooms and sauté until the mushrooms are tender. Add the frog meat and pour in the eggs and stir well. When the omelet is cooked on the bottom flip it over. When top is done pour the sauce over it and serve. Serves 5.

CREPE DE MAIS AUX CUISSES DE GRENOUILLES

6 frog legs
Corn Pancake Batter
1 tsp paprika
⅛ tsp black pepper
1¼ tsp oil (or more)
1¼ tsp garlic, minced
salt and white pepper to taste
1 leek, green part only
milk
plain flour
¼ tsp salt
3 tbsp butter (or more)
1½ tbsp shallot, minced
1 tbsp parsley, minced
clarified butter

Combine frog legs with enough milk to cover and refrigerate 2-3 hours. Prepare pancake batter. Drain the frog legs thoroughly and toss in the flour seasoned with paprika, salt and pepper until well coated. Melt 2 tbsp butter with 1 tbsp oil in a large skillet over medium heat. Add shallot, 1 tsp garlic and frog legs and sauté, stirring frequently, until legs are golden and fork tender adding more butter and oil if necessary. Transfer to a platter with a slotted spoon and bone and cut meat into 1 inch julienne strips. Then melt 1 tbsp butter with 1½ tsp oil in same skillet over a medium heat. Add frog meat, parsley and remaining shallot and garlic and sauté lightly. Reduce heat, cover, and slowly cook until very tender. Season with salt and pepper. Remove frog legs from the skillet using a slotted spoon and reserve the seasonings. Now heat a crepe pan over a medium-high heat and brush with clarified butter. Sprinkle with a small amount of water and if beads "dance" on the pan it is ready. Remove pan from heat and working quickly add 3 tbsp batter to one edge of pan tilting and swirling until bottom is covered with a thin layer. Pour excess back into the bowl. Return pan to heat and loosen edges of crepe with a small spatula or knife discarding any pieces clinging to sides of pan. Cook crepe until bottom is brown. Turn or flip crepe over and cook the other side until brown. Slide out onto a plate and top with a sheet of waxed paper or foil. Repeat with remaining batter. Preheat the oven to 200°F. Butter a large baking dish and place about ½ cup frog legs in center of each crepe and top with a little reserved seasonings. Bring edges together and tie with leek (which has been cut into ribbons) to resemble a small bag. Arrange in a single layer in the prepared dish and cover with foil. Bake until warmed through and serve hot. Serves 6.

CORN PANCAKE BATTER

- 4 eggs
- ¾ c self rising flour
- ¼ c oil or melted butter
- 1 8¾ oz can cream corn, pureed
- ½ c water
- ½ tsp salt

Whisk the eggs in a medium bowl until light and lemon colored. Add the remaining ingredients and mix well. Refrigerate for 1-2 hours.

FROG LEGS CALVADOS

- 4 pr frog legs, separated
- clarified butter
- plain flour
- salt
- 1 lge granny smith apple
- 2 tbsp sour cream
- 2 tbsp worcestershire sauce
- 2 tsp capers
- 1 tbsp fresh chives, minced
- 2 ozs calvados
- 1 tbsp fresh tarragon, minced
- 2 ozs + 2 tbsp dry white wine
- freshly ground black pepper
- 2 tbsp parsley, chopped
- 2 cloves garlic, minced
- 1 tbsp fresh basil, minced

Dry frog legs thoroughly and lightly flour. If dried chives, tarragon, parsley or basil is used reduce to 1 tsp each and tie in a small muslin bag. Place enough clarified butter in a heated fry pan to just cover the bottom. Add frog legs and fry gently. Smother frog legs with the garlic and turn the legs after 2 minutes. Add the herbs to pan and then pour over heated Calvados and set alight but if using a small bag do not allow to burn. Give the pan a good shake and add the 2 ozs of wine and reduce the heat, cover, and allow to cook gently for 4 minutes. Cut the apple in half and scoop out the center leaving ½ inch of flesh around the outside. Heat the broiler unit and place apples under broiler for 10 minutes. To the frog legs add the sour cream, worcestershire sauce, capers and the 2 tbsp wine and stir gently to combine. Season with the salt and pepper. Place the frogs on a heated serving dish and coat with a little of the sauce and serve the rest of the sauce in the broiled apple halves. Serves 4.

FROG LEGS SAUTÉ

- 8 pr sm frog legs
- 2 tbsp dry sherry
- 2 cloves garlic, minced
- 2 tbsp orange juice
- 2 tbsp parsley, minced
- ¼ tsp marjoram
- 3 tbsp peanut oil
- 3 tbsp butter
- 2 shallots, minced
- 2 tbsp orange zest, grated
- ¼ tsp basil
- salt and pepper to taste

Separate the legs. Heat the oil in a pan over high heat. Quickly sauté the legs for 2 minutes. Pour off the oil and carefully add the sherry and allow to boil for 15 seconds. Remove legs and keep warm and add the butter, garlic and onions to the pan. Cook for 1 minute and add the orange juice, zest and herbs. Cook for 30 seconds and pour over the frog legs. Serves 4 as an appetizer.

SAUTÉ FROGS, CRAB MEAT AND OYSTERS

6 frog legs
1 c white lump crab meat
salt and white pepper to taste
oil
1 c mushrooms
1 pt oysters
dash of brandy

Fry legs in oil and then cut off the meat from the bones. Sauté the meat with the crab meat for 5 minutes. Add mushrooms and let brown. Add oysters and seasonings and cook until the oysters curl. Add the brandy and let steep for a few minutes. Serves 6.

FROG LEGS AND CRAB MEAT

18 pr med frog legs
2 whole cloves
½ lb crab meat
2½ tbsp + butter
1½ c heavy cream
1 tbsp plain flour
2 egg yolks
cayenne pepper to taste
Melba toast
dry white wine
1 lge bay leaf
½ c sherry
1 c button mushrooms
1½ tbsp chives, chopped
1 tbsp water
salt to taste
1 tsp brandy

Bone the frog legs and place into a saucepan with enough wine to cover. Add the cloves and bay leaf and cook over a medium heat for about 8 minutes or until tender. Drain and combine with the crab meat, sherry and 2½ tbsp melted butter. Cover and let mixture stand for 40 minutes. Transfer the mixture to a shallow flameproof casserole and cook it over a low heat for 8-10 minutes stirring occasionally. Drain off half the liquid and add to the casserole the mushrooms, sautéed in a little butter and well drained. Heat the cream with the chives and thicken with the flour mixed with the tbsp water. Add the cream to the frog legs. Simmer the casserole gently, covered, for 15-20 minutes. When ready to serve beat egg yolks with salt and cayenne to taste and add brandy, a little of the sauce, and stir the mixture into the casserole. Serve immediately with heated Melba toast.

FROG LEGS WITH SCALLOPS

18 pr med frog legs
milk
seasoned flour*
scallops, poached
croutons, seasoned
parsley, chopped
cold water
brandy
sweet butter
bechamel sauce*, hot
paprika

Soak the frog legs in the cold water for 2 hours and drain and dry thoroughly. Then soak for 15 minutes in barely enough milk to cover adding 2 tbsp brandy for each cup of milk used. Roll the legs in seasoned flour and sauté them in sweet butter on all sides until crisp and golden, about 5-6 minutes. Arrange the legs in a wreath on a heated platter and fill the center with poached scallops mixed with hot bechamel sauce. Garnish with sautéed croutons dipped in paprika and sprinkled with parsley.

BROCHETONS "TATAN NANO"

4 sm pike, 4½-5 ozs each
24 sm frog legs
¼ recipe court bouillon with vinegar
1½ tbsp scallion, chopped
¼ lb spinach, blanched, seasoned
½ c white wine
¼ lb butter
1 egg white + 2 yolks
salt and pepper to taste
3 tbsp mushrooms, chopped
1 c creme fraiche or heavy cream
velouté sauce 2

Scale, trim, clean and remove the gills of the young pike and make an incision in the backs to remove the bones. Use the back of a tbsp to push the flesh aside and remove the very fine bones and gut the fish through the incision. Wash thoroughly. Bone the frog legs. Take the pike and frog bones to make the court bouillon. Pound the frog meat together with the egg whites and rub through a fine sieve to eliminate the tendons. Beat over ice with the cream and season highly. Stuff each pike with this filling. To close the incision press the edges together as they are gelatinous and will stick easily. Place the stuffed pike in a buttered pan and on a bed of chopped scallions and half the chopped mushrooms. Pour over ½ cup white wine and the court bouillon which has been reduced to 1 cup. Season again, cover with buttered wax paper and braise over a low flame. When cooked drain the pike on a cloth and remove the skins. Trim again and envelop each pike (except the head) in the spinach. Arrange the pike on a serving platter and on top of each pike make a stripe of chopped mushrooms. Strain the velouté sauce 2 through a double cheesecloth and coat the fish with it. Glaze in a 475°F oven until golden brown. Serves 4.

COURT BOUILLON WITH VINEGAR

3½ qt water
3 tbsp salt
2 lge onions, thinly sliced
1 sprig thyme
pinch crushed peppercorns
1 tbsp butter
1 c vinegar
2 med carrots
some sprigs of parsley
½ bay leaf
fish and frog bones

Cook the flavoring vegetables in the butter. Add the remaining ingredients except peppercorns and bring to a boil and then simmer for 30 minutes. Add peppercorns and simmer 10 minutes longer. Strain through a fine sieve.

VELOUTÉ SAUCE 2

butter
heavy cream
egg yolks
court-bouillon with vinegar

Beat court-bouillon with butter and thicken with egg yolks and cream to produce a smooth sauce to glaze the fish.

FROG LEGS WITH MELON

6 pr frog legs
1 tsp sherry
2 ozs sliced mushrooms
½ tsp sugar
salt to taste
1 tbsp water
chinese stock*
1 tsp sesame oil
1 tsp ginger root juice
2 lb winter melon, diced
1 tsp cornstarch
10 scallions, 3 inch lengths

Put frog legs in hot oiled pan and add sugar, ginger root juice, sherry and salt and cook for 2 minutes. Remove from pan and cook the melon for 2 minutes. Add the frog legs, mushrooms and enough chinese stock to cover and simmer gently for 20 minutes. Add the scallions and cook for 1 minute. Add the cornstarch mixed with the water and sesame oil and cook for 1 minute and then serve. Serves 6.

FROG LEGS BRAISED WITH WINTER MELON

8 pr frog legs
½ c chicken stock
cornstarch
½ lb winter melon
2 tbsp rice wine or sherry
sesame oil
1 clove garlic, minced
½ c oil
2 lge dried chinese mushrooms*

Peel melon and cut in ½ inch cubes. Soak the dried mushrooms for a while in warm water and then cut into strips. Sprinkle the frog legs with salt and then dust with cornstarch. Fry in oil until light golden; drain and keep warm. Discard all but a little of the oil. Sauté the garlic, add the winter melon and mushrooms. Sauté 2 minutes and then add the wine and stock. Cover and braise 5 minutes. Add the frog legs and heat through. Transfer to serving dish and sprinkle with a few drops of sesame oil.

BRAISED FROGS LEGS WITH VEGETARIAN STEAK

1 lb med frog legs
¼ c vegetable oil
5 oz fried vegetable steak*
2 scallions, 1½ inch pieces
2 tsp cornstarch
3 tbsp sherry
½ tsp salt
¼ tbsp light soy sauce
5 slices ginger
1 tsp sugar
1 c chicken soup
½ tbsp cornstarch + mixed in 2 tbsp water
1 tsp sesame oil (optional)

Mix well the cornstarch, sherry, salt, soy sauce and sesame oil and separate into 2 portions of ⅔ and ⅓. Mix the ⅔ portion with the frog legs and add the ginger and sugar and let stand. Cut the vegetable steak into ½ x 1½ inch pieces and mix with the ⅓ portion. Heat the vegetable oil in a frying pan and add the vegetable steak and scallions and stir fry 1-2 minutes. Add the frog mixture and stir fry for 2 minutes. Add small amounts of soup if pan dries out. Then add ¼–½ cup soup and simmer for 2 minutes. Thicken with the cornstarch and water mixture and then serve.

FROG LEGS WITH PEAS

1 lb med frog legs
¼ c vegetable oil
½ tsp sugar
1 tsp sesame oil (optional)
2 tsp sherry
1 tsp salt
¼ tsp ginger powder
2 tsp cornstarch
1 tbsp light soy sauce
1 10 oz pkg frozen peas
2 scallions, shredded
boiling water

Cover the peas with enough boiling water until the peas are completely thawed. Drain and set aside. Mix the frog legs, sugar, sesame oil, sherry, salt, ginger, cornstarch and soy sauce and mix well. Heat the vegetable oil and add the frog leg mixture and stir fry for 2 minutes. Add the peas and some water (if needed) and simmer for 1-2 minutes or until legs and peas are cooked. Remove to a serving dish and garnish with the scallions.

FROG LEGS WITH BLACK BEANS

6 pr frog legs
2 c and 2 tbsp peanut oil
¼ tsp msg
salt to taste
⅓ c fermented salted black beans*
2 tbsp garlic, chopped
4 sm hot peppers
1 tbsp cornstarch
4 tbsp dry sherry
½ tsp sugar

Split legs in half and each leg into pieces at the joints. Place the pieces in a bowl and add 1 tbsp sherry. Using a sieve, sprinkle the pieces with the cornstarch, stir to coat, and then set aside. In a wok or skillet, heat 2 cups oil until smoking, then turn off the heat. Add the frog legs and stir. Immediately turn on the heat and cook about 45 seconds, stirring. Add the hot peppers and cook stirring about 30 seconds. Drain well. In a clean wok or skillet add 2 tbsp oil and when hot add the beans. Stir, mashing with a spoon until the beans are dry. Add the garlic and stir. Add the frog legs and hot peppers, then the sherry, msg, salt to taste and sugar. Cook briefly and then serve. Serves 3-6.

RICE WINE FROG LEGS

16 pr sm frog legs
2 cloves garlic, minced
1 tbsp soy sauce
3 tbsp peanut oil
8 asparagus spears, 2 inch lengths
2 c chicken stock
2 tbsp ginger, minced
2 tbsp chinese rice wine
1½ tsp sesame oil
3 tbsp unsalted butter
1 med leek, white part

In a medium bowl marinate the frog legs with the ginger, garlic, wine, soy sauce and sesame oil at room temperature for 1 hour. Drain and reserve the marinade. In a wok or large heavy skillet melt 1 tbsp butter with 1 tbsp peanut oil over a moderately high heat. Add the asparagus and leek (cut into ½ inch pieces) and stir fry until just tender, about 5 minutes. Remove the vegetables. Wipe out the inside of the wok or skillet with paper towels and add the remaining peanut oil and place over a moderately high heat until almost smoking. Add the frog legs and stir fry until a light golden brown, about 3-5 minutes. Pour in the stock and the reserved marinade and reduce the heat to low and simmer for 5 minutes. Remove the frog legs. Increase the heat to high and boil the stock uncovered until reduced to 1 cup, about 5 minutes. Slowly swirl in the remaining butter into the reduced cooking liquid and return the frog legs and vegetables to the wok and cook briefly over a moderate heat until just heated through, about 1 minute.

STIR-FRY CHICKEN WINGS WITH FROG LEGS

6 chicken wings
8 pr frog legs
¼ tsp freshly ground black pepper
2 tsp + 1 tbsp cornstarch
8 slices peeled ginger, thin sliced
3 scallions, white only, in 3 inch lengths
1-2 sm red peppers
1 egg white
1 tbsp rendered chicken fat
3 cloves garlic, crushed
½ c peanut oil
1 tbsp dry sherry
5 dried black mushrooms*
½ c chicken broth
1 tsp sesame oil
2 tbsp water

Cut the wings into 3 parts at the joints and discard the wing tip (the wing tip can be used to make the chicken stock). Score each piece lightly. Cut each pair of frog legs in half and separate each leg at the joints. Combine. (If you wish you can bone wings and legs.) Beat the egg white lightly and use half of it discarding the remainder. Blend the egg white and 2 tsp cornstarch and stir into the chicken and frog mixture. Refrigerate for 30 minutes. Put the mushrooms into a bowl and add boiling water to cover. Let stand 15-20 minutes. Drain, and squeeze to extract most of the moisture. Cut off and discard the tough stems. Slice. In a wok or deep fryer heat the oil for deep frying and when hot but not smoking add the chicken and frog mixture. Cook briefly, about 1 minute, stirring, then drain thoroughly. Do not cook through. Pour off all but 1 tbsp oil from the pan. Add the garlic, scallions, peppers, ginger and mushrooms. Cook about 1 minute, then add the chicken and frog mixture. When thoroughly hot add the chicken broth, sherry, pepper and sesame oil. Cook about 30 seconds. Blend the remaining tbsp of cornstarch and the water and stir in. When slightly thickened stir in the chicken fat. Serves 4-6.

CHINESE FROG LEGS

9 pr med frog legs
2 scallions, 1 inch lengths
½ bell pepper, diamond cut
5 tbsp lard
2 slices fresh ginger
dash white pepper
⅓ tsp sugar
1 tsp cornstarch
3 ozs dried mushrooms
hot water
3 ozs water chestnuts, sliced
2 cloves garlic, crushed
1 tbsp sherry
¼ c chicken stock
pinch of salt
2 tbsp cold water

Soak mushrooms in hot water for 30 minutes, wash them, discard the stems and cut into small squares. Wipe frog legs with a damp cloth and cut each pair into 4 pieces. In a skillet sauté the frog legs in the lard with the garlic and ginger for 2 minutes. Add the scallions, sherry and white pepper and sauté 3 minutes longer. Remove the legs and vegetables leaving the oil. In the oil sauté the bell pepper, mushrooms and water chestnuts for 2 minutes. Remove the vegetables. To the juices in the pan add chicken stock, sugar, salt and the cornstarch which has been mixed with the cold water. Cook, stirring, until the sauce is smooth and well blended. Add the frogs and vegetables and cook over high heat for 1½ minutes. Serves 4.

FROG LEGS HOPALONG CASSIDY

4 lbs frog legs
1 tbsp + 1 tsp cayenne pepper
1 c oil
1½ c onions, chopped
7 c rich chicken stock
2 c scallions, chopped
4 c hot cooked rice
1 tbsp + 2½ tsp salt
1 tbsp black pepper
1 c plain flour
2 c bell peppers, chopped
6 tbsp garlic, minced
1 c parsley, chopped

Separate the frog legs into 2 and place in a large bowl. Mix the salt, cayenne pepper and black pepper in a small bowl. Sprinkle 1 tbsp plus ½ tsp evenly over the frog legs working it in with your hands. Cover and refrigerate overnight. Reserve the remaining seasonings mix. In a 6-quart dutch oven heat oil ⅛ inch-deep over high heat and heat until a drop of water sizzles when dropped in. Meanwhile place the flour in a pan and mix well with 1 tbsp of the seasoning mix and mix well. Just before frying dredge each leg in the seasoned flour shaking off the excess. Reserve the leftover flour. Sauté the frogs in a single layer in the hot oil until very lightly browned, about 1-2 minutes. Turn over and lightly brown the other side, about 1 minute more. Do not overcook. Transfer to a large bowl and fry remaining legs and transfer. Pour the hot oil remaining into a glass cup and include as many of the brown particles as possible to make 1 cup, add more oil if needed. Return to the dutch oven. Measure out the leftover flour and add more flour if necessary to get ½ cup. Add the flour to the oil and cook until light to medium brown scraping the bottom of the pan so roux doesn't burn. Add ½ cup onions and cook about 2 minutes stirring constantly. Add the remaining seasoning mix and 1 cup of the bell peppers and cook about 2 minutes stirring frequently. Gradually stir in the stock stirring constantly until well blended. Bring to a boil stirring occasionally. Add the frog legs, garlic, remaining onions and bell peppers stirring well. Return to a boil and cover oven. Reduce heat and simmer about 20 minutes stirring frequently and scraping oven bottom to make sure mixture doesn't stick. Now add the scallions and parsley stirring well and simmer uncovered for about 15 minutes, skimming any oil from the surface as it develops and stirring and scraping bottom frequently. Add about 1 cup more stock or water if gravy gets too thick. Remove from heat and let sit covered about 10 minutes before serving. Serve over rice allowing about ½ cup rice, ¾ cup sauce and a portion of the legs for each serving.

RICH CHICKEN STOCK

Take a basic chicken stock and simmer it down until it is reduced by half or more.

SAUTÉED FROG LEGS WITH GARLIC

18 pr frog legs
2 tbsp tamari soy sauce
¾ tsp black pepper
¾ c butter
juice of ½ lemon
4 c milk
¾ c whole wheat flour
4 tbsp olive oil
6 lge cloves garlic, minced
½ c parsley, minced

Place the frog legs in a bowl with the milk and soy sauce. If legs are frozen let them defrost in the milk either overnight in the refrigerator or at room temperature for several hours. About 20 minutes before serving remove the legs and place them in a pie pan with the flour and pepper and coat evenly. Heat the olive oil until very hot and add the legs a few at a time and sauté 4-5 minutes turning so that they brown evenly and become crisp without drying out. Drain on paper towels and place on a heated platter and keep warm in the oven. More oil can be added if necessary to sauté all the legs. While the legs are cooking cut the butter into pieces and melt it in a second skillet over low heat and then add the garlic. Cook just long enough to turn a golden color. Add the lemon juice and pour sauce over the frog legs and sprinkle with the parsley. Serves 6.

FROG LEGS WITH GARLIC

24 pr med frog legs
2 bay leaves
salt and pepper to taste
1 tbsp hot red chili, minced
2 tsp paprika
water
¼ c onion, minced
½ c wine vinegar
8 cloves garlic, minced
¼ c olive oil
2 eggs, hard boiled, chopped

Place the frog legs in a heavy pot with the onion, bay leaves and vinegar and add just enough water to cover and then add salt and pepper and bring the liquid to a boil. Reduce the heat to low and simmer for 5 minutes. In a skillet fry the garlic and the red chili in the oil. When they begin to brown add the paprika (taking care not to let it burn) and pour the mixture over the frog legs. Scatter the eggs over the mixture. Cover the pot and simmer the legs for another 3 minutes. Season to taste and serve very hot. Serves 4-6.

MISSOURI LEVEE FROG LEGS

2 lbs frog legs
½ c butter
½ c sliced fresh mushrooms
2 tbsp fresh lemon juice
salt and pepper to taste
½ c shallots, chopped
½ c white wine
2 tbsp snipped fresh parsley

Sprinkle frog legs with salt and pepper. In a 12 inch skillet melt the butter over a medium heat. Add the legs to the skillet and turn to coat. Fry for 8-10 minutes, or until browned on both sides turning at least once. Add the shallots and mushrooms and reduce the heat to medium-low. Simmer 5 minutes. Add the wine and juice and simmer 5-6 minutes until legs are tender. Arrange legs on a serving platter and spoon the shallot-mushroom mixture over the legs. Just before serving sprinkle with the parsley.

FROG LEGS WITH GARLIC AND PINE NUTS

18 pr frog legs
3 sprigs parsley
2 tbsp lemon juice
1 tbsp scallions, minced
¼ c parsley, minced
lemon slices for garnish
1 tbsp scallions, thinly sliced
½ c water
¾ c butter, softened
2 tsp garlic, minced
¼ c pine nuts, toasted

In a skillet mix the sliced scallion, sprigs of parsley, water and lemon juice and bring to a boil. Add the frog legs, cover, and simmer for 5 minutes. Lift legs out with a slotted spoon and remove to a broiler plate. Reduce the liquid to 1 tbsp and strain and reserve. In a separate bowl mash the butter, minced scallions, garlic and minced parsley with the reserved liquid. Smear lavishly over the top of the frog legs and broil at least 6 inches from the heat for 4 minutes or less, until the butter melts into a sauce. Sprinkle with toasted pine nuts (sunflower seeds may be used if pine nuts are not available) and garnish with lemon slices. Serves 6.

RO-TEL FROG LEGS

12 frog legs
4 tbsp oil
1 med white onion, chopped
1 med bell pepper, chopped
1 8 oz can tomato sauce
juice of 1 lime
cooked wild rice
2 tbsp flour
6 scallions, chopped
2 cloves garlic, chopped
1 stalk celery, chopped
1 10 oz can Ro-Tel Tomatoes
salt

Salt frog legs generously and brown in skillet and remove. Blend flour and oil in skillet and brown and then add onion, garlic, bell pepper and celery and mix thoroughly. Add tomato sauce and Ro-Tel Tomatoes. Stir and add frogs and cover and simmer until legs are tender. Just before removing, add scallions and the lime juice and pour over wild rice. Serves 4.

GOLD IN VEAL

Veal Clifton

3 lb baby veal rib eye roast
½ tsp white pepper
½ tsp salt
½ tsp garlic powder

Frog Leg Dressing

2 lb frog legs
½ tsp salt
¼ tsp cayenne pepper
½ c plain bread crumbs
brown veal demi glaze sauce
½ stick butter
½ tsp garlic powder
½ tsp white pepper
¼ c shallots, sliced

Bake the frog legs at 350°F for 45 minutes. Remove meat from bones and grind up fine. Add the ground frog meat to the frog leg dressing ingredients and cook down for 10 minutes. Carve a hole in the center of the veal roast and stuff with the dressing. Season the veal with the veal clifton seasonings and bake at 250°F for 45 minutes. Remove from the oven and slice about ½ inch thick. Serve with the brown veal demi glaze sauce poured over. Serves 6.

BROWN VEAL DEMI GLAZE SAUCE

2 lbs veal bones
1 c white wine
2 cloves garlic
1 carrot, halved
2 bay leaves
½ tsp thyme
½ gal + 2 c water
½ lb butter
1 c red wine
2 whole tomatoes
1 lge onion, halved
1 rib celery
1 tbsp peppercorns
2 tbsp beef bouillon granules
½ tbsp Kitchen Bouquet
1 c plain flour

Brown bones in a 350°F oven for 45 minutes. Put the baked bones in a pot and add all the other ingredients except butter, flour, Kitchen Bouquet and the 2 cups water and bring to a boil. Reduce to slow and simmer for 4 hours. Add the 2 cups of water and again simmer for 30 minutes. Strain the stock with a fine sieve and put stock in a clean pot on medium heat and simmer for 75 minutes. Make a stiff roux with the butter and flour and check for thickness. (Thicken to taste.) Add to the stock and add the Kitchen Bouquet for color.

LEMON DEVILED FROG LEGS

8 pr med frog legs
½ c olive oil
milk
salt and pepper to taste
plain flour
juice of 1 lemon
2 eggs, beaten
1 tsp dry mustard
bread crumbs, plain
1 tbsp worcestershire sauce
2 ozs butter
dash of brandy

Soak the frog legs in milk at least 30 minutes. Drain and dry. Roll them in flour. Dip them in egg, and roll in crumbs. Sauté in butter and olive oil. When nicely browned, remove frog legs from pan. Add salt, pepper, lemon juice, mustard, worcestershire sauce and brandy to the pan. Mix well and pour this sauce over the frog legs. Serves 4.

SCALLOPED FROG LEGS

8 pr frog legs
salt and pepper to taste
3 c milk
½ c plain flour
½ c butter

Separate and dredge frog legs in seasoned flour*. Arrange the legs in a baking dish in which the butter has been melted turning each leg in butter. Pour milk into the bottom of dish and bake at 350°F for 60-90 minutes until legs are browned and gravy has formed in bottom of dish. Serves 4.

DEVILED FROG LEGS

18 pr med frog legs
milk
plain bread crumbs, fine
¼ c butter
¼ tsp dry mustard
dash of whiskey
eggs, beaten
plain flour
¼ c olive oil
salt and pepper to taste
dash worcestershire sauce

Soak the legs in the milk for 1-2 hours. Wipe dry and roll in flour and dip in eggs and then roll in bread crumbs. Sauté legs in olive oil and butter for about 5 minutes or until nicely browned. Arrange the legs on a hot platter and to the juices in the pan add salt and pepper to taste, mustard and generous dashes of worcestershire sauce and whiskey. Heat the sauce well blending thoroughly and pour it over the frog legs. Serve.

BRAISED FROG LEGS WITH SHERRY

18 pr frog legs
plain flour
2 tbsp shortening
¼ cup sherry
8 tbsp butter
salt and pepper to taste
1¼ c beef broth
1 tbsp parsley, chopped

Preheat an oven to 400°F. Place the legs in a mixing bowl and add cold water. Drain well and sprinkle with salt and pepper. Dredge the legs 1 pair at a time in the flour and shake off excess. Heat the butter and shortening in a heavy skillet and when very hot add the legs. Cook until golden on all sides. Transfer the legs to a baking dish and add the broth and sherry to the fat remaining in the skillet and bring to a boil. Pour this over the frog legs and bake 20 minutes. Sprinkle with parsley and serve. Serves 6.

FROG LEGS WITH MUSHROOMS

12 frog legs
3 tbsp butter
½ tsp salt
1 c milk
2 egg yolks
bread crumbs, seasoned*
1 4 oz can mushrooms, sliced
1 tbsp plain flour
⅛ tsp black pepper
½ c meat or chicken stock
1 tbsp cream
hot buttered toast

Dredge frog legs in seasoned crumbs. Melt butter in pan and put in the frog legs and mushrooms and sauté to a light brown. Sprinkle with flour and salt and pepper. Add milk and stock and bring to a boil, cover and reduce heat and let simmer for 10 minutes. Beat egg yolks with cream and add to the pan. Stir well but do not boil. Serve hot with buttered toast garnished as desired.

FRICASSEE FROGS AND OYSTERS

1 doz frog legs
1 sprig thyme, minced
1 egg yolk
½ pt water
1 bay leaf
croutons
1 tbsp plain flour
1 sprig marjoram, chopped
cayenne pepper to taste
2 doz oysters
1 clove
½ pt oyster water
10 allspice
3 tbsp butter
1 sprig parsley, chopped
salt and pepper to taste

Take 1 tbsp butter and put in frying pan. When it begins to melt, add flour and stir constantly. When it begins to brown nicely, add water and oyster water. Throw in the frogs legs as it begins to boil, add salt and pepper, a little cayenne, thyme, bay leaf, marjoram, allspice and clove. Let simmer about 15 minutes and take off fire. Sauté oysters in 1 tbsp butter until curled. Beat the egg yolk and add to the frog mixture, blending well, and serve immediately with croutons fried in a little butter with the sautéed oysters laid upon them.

FRICASSEE FROG LEG PLATTER

12 frog legs
1 tsp salt
3 tbsp butter
1 tbsp parsley, chopped
¼ c plain flour
⅛ tsp black pepper
½ c heavy cream

Sprinkle frog legs with some flour, salt and pepper and brown in melted butter for 3 minutes, turning frequently. Remove the legs and blend in 1 tbsp flour with the remaining butter. Add cream slowly and cook over low heat for 5 minutes stirring constantly. Add legs and parsley. Serve on a hot platter.

FROG FRICASSEE AND FRENCH BREAD

12 whole frogs, dressed
1 sm onion, chipped
¼ c white wine
¼ c dried mushrooms, soaked ½ hr
2 tbsp plain flour
2 egg yolks, lightly beaten
2 tbsp olive oil
1 clove garlic, chopped
½ tsp pepper
1 tbsp parsley, chopped
1 tbsp lemon juice
12 thin slices french bread, toasted

Remove the legs from the dressed frogs and place the legs in cold water and let stand for 2 hours. Brown the garlic and onion in oil and add the wine. When wine has evaporated add the frog bodies, salt, pepper and dried mushrooms. Add just enough water to cover and simmer for 1 hour, covered. This will make a delicious broth. Strain. Flour the legs and add to the broth. Cook slowly about 30 minutes, mixing well occasionally. When the legs are tender add the parsley. Remove from fire and add the egg yolks and lemon juice. Mix well and serve on toasted bread slices. Serves 4.

FROG LEGS IN RIESLING WITH FRESH HERBS

40 pr med frog legs
2 shallots, minced
1 c Alsatian Riesling
½ tsp salt
½ c creme fraiche or heavy cream
1 tbsp parsley, chopped
8 tbsp unsalted butter
1 clove garlic, minced
1 c chicken stock
¼ tsp black pepper
1 tbsp lemon juice
1 tbsp chives, minced

In a large flameproof casserole melt 3 tbsp butter over a moderate heat and add the shallots and garlic and sauté until tender, about 5 minutes. Add the frog legs, Riesling, stock, salt and pepper and increase the heat to moderately high and stir occasionally until the legs are tender, about 10 minutes. Transfer the legs to a platter. Bring the liquid in the casserole to a boil over high heat and then stir in the creme fraiche and cook until the liquid has reduced to 1½ cups, about 10 minutes. Reduce the heat to moderate and cut the remaining butter into small pieces and whisk into the sauce a few pieces at a time. Stir in the lemon juice and more salt and pepper to taste. Return the legs to the casserole and toss gently until heated through. Add the parsley and chives and toss. Serves 4.

CREOLE FROG LEGS

8 pr med frog legs
1 tbsp parsley, chopped
⅛ tsp cayenne pepper
½ c butter
2 tbsp scallion tops, chopped
1 egg, slightly beaten
¼ c bell pepper, chopped
2 c stewed tomatoes
1 tsp soy sauce
½ c onions, chopped
1 tsp msg
1 c milk
2 cloves garlic, minced
½ tsp salt
plain flour
1 8 oz can mushrooms, sliced
1 tsp sugar

Melt half the butter in a skillet and sauté bell pepper, onion, garlic, parsley, mushrooms and scallion tops. Add stewed tomatoes and seasonings, sugar and soy sauce. Simmer over low heat 10 minutes. While the sauce cooks mix the egg and milk. Dip frog legs in the egg wash and drain. Dredge in flour; fry in remaining butter about 10 minutes. Add legs to sauce and simmer 5 minutes. Arrange frog legs in center of dish surrounded by hot cooked rice; pour sauce over and between frog legs. Garnish. Serves 4.

FROG LEGS SAUTÉ A LA CREOLE

6 frog legs
2 tbsp butter
3 lge onions, sliced
6 fresh tomatoes, chopped
1 sprig thyme
1 bay leaf
2 cloves garlic, minced
6 bell peppers, minced
1 c consommé or boiling water
salt and pepper to taste

Brown washed frog legs in butter being careful not to burn. After 10 minutes take onion and brown with the legs. Add tomatoes, cover, and let brown. Cook very slowly adding salt and pepper to taste, thyme, bay leaf and garlic. Smother mixture slowly for 20-30 minutes, stirring frequently. Add bell pepper and continue to smother until legs are tender. Add consommé and continue to cook slowly, covered, for 30 minutes. Serve hot. Serves 4.

ONION CREAM FROG LEGS

16 frog legs
juice of ½ lemon
2 eggs, well beaten
oil
boiling water
salt and pepper to taste
bread crumbs, fine
onion cream sauce

Put the legs in the boiling water with the lemon juice, salt and pepper and scald for 4 minutes. Drain and pat dry. Dip legs into eggs and roll in bread crumbs and fry in oil for 4-5 minutes until legs are tender. Serve with onion cream sauce.

ONION CREAM SAUCE

2 tbsp butter
1½ c light cream or evaporated milk
2 tbsp onion, minced
1 egg, well beaten
2 tbsp plain flour
½ tsp salt
1 tbsp parsley, minced

Melt butter and stir in the flour blending well and gradually stir in the cream. Add salt, onion and parsley and simmer over low heat, stirring constantly, until smooth and thickened. Beat some of the sauce in the egg. Add to the remainder of the sauce and cook for 2 minutes.

FROG LEGS N' RICE CREOLE

12 lge frog legs
4 tbsp plain flour
1 tbsp lemon juice
2 tbsp soy sauce
½ tsp Accent
⅛ tsp cayenne pepper
6 tbsp butter
3 cloves garlic, crushed
1 18 oz can stewed tomatoes
2 tbsp brown sugar
1 tsp salt

Roll the legs in the flour. Melt the butter in a skillet or dutch oven and add the garlic and lemon juice. Add the legs and slowly brown on all sides, about 6-8 minutes. Shake skillet to prevent sticking and add the tomatoes, soy sauce, sugar and seasonings and slowly cook for 8 minutes. Serve over rice. Serves 4.

PICAYUNE FROGS A LA CREOLE

½ doz frog legs
½ tsp salt
5 eggs
½ onion, finely chopped
1 c bread crumbs, plain
salt and pepper to taste
1 tbsp butter
oil
1 tbsp plain flour
1 tsp parsley, chopped
8 ozs fresh cream

Beat 3 eggs and mix in the bread crumbs. Have ready heated oil. Rub the frog legs well with salt and pepper and dip in egg mixture and place into hot oil. Let them fry to a nice golden brown. Remove, drain and keep warm. Make a sauce of the flour, butter, cream and salt putting the butter into a saucepan over the fire and as soon as it melts add the flour, stirring gradually, and when blended add by degrees the milk, stirring constantly, and then add the parsley and onion. Have ready the remaining eggs beaten in a bowl. Warm the frog legs in the sauce and when it begins to simmer stir the eggs briskly into the sauce. Return to the back of the stove a minute or two being careful not to let the sauce boil after the eggs have been added as they will curdle. Serve hot.

FROG LEGS A LA SQUANOCOCK

2 lbs frog legs
½ c plain flour
2 tsp lemon juice
salt and pepper to taste
½ c milk
1 c butter, melted
1 tsp soy sauce
dash of garlic salt

Coat the legs with the milk and then dredge in flour coating completely and shake off excess. Place ½ cup melted butter in a hot frying pan and add the legs and brown evenly. Combine the remaining butter, lemon juice, soy sauce, salt, pepper and garlic salt and heat. Pour over the browned frog legs and serve hot. Serves 5-6.

FROG LEGS JESSE

4 pr med frog legs, separated
salt and pepper to taste
½ c tbsp plain flour
½ stick butter
brown sauce
½ c mushrooms, sliced
¼ c scallions, chopped
juice of ½ lemon
2½ tbsp garlic, minced

Season legs with salt and pepper and dredge in the flour and sauté in butter until golden brown on all sides. Add the garlic and mushrooms and cook approximately 5 minutes. Add scallions and cook 3 minutes. Add lemon juice and brown sauce and cover and cook over low heat for 10-15 minutes. Serve.

BROWN SAUCE

1½ tsp clarified butter
1½ tbsp plain flour
2 c beef consommé

Melt the butter in a saucepan and blend in a generous 1½ tbsp flour and cook slowly over a low heat, stirring, until thoroughly blended and the color of brown wrapping paper. Moisten gradually with consommé, bring to a boil, and simmer 3-5 minutes, stirring constantly. Lower heat and simmer gently for 30 minutes, stirring occasionally. Skim off the fat and strain the sauce through a fine sieve. Makes about 2 cups.

FROG LEGS SAUCE PIQUANT

4 lbs individual frog legs
2 tbsp bell pepper, chopped
1 lemon, thinly sliced, seeded
½ c onions, minced
2 tbsp parsley, chopped
1½ tsp salt
1 lb can tomatoes, chopped
1 tsp cayenne pepper
2 tbsp celery, chopped
4 tbsp brown roux*
2 tbsp scallions, chopped
½ c oil
2 tsp garlic, minced

Pat the frog legs completely dry and season them evenly on all sides with the cayenne pepper and 1 tsp salt. In a skillet heat half the oil over moderate heat until a light haze forms above it. Add frog legs and fry them for 4-5 minutes, turning them once or twice and regulating the heat so that they color richly and evenly without burning. As they are cooked, transfer the frogs to a platter. Pour the remaining oil into the skillet and when it is hot fry the other frogs in the same fashion. Stirring constantly, add the brown roux to the fat remaining in the skillet. Drop in the onions, bell pepper and celery and stirring frequently cook over moderate heat for 5 minutes or until the vegetables are soft. Stir in the tomatoes, add the garlic and the remaining ½ tsp salt and boil briskly, uncovered, for about 5 minutes. When the vegetable sauce mixture is thick enough to hold its shape almost solidly in the spoon, return the frogs and the liquid that has accumulated around them to the skillet. Stirring and basting the frogs with the sauce, cook for a minute or so longer to heat the legs through. Then mix in the scallions and parsley and taste for seasonings. If more is needed, add. Arrange the frogs attractively on a large heated platter. Pour the sauce piquant over the frog legs, garnish the platter with the lemon slices and serve at once over hot cooked rice. Serves 4-6.

SAUCE PIQUANT A LA RANA

16 pr sm frog legs
cornstarch
4 ozs margarine
parmesan cheese (optional)
milk
4 ozs butter
4 tomatoes, broiled, (optional)

Sauce

4 ozs butter
1 med onion, chopped
½ bell pepper, chopped
2 scallions, thinly sliced
1 16 oz can tomatoes, chopped
3 c bouillon (or water)
3 tbsp plain flour
4 cloves garlic, chopped
1 rib celery, chopped
2 tbsp parsley, chopped
1 tsp cayenne pepper (or less)
salt and black pepper to taste

To make the sauce melt the butter and add the flour and cook slowly while stirring constantly for about 15 minutes. Add the onion, garlic, bell pepper, celery, scallions and parsley and cook until soft and transparent. Add the bouillon, tomatoes, cayenne pepper, salt and pepper and simmer over low heat for 1 hour adding more bouillon if necessary. While the sauce is simmering cook the frog legs. Melt the margarine in a skillet. Dip legs in milk and then the cornstarch shaking off the excess and cook gently in margarine until golden brown on both sides. Keep warm until all legs are cooked. Place legs in the sauce and cook 5-10 minutes. Use preheated round plates and arrange the legs on the plate like the spokes on a wheel and cover them with sauce. Although optional a fresh tomato with its top sliced off and broiled or baked with a cap of parmesan cheese makes a nice hub. Serves 4.

CAJUN SAUCE PIQUANT A LA FROG LEGS

18 med frog legs
½ c plain flour
4 cloves garlic, minced
1 lge bell pepper, chopped
3 15 oz cans stewed tomatoes
½ tsp garlic powder
½ tsp black pepper
½ tsp tabasco sauce
1 tbsp lemon juice
cooked white rice
1 c + 2 tbsp peanut oil
4 med onions, chopped
1 c celery, chopped
1 cayenne pepper, minced
1½ c chicken stock
½ tsp onion powder
¼ tsp white pepper
1 tbsp worcestershire sauce
2 tsp seafood seasoning mix

In a large saucepan, over medium heat, heat the cup of oil until it is hot. Add the flour and heat, stirring constantly, until a dark brown. Add the onions, garlic, celery, bell pepper and cayenne pepper and sauté for 5 minutes over a medium heat. Add the solid pieces from the stewed tomatoes, reserving the liquid, and sauté for 5 minutes. Add the stock, garlic powder, onion powder, peppers, tabasco sauce, reserved tomato liquid, worcestershire sauce and lemon juice. Bring to a boil and then reduce the heat to a low simmer and simmer the sauce for 1 hour, stirring frequently. In a heavy skillet over medium heat, heat the remaining oil until hot. Season the frog legs evenly with the seafood seasoning Mix rubbing the seasoning in with your hands. Fry the legs 4-5 at a time until they are a nice golden brown on all sides. Set them aside. When the sauce is cooked add the legs and stir them in and continue to simmer for 3 more hours over low heat. Serve hot over cooked white rice. Serves 6-8.

HAM STUFFED FROG LEGS

1 lb lge frog legs
oil
¼ tsp pepper
1-2 egg whites
¼ lb ham
½ tsp salt
2-3 tbsp plain flour

Remove bones of each leg by snipping each end to remove center part of bone. Cut ham into very slender sticks and insert in each leg cavity replacing the bone. Beat egg whites and mix with flour and seasonings. Dip frog legs in egg mixture and fry in hot oil until light golden and crusty. Serve hot with soy sauce or Plum sauce. Serves 4.

PLUM SAUCE

1 c plum jelly
1 tbsp sugar
½ c chutney, finely chopped
1 tbsp vinegar

Mix together thoroughly.

SEAFOOD STUFFED FROG LEGS

12 pr large frog legs
1 c fresh crab meat
fish bake
2 c fresh ground shrimp
milk
1½ c onion, minced
1 can oyster soup
3 tbsp butter
1½ c cooked rice
¼ c parsley, minced
salt and pepper to taste
¼ c bell pepper, minced
dash tabasco sauce

Make pockets in thighs of the frog legs by removing center part of bone as in Stuffed Frog Legs. Season with salt and pepper and set aside. Sauté onion and bell pepper in butter until brown. Add shrimp, crab meat, parsley, soup, and tabasco sauce and simmer about 30 minutes. Add rice and mix well. Stuff mixture into pockets of frog legs. Moisten with milk and coat with fish bake. Arrange in a shallow pan and bake at 375°F for 40-50 minutes. Serves 6.

STUFFED FROG LEGS

12 lge frog legs
1½ tbsp butter
¾ c onions, chopped
⅛ c celery, chopped
1 c shrimp. ground or chopped
milk for dipping
fish fry for coating
½ c crabmeat, fresh
¾ c rice, cooked
⅛ c bell pepper, chopped
⅛ tsp tabasco sauce
½ tsp creole seasoning
½ can (10.5 oz size) oyster soup

Preheat oven to 375°F. Make pockets in largest part of legs and set aside. Sauté onions, bell pepper and celery in the butter until brown. Add the shrimp and crabmeat and simmer about 30 minutes. Add the rice and mix well. Season with the tabasco and creole seasoning. Add the soup as needed to moisten. Stuff the mixture into the frog leg pockets. Dip legs into milk and coat with fish Fry. Arrange in a shallow pan and bake for 40-45 minutes.

FROG LEGS SUPREME

12 lge frog legs
½ c yellow cornmeal
½ c plain flour or Bisquick
oil
⅓ c Wish Bone California Onion Dressing
1 tsp msg
¼ tsp cayenne pepper
1 tsp salt
1 6 oz pkg. Good Seasons
Salad Dressing Mix

Combine salt, pepper, msg, salad dressing mix and onion dressing. Pour over frog legs and coat well. Let stand in refrigerator for 20-30 minutes or overnight. Combine the flour and cornmeal in a paper bag. Add frog legs a few at a time and coat well. Fry in hot oil at 375°F a few at a time until golden brown. Drain. Serves 4-6.

FROG LEGS SUPREME CHABLIS

12 frog legs
2 cloves garlic, chopped
1 tsp salt
1 tbsp tarragon
2 tbsp parsley flakes
2 ozs chablis
plain flour
5 tbsp butter
2 tbsp lemon juice
1 tsp black pepper
2 tbsp chives, minced
1 oz brandy
water

Wash legs with 1 tbsp lemon juice and water. Roll in flour. Sauté garlic in butter and 1 tbsp lemon juice and add the frogs and brown. Add spices and cook for 2 minutes and then add the brandy and chablis. Simmer uncovered for 1 minute. Serve immediately. Serves 4.

SWEET AND SOUR FRIED FROG LEGS

1 lb frog legs
¾ tsp salt
¼ tsp msg
5 tbsp plain flour
1 tsp soy sauce
3 tsp cornstarch mixed with 6 tbsp water
¼ c vinegar
½ c oil
½ bell pepper, chopped
5 slices ginger root
dash of pepper
1 egg
5 tbsp sugar
1 carrot, chopped
1 slice pineapple, chopped
1½ c water

Mix the egg and flour. Dip frog legs in the batter. Deep fry in boiling oil 4 minutes on each side. Remove and place in serving dish. Using a high flame, heat a well-greased pan and add the salt, ginger, water, vinegar, bell pepper, carrot, sugar, pineapple, soy sauce, msg and pepper. Bring the mixture to a boil. Stir in cornstarch paste and cook until thickened and pour over frog legs. Serve.

SWEET AND SOUR SAUTÉED FROG LEGS

frog legs
3 tbsp cornstarch
⅔ c water
2 ozs butter
¾ c sugar
1 tbsp tomato sauce
⅓ c vinegar
¼ c oil

Sauté the frog legs in the butter about 5 minutes browning evenly. Cook all other ingredients over low heat until thickened, stirring constantly. After the frogs are sautéed place them in the sauce. Serve in a chafing dish.

SWEET AND SOUR FROG MEAT

½ lb frog meat
1 bell pepper, chopped
1 carrot, sliced thin
½ onion, sliced thin
1½ tsp water
1 cup egg whites
4 c oil
½ c pineapple pieces
1½ tbsp cornstarch
seasoning A & seasoning B

Cut frog meat into small pieces and then place in seasoning A for 20 minutes. Rinse with egg white and deep fry in oil until it turns orange. Remove. In a prewarmed frying pan add 2 tbsp oil and when warmed add the onions and sauté until the flavor comes out, then add the bell pepper, carrot and pineapple, and sauté. Add seasoning B and fried meat. Sauté and let it boil a few minutes and then add the cornstarch mixed with the water. Sauté a few seconds more and serve.

Seasoning A

6 tbsp cornstarch
½ tsp sugar
1 tbsp soy sauce

Seasoning B

3 tbsp sugar
3 tbsp water
⅓ tsp salt
3 tbsp vinegar
3 tbsp ketchup

FROG LEGS FORESTIERE

12 pr sm frog legs
2 tbsp plain flour
3 scallions, finely sliced
juice of 1 lemon
1 c + 1 oz rum
1 c water
8 ozs butter
1 sm bell pepper, chopped
2 c mushrooms, sliced
2 tbsp parsley, chopped
salt and pepper to taste
cayenne pepper to taste

Divide legs into 2 and parboil them for 3 minutes and then let them cool. Mix the cup of rum with the water in a glass dish. Add salt and pepper and marinate for 1 hour. Remove, drain and dry them. Rub with salt and peppers. Cook them in butter until golden brown on both sides and set aside. Add flour to the butter in the skillet and blend well. Add the bell pepper, scallions, mushrooms, parsley and the ounce of rum and cook until the vegetables are soft. Add the lemon juice and season with salt and pepper. If the sauce is too thick add a little hot water. Place the frog legs in the sauce and cook until heated through. Arrange on warm serving plates and pour the sauce over them. Serves 4.

FROG LEGS WITH PEARS

6 pr medium frog legs
1 tsp sherry
1½ lbs hard pears
salt
1 tsp ginger root juice
1 tsp cornstarch
½ tsp sugar
few drops sesame oil
½ pt veal or chicken stock
2 tsp water
3 tsp oil

Peel, core and slice the pears. Cut each frog leg into 2 pieces and put in hot oiled pan together with ginger root juice, sugar, sherry and salt to taste. Cook for 2 minutes. Remove legs and put in a dish. Add pears to the pan and cook for 2 minutes. Add frog legs and stock together and cook for 2 minutes. Add the cornstarch diluted with the water to make a thin paste and the sesame oil. Cook for 1 minute and serve. Serves 6.

FROG LEGS WITH TOMATOES & GREEN PIMENTOS

6 pr frog legs, boneless
¾ pt hot chicken stock
¾ scallion, chopped
1-2 slices green ginger root, pounded
4 tbsp oil
2 tbsp cold water
2 lge green pimentos
2 tbsp soy sauce
¼ tsp pepper
2 tbsp cornstarch
1 clove garlic, crushed
2 tomatoes, quartered

Toss the legs in a hot oiled frying pan for 1 minute and remove. Cut pimento into 6-8 pieces discarding seeds and fibers. Heat the garlic, oil, salt, pepper, pimentos and ginger in the frying pan. Add the frog legs and cook for 3 minutes. Add pimentos mix, then stock and stir well. Cover and allow to come up to a boil and then simmer 3-4 minutes shaking the pan from time to time. Add tomatoes and mix well and cook for 2-3 minutes. Mix soy sauce, cornstarch and cold water and pour into the pan. Cook awhile stirring all the time for 1-2 minutes until the sauce thickens. Add scallions and serve. Serves 6.

FROG LEGS TERIYAKI

2 lbs frog legs
2 tsp cornstarch
¼ c dry sherry
4 sprigs parsley, chopped
¼ c soy sauce
½ c chicken stock
1 tbsp water
1 tbsp sugar
1 tbsp dry mustard
1 tbsp oil

Preheat the broiler to its highest setting. In a small bowl mix the mustard with just enough hot water to make a thick paste. Set aside for 15 minutes. Combine the soy sauce, sherry, chicken stock and sugar in a small saucepan and bring to a simmer over moderate heat. Stir the combined cornstarch and water into the sauce and cook over low heat stirring constantly until the mixture thickens to a clear syrupy glaze. Immediately pour into a dish and set aside. Pat the frog legs completely dry. With a pastry brush, spread the oil over the rack of the broiler pan. Place the frog legs on the rack and brush them with about 2 tbsp of the glaze. Broil 4 inches from the heat for 6-8 minutes while brushing the frog legs 3-4 times with the remaining glaze and spoon a little over each serving of frogs. Sprinkle with parsley. Serve.

FROG LEGS & MUSHROOMS

1 lb sm frog legs
4 mushrooms
1 tbsp sherry
salt to taste
1 tsp cornstarch
2 scallions, chopped
¼ c chicken stock
6 water chestnuts
1 sm bell pepper
½ tsp sugar
4 tbsp oil
2-3 slices fresh ginger
1 clove garlic
1 tsp water

Separate the frog legs and if they are large cut each leg in two. Heat the oil in a frying pan and sauté the frog legs with garlic and ginger for 2 minutes. Then add the scallions and the sherry and cook for another 3 minutes. Remove the frog legs, garlic and ginger and discard the garlic and ginger and keep the legs warm. Slice the mushrooms and water chestnuts and cut the bell pepper into 1 inch square pieces and add all to the hot oil and cook for 2 minutes. Remove the vegetables and put them with the frog legs. To the oil in the pan add the chicken stock, sugar and cornstarch mixed with the water. Cook this sauce over a moderate flame until it is smooth and slightly thickened. Add the frog legs and vegetables and cook briskly for 2 minutes stirring frequently. Serve immediately.

SWEET SESAME FROG LEGS

12 frog legs
¼ cup butter
¼ tsp salt
½ c soy sauce
⅔ c honey
sesame seeds, toasted

Put frog legs in soy sauce for 4-5 hours. Remove and pat dry reserving ¼ cup soy sauce. Melt butter and combine with honey, salt and soy sauce. Dip legs in honey mixture coating each well. Roll legs in toasted sesame seeds and place directly on the smoker grid or a sheet of greased heavy duty foil. Use 5 pounds charcoal, 3 quarts hot water, 1 wood stick and smoke for 1-2 hours. Serves 6.

BARBECUED FROG LEGS

12 frog legs
⅓ c catsup
1 tbsp brown sugar
1 tsp prepared mustard
⅔ c seafood cocktail sauce
2 tbsp worcestershire sauce
¼ c bell pepper, minced

Rinse frog legs and dry. Place in a shallow heat-proof dish or foil pan. Combine all other ingredients and pour over frog legs. Let stand 3-4 hours in a refrigerator. Place container on smoker grid and water smoke as above recipe. Serves 6.

GRILLED MARINATED FROG LEGS

6 pr frog legs
1 clove garlic, minced
¼ c butter

MARINADE

½ c vegetable oil
2 tbsp snipped fresh parsley
1 tbsp grated lemon peel
1 tsp salt
3 tbsp finely chopped red onion
1 tbsp + 1½ tsp lemon juice
1 tsp dry mustard
1 tsp dried basil leaves

Arrange the frog legs in a single layer in an 11 x 7 inch baking dish. In a small mixing bowl combine the marinade ingredients and reserve ⅓ cup and cover with plastic wrap and chill. Pour remaining marinade over frog legs turning to coat. Cover with plastic wrap and chill for 3 hours, turning occasionally. Spray cooking grate with nonstick vegetable cooking spray and prepare the grill for medium direct heat. Drain and discard the marinade from the frog legs. Arrange the legs on the grill and grill uncovered for 3 minutes. Turn legs over and grill for 2-5 minutes longer or until meat is no longer pink and begins to pull away from the bones. In a 1 qt saucepan combine the reserved marinade with the butter and garlic and cook over a medium heat for 1-2 minutes or until the mixture is hot and butter is melted, stirring frequently. Before serving pour the garlic butter mixture over the legs. Serves 3-6.

SMOKED FROG LEGS

frog legs
salt and pepper
butter

Apply salt and pepper liberally. Cold smoke at 75-85°F for about 1 hour, or until the meat takes on a rich, golden color. Then raise the smoker temperature to 225-250°F and roast until done, basting carefully with butter.

FROG LEGS OVER CHARCOAL

frog legs, 2 pr/person
salt and pepper to taste
msg
oil
milk
cayenne pepper to taste
plain flour
sauce

If legs are frozen thaw in milk almost to cover. Drain, shake well in a paper bag with flour, salt, peppers and msg. Shake off excess. Make a good hot fire with charcoal and fill a kettle half full with oil and fry legs until golden. Combine the sauce ingredients and heat through. Serve the legs on a large platter with the sauce drizzled over them.

SAUCE

1 stick butter
2 tbsp lemon juice
1 clove garlic, crushed
½ tsp tabasco sauce
1 tsp worcestershire sauce

Mix ingredients thoroughly.

ROSEMARY BARBECUED FROG LEGS

frog legs
worcestershire sauce
lemon juice
rosemary leaves
butter, melted
garlic, minced
lemon slices
french bread

In a large rectangular pan lay the legs flat across the bottom. Pour your sauce over the frog legs. For the sauce combine all the ingredients using lots of melted butter. Place in the oven at 400°F and cook until the meat starts to fall off the bones. Serve in a bowl and use french bread for dipping in the sauce.

E. D.'S BARBECUED FROG LEGS

12 frog legs
1 c italian salad dressing
1 c barbecue sauce, spicy
garlic powder
cajun seasoning

Season legs with seasoning and garlic and marinate in the dressing and barbecue sauce overnight. Place on a low to medium grill and grill for 30-40 minutes depending on heat and size of legs. Add more barbecue sauce 5 minutes before removing from grill. Serve hot.

GLORIFIED FROG LEGS

10-12 frog legs
½ c oil
4 tbsp salt
1 lge onion, chopped
¼ c sour cream
3 pts water
1 4 oz can mushrooms, sliced
pepper to taste
1 10¾ oz can clam chowder soup
1 c plain flour
3 tbsp white wine
2 ozs butter

Mix salt with water and soak frog legs for 2 hours or longer. Drain and add pepper and roll in the flour. In a skillet heat the butter and oil. Brown legs quickly, about 5 minutes each side. Pour off drippings and place the legs in a baking dish. In same skillet sauté the onion and drained mushrooms. Add the soup, wine and sour cream and mix well. Pour over the frog legs. Bake in 250°F oven for 30 minutes.

FROG LEGS IN CASSEROLE

18 pr med frog legs
12 blades chives
2 bay leaves
1 whole clove, bruised
2 tbsp brandy, warmed
salt and pepper to taste
pinch nutmeg, grated
2 tbsp truffles, chopped
1 c white wine
6 sprigs parsley
1 sprig thyme
¼ c butter
1 c cream, scalded
pinch cayenne pepper
2 tbsp sherry
toast

Bone the legs and marinate the meat 1 hour in white wine, chives, parsley, bay leaves, thyme and clove stirring frequently. Drain and dry the meat and in a skillet with the butter sauté the meat for 10 minutes. Transfer to a heated casserole and pour over it the brandy and ignite. When the flame dies add the cream seasoned with salt, peppers, nutmeg and sherry. Add truffles if available and cover the casserole and bake in a 350°F oven for 25 minutes. Serve in the casserole with freshly made toast.

FROG LEGS SAUTÉ DEMI BORDELAISE

2 lbs frog legs
salt and pepper to taste
1 tsp garlic, minced
½ c plain flour
¼ c olive oil
2 tbsp parsley, minced
2 tbsp onions, grated

Coat frog legs in seasoned flour. Sauté in olive oil until tender, about 15-20 minutes, depending on size. Place frog legs on heated serving platter. In the same pan used for cooking frogs, sprinkle the onion, parsley and garlic. Cook a minute or so and pour over the frog legs. Serve piping hot with apricot brandy Pears for a compliment. Serves 4-6.

APRICOT BRANDY PEARS

4 fresh pears
¼ c apricot brandy
⅓ c slivered almonds
1¼ c water
¾ c sugar
toasted chocolate shavings
½ c apricot jam
few dashes salt

Peel and cut in halves and core the pears. In saucepan combine sugar, water and salt and bring to a boil. Add the pears and simmer uncovered for 8 minutes turning twice. Chill fruit in syrup. Sieve the jam and mix with the brandy. Serve pear halves topped with the jam mixture, slivered almonds and chocolate shavings.

FROG LEGS SAUTÉED BORDELAISE

frog legs
salt and pepper to taste
mushrooms, sliced
bordelaise sauce
parsley, chopped
butter
shallots, sliced
1 glass claret
1 clove garlic, crushed

Season frog legs with salt and pepper and fry lightly in butter. Add some shallots and mushrooms and moisten with the claret and some bordelaise sauce. Cover and let simmer about 5 minutes. Then add the garlic, a piece of fresh butter and some parsley. Dress the frog legs in a chafing dish and serve very hot.

BORDELAISE SAUCE

1½ c red wine
thyme
3 oz espagnole sauce*
1 tbsp melted meat glaze
2 ozs beef marrow
2 tbsp shallots, chopped
½ bay leaf
coarsely ground pepper
juice ¼ lemon

Combine red wine. shallots, thyme, bay leaf, and pepper and reduce to cup. Add the espagnole sauce and simmer 15 minutes, skimming from time to time, and strain through a cloth. Season to taste with the meat glaze, lemon juice and marrow which has been diced and poached.

MEAT GLAZE

Reduce clear and greaseless brown stock* by boiling and straining through cheesecloth several times. When the glaze is reduced (when it coats a spoon) pour into small jars and keep in the refrigerator.

CREAMED FROG LEGS

2 lbs frog legs
1 c sherry
2 c white sauce
hot water
toast or patty shells

Steam legs until tender placing them on a rack in a pan with hot water covering the bottom of the pan but not touching legs. Cover and steam in oven at 350°F. Remove frog legs from the oven and place in a heat proof casserole with the white sauce that has the sherry added. Heat through in 350°F oven. Serve on the toast or in patty shells. Serves 6.

WHITE SAUCE

2 tbsp butter
1 c scalded milk
few grains black pepper
2 tbsp plain flour
¼ tsp salt

Melt the butter and add the flour mixed with the seasonings and stir until well blended. Pour the milk in gradually, stirring constantly. Bring to a boil and cook for 2 minutes.

FROG LEGS IN CREAM SAUCE

8 pr frog legs
2 tbsp scallions, minced
salt and pepper to taste
1 c heavy cream
1 egg yolk, lightly beaten
⅛ tsp cayenne pepper
4 tbsp butter
1 lb mushrooms, quartered
½ c dry white wine
2 tsp plain flour
juice of ½ lemon
2 tbsp chives, minced

Pat the frog legs dry and cut off and discard the feet. Heat 2 tbsp butter in a skillet large enough to hold the legs in 1 layer. Add the scallions and cook briefly. Add the legs, mushrooms, salt and pepper; cook, stirring, about 1 minute. Add the wine and cook for 5 minutes. Transfer the legs to a warm serving platter. Bring the cooking liquid to a boil over high heat and cook down to a cup. Add the cream and cook about 1 minute. Blend the flour with the remaining butter and, when thoroughly mixed add gradually to mixture. Add the egg yolk. Remove the sauce from the heat and stir rapidly. Add the lemon juice and the cayenne pepper. Pour and scrape the sauce over the legs. Serve with sprinkled chives. Serves 4.

FROG LEGS WITH TOMATO SAUCE

8 pr frog legs
2 tbsp olive oil
1 bay leaf
½ c plain flour
juice of ½ lemon
1 tbsp garlic, minced
6 tomatoes, ripe
salt and pepper to taste
½ c milk
¾ c peanut or corn oil
4-6 tbsp butter
2 tbsp parsley, minced

Pat the frog legs dry and cut off the feet. Peel the tomatoes, cut in half, and remove the seeds, and chop (there should be about 3 cups). Heat the olive oil in a heavy skillet and add the tomatoes, salt, pepper and bay leaf. Cook while stirring about 20 minutes or until the sauce is thick. Pour the milk on the frog legs. Season the flour with salt and pepper; drain legs and dredge in flour, shaking off excess. Heat peanut or corn oil in 1 or 2 large skillets, adding more oil if necessary. Add the frog legs and cook over moderately high heat about 3-4 minutes on each side or until golden brown. As the legs cook transfer them to a warm serving platter. Spoon tomato sauce in the center and over the frog legs. Sprinkle with lemon juice. Heat butter in a skillet and when it is hot add garlic. Cook briefly, swirling it around. Pour the butter over the legs and serve sprinkled with parsley. Serves 4.

SAVORY FROG LEGS IN OYSTER SAUCE

8 frog legs
4½ tsp salt
2-3 c boiling water
2 tsp sugar
½ lemon
1 tbsp powdered cinnamon
2 tbsp soy sauce
½ tbsp powdered ginger
¼ c oyster sauce*
¼ tsp pepper
1 c chicken broth

Rinse the legs and place in a saucepan; barely cover with boiling water. Squeeze lemon into the pan and then drop it in. Cover and bring to the boiling point. Lower heat and simmer 5 minutes, then drain. Cover frogs with mixed soy sauce, oyster sauce, chicken broth, ½ tsp salt and sugar; bring to a boil, then lower heat and cook 20 minutes or until done. Drain and cool. Mix remaining 4 tsp salt, cinnamon, ginger, and pepper in a small frying pan or chafing dish. Stir or shake until hot, then dip the legs until coated evenly. Serves 4.

SAVORY FROG LEGS

4 pr lge frog legs
¾ c cream
3 tbsp butter
2 tbsp plain flour
1 c milk
1 tbsp parsley, minced
dash of pepper
⅓ tsp salt

Soak frog legs in milk for 30 minutes, drain, pat dry and then roll in seasoned flour*. Sauté in butter about 10 minutes turning to brown both sides. Lift out legs and to the butter remaining in the pan add leftover seasoned flour. When smooth pour in cream gradually; bring to boiling point, stirring constantly. Return frog legs to this sauce and cook for 1 minute and then add the parsley. Serves 4.

LOUISIANA FROG LEGS

12 frog legs
2 tbsp apple vinegar
4 cloves
½ tsp salt
1 tbsp cornstarch
½ c canned tomatoes
2 tbsp brown sugar
2 c mushrooms
1½ c water
few grains mace
3 tbsp butter
1 tbsp cold water
1 tbsp lemon juice
salt and pepper to taste

Broil the legs with 1 tbsp of the butter until slightly browned. Cook the mushrooms 5 minutes in the vinegar and water with the cloves, mace, and the ½ tsp salt and slice them and put in a sauce pan with the remaining butter and sugar and sauté for 5 minutes. Add the cornstarch blended with the tbsp of water and then the tomatoes, lemon juice and season to taste with salt and pepper. Put the legs on a platter and pour the sauce over them. Serve with rice.

MISSISSIPPI FROG LEGS

8 pr frog legs
½ bottle white maconnais
garlic salt
lemon pepper seasoning
dash lemon juice
1 4 oz can mushrooms, sliced
1 stick butter
½ c olive oil
2-4 cloves garlic, chopped
pimento strips
parsley, chopped

Place the frog legs in a large sauté pan so that they are in 1 layer. Add the wine and sprinkle with garlic salt, lemon pepper seasoning and a dash of lemon juice. Cover and bring to a boil and then simmer gently for about 5 minutes. Remove the legs and arrange them carefully in an ovenproof serving dish. Discard the marinade. Melt butter and add the olive oil and heat. Sauté the garlic for a minute or so. Pour over the frog legs and garnish with pimento, parsley, and mushrooms. Place in a 350°F oven just long enough to heat well. Serve.

FROG LEGS A LA KING

12 pr large frog legs
1 tsp msg
2 c milk
⅛ tsp cayenne pepper
1 tsp prepared mustard
½ c butter
1 tsp worcestershire sauce
¾ c heavy cream
6 tbsp plain flour
¼ c parsley, chopped
1 tsp salt

Cover frog legs with combined milk, worcestershire sauce, and mustard. Soak 15 minutes. Combine 2 tbsp flour with salt, msg and pepper. Roll frog legs in flour mixture. Sauté in butter until browned, turning often, about 5 minutes. Remove legs from pan. Stir in remaining flour. Add cream gradually stirring on low heat until smooth and thickened for about 5 minutes. Add legs to sauce and cook slowly about 7 minutes longer. Add the parsley. Serves 4-6.

FROGS-RAYNE STYLE

12 frog legs
3 tbsp chives, chopped
¼ c dry white wine
1 tbsp worcestershire sauce
⅛ tsp cayenne pepper
6 tbsp butter
2 tbsp parsley, chopped
½ c plain flour
2 tbsp tarragon, chopped
1 tbsp lemon juice
1 tsp salt
3 cloves garlic, minced
1 tsp msg

Melt butter in skillet and add the garlic, lemon juice and worcestershire sauce. Add frog legs coated with flour, turning to prevent sticking. Slowly brown on all sides 6-8 minutes. Add chives, tarragon, parsley and seasonings and cook 2 minutes. Add wine and cook 1 minute. Serve immediately. Serves 4.

FROG LEGS IN CREOLE SAUCE

12 frog legs
dash of cayenne pepper
6 ozs fine bread crumbs, plain
1 tsp msg
6 ozs white wine
½ tsp tabasco sauce
salt and pepper to taste
4 ozs butter
2 eggs, well beaten
1 clove garlic, crushed

Marinate the legs overnight in a refrigerator. Lift out and dry. Dip in white wine and dry again. Melt the butter in a saucepan and add the tabasco sauce and garlic. Blend the dry ingredients into the bread crumbs. Dip legs in the beaten eggs and then into the bread crumbs. Repeat. Sauté the legs in butter, turning so all sides get brown, 5 minutes on each side. Serve on a dish with the creole sauce poured over them.

MARINADE

⅓ pt white wine
2 bay leaves
1 sprig thyme
6 sprigs parsley
12 sprigs chives
1 clove garlic, bruised

Mix all ingredients thoroughly.

CREOLE SAUCE

2 ozs butter
1 clove garlic, minced
½ lb tomatoes, peeled
2 red pimentos, sliced
salt and pepper to taste
1 8 oz can sliced mushrooms
2 bell peppers, sliced
2 med onions, chopped
2 tbsp tomato paste

Melt the butter in a saucepan and sauté the onions and peppers until the onions are transparent. Blend all the remaining ingredients in the top of a double saucepan. Add the onions and pepper mixture. Cook covered for 1 hour. Season to taste before serving.

BROILED FROGS

12 whole frogs, dressed
1 c milk
8 oz butter
2 tbsp parsley, chopped
salt and pepper to taste

Dip frogs in milk. Season with salt and pepper and place in a baking dish. Cover frogs with cut up chips of butter and place in 400°F broiler. Turn over when brown, baste and cook 10-20 minutes until brown on the other side. Garnish with chopped parsley and serve with lemon sauce or creole tartar sauce and toast. Serves 6.

CREOLE TARTAR SAUCE

1½ c olive oil
3 egg yolks
½ c scallions, chopped
1½ tsp salt
½ c dill pickles, chopped
½ c parsley, chopped
1 tbsp brown mustard
¼ tsp cayenne pepper

With a wire whisk beat the egg yolks vigorously in a deep bowl for about 2 minutes until they thicken and cling to the whisk. Beat in ½ cup of the oil, ½ tsp at a time, making sure each addition is absorbed before adding more. By the time ½ cup has been beaten in, the sauce should have the consistency of thick cream. Pour in the remaining oil in a slow thin stream beating constantly. Add the mustard, pepper, and salt and continue to beat until the sauce is smooth. Then stir in the scallions, including some of the top, parsley, and pickles and taste for seasoning. The creole Tarter sauce may be served immediately or, if you prefer, it may be covered tightly and refrigerated 2-3 days before serving. Makes 2-3 cups.

LEMON SAUCE

3 tbsp butter
1 c hot water
2 tbsp lemon juice
1 tbsp plain flour
salt and pepper to taste

Melt 2 tbsp butter in a saucepan but do not brown. Add flour and mix well. Slowly add the water, stirring thoroughly. Then smooth and add remaining butter. When smooth again add seasonings and lemon juice. Serve over frogs.

BROILED FROG LEGS

6 frog legs
1 tbsp salad oil
lettuce leaves or parsley
salt
olives
1 c boiling lemon juice
sliced lemon
salt and pepper to taste

Scald legs well in the lemon juice with some salt. Dry. Mix salt, pepper and salad oil together. Rub legs thoroughly with this mixture rolling them over and over. Place on a double-wide broiler being careful to turn legs to prevent scorching. When done place on a platter of lettuce leaves or parsley and garnish with sliced lemon and olives. Serves 3.

MAÎTRE D'HÔTEL FROG LEGS

18 pr med frog legs
salt and pepper to taste
parsley, chopped
juice of 2 lemons
oil
maître d'hôtel butter*

Wash legs in cold water and wipe dry and put them in a bowl, sprinkle with salt and pepper and pour over them the lemon juice and marinate for 30 minutes turning occasionally. Drain and dry them again. Brush with oil and broil them 2 inches from broiler 4-5 minutes on each side, or until golden brown. Arrange them on a heated serving dish and pour over them a little maître d'hôtel butter. Sprinkle with a little parsley.

FROG LEGS FEAST

2 pr lge frog legs
lemon butter sauce

Marinate the frog legs in lemon butter sauce for 30 minutes, turning several times. Drain and reserve the marinade. Place frog legs on a greased steak broiler and grill about 3 inches from the coals, about 5 minutes each side, or until tender. Baste often with the marinade. Serve with hot lemon butter sauce.

LEMON BUTTER SAUCE

¼ c butter
¼ tsp salt
¼ tsp paprika
¼ c lemon juice
⅛ tsp pepper
¼ c parsley, chopped (optional)

Combine all ingredients in a small saucepan and place over a medium heat, stirring to blend, and heat through. If desired add parsley. Makes about 1 cup sauce.

LEMON LIME BROILED FROG LEGS

1 doz frog legs
1 tbsp soy sauce
salt to taste
juice of 2 lemons
1 tbsp parsley, chopped
1 lb butter
1 tbsp worcestershire sauce
1 tsp tabasco sauce
juice of 2 limes

Broil frog legs plain. Combine all other ingredients and heat but do not boil. Pour sauce over frogs when they are done. Serve.

STEAMED FROG LEGS

24 frog legs
¼ c scallions, minced
salt and pepper to taste
dry white wine
1 c fresh mushrooms, minced
¼ c parsley, minced
plain flour

Flour the frog legs and sauté in butter until delicately brown on all sides. Season. Sprinkle with parsley and brown thoroughly. In a separate skillet, sauté the mushrooms and shallots, and sprinkle with wine. When the vegetables are tender add the legs, cover, and let them steam for a few minutes. Serve immediately.

SAUTÉED FROG LEGS IN LEMON BUTTER

8 lge or 16 sm to med frog legs
½ c plain flour
1 tsp lemon rind, grated
1 tbsp onion, minced
cayenne pepper
salt and pepper to taste
7 tbsp butter
1 tbsp lemon juice
1 lemon, quartered

Dry and sprinkle the legs with salt and pepper and coat evenly with flour. Heat 4 tbsp butter in a large skillet and add the frog legs and sauté gently 5-6 minutes on each side. In a small skillet melt the remaining butter and stir in the rind, lemon juice, onion and a pinch of cayenne pepper mixing well and then spoon over the legs. Cover the skillet and simmer 6-8 minutes or until the legs are tender. Serve at once with lemon quarters. Serves 4.

FROG LEGS WESTFIELD

6 pr frog legs
1 c white wine
garlic powder to taste
2 tbsp parsley, minced
paprika
olive oil
salt and pepper to taste
2 tbsp lemon juice
1 4 oz can mushrooms, sliced
cornstarch

Coat the skillet with oil and add the wine. Sprinkle the legs with salt, pepper and garlic powder and place in the skillet. Sprinkle with lemon juice and parsley and cook at low heat (250°F), covered, for about 20 minutes or until tender. Remove to a warm platter. Add mushrooms to remaining juices and thicken with cornstarch. Pour sauce over the legs and sprinkle with paprika. Serve.

PEPPERGRASS FROG LEGS

2 lbs frog legs
½ cup cornmeal
½ c safflower oil
1 egg, lightly beaten
¼ tsp peppergrass seeds
1 lemon, cut in wedges (optional)

Combine cornmeal and peppergrass seeds with the egg to make a batter. Coat each leg in the batter and drain off excess. Heat oil in a large skillet over a medium-high heat but do not allow to smoke. Carefully place the legs in the oil and sauté 20 minutes, turning to brown all sides. Serve with the lemon wedges. Serves 2-4.

FROG LEG SAUTÉ WITH VERMOUTH

4 pr frog legs, separated
4 tbsp butter
½ c shallots, chopped
½ c beef stock
seasoned flour*
2 tbsp oil
¼ c wine or dry vermouth
1 tsp lemon juice

Dry frog legs and dredge in flour. Sauté in butter and oil until brown on all sides. Remove to a heated platter and keep warm. Sauté shallots in the same skillet until tender. Add wine and stock to make a glaze. Cook over medium heat for several minutes; add the lemon juice and spoon over the warm frog legs. Serves 2.

FROG LEGS SAUTÉ ITALIENNE

8 pr med frog legs
3 tbsp parsley, chopped
1 onion, minced
plain flour
4 ozs mushrooms, sliced
milk
salt and pepper to taste
½ c olive oil

Soak the frog legs in milk. Drain, dry, and roll them in flour. Sauté in olive oil. Just before they are done, add mushrooms and onions to the pan and cook until done. Salt and pepper to taste and serve sprinkled with parsley. Serves 4.

FROG LEGS SAUTÉ FINES HERBES

8 pr med frog legs
6 ozs butter
milk
1 c plain bread crumbs
plain flour
½ c olive oil
2 tbsp parsley, chopped
salt and pepper to taste
2 tbsp chives, chopped
lemon wedges
2 tbsp tarragon, chopped

Soak the frog legs in milk. Drain, dry and roll in flour. Add herbs to bread crumbs which has been toasted with 2 ozs butter. Place remaining butter and olive oil in a skillet. Heat and sauté the legs very quickly and, when they are nicely browned, add the herbed crumbs. Mix these well with the frog legs, salt and pepper to taste, and serve with lemon wedges. Serves 4.

FROG LEGS SAUTÉ BOURGEOISE

6 frog legs
white pepper
salt
1 8 oz can sliced mushrooms
2 ozs butter
dash of brandy
1 c crab meat
1 pt oysters, drained

Fry frog legs in the butter and cut off meat when done. Dice. Sauté frog meat with crab meat (preferably white lump meat) for 5 minutes. Add drained mushrooms and brown. Add oysters and season with salt and pepper to taste. Cook until the oysters curl. Add a dash of brandy and let steep for a few minutes. Serves 6.

FROG LEGS VINAIGRETTE

12 pr small frog legs
½ c white wine
½ c water
green olives
pickles
salt and pepper to taste
romaine or boston lettuce
vinaigrette sauce
2 hard boiled eggs, quartered

Poach frog legs in a mixture of the wine and water for 5 minutes. Season to taste with salt and freshly ground pepper. Drain and chill. Serve on lettuce leaves and top with vinaigrette sauce. Garnish with the eggs, olives and pickles. Serve.

VINAIGRETTE SAUCE

1 c olive oil
½ tsp freshly ground pepper
1 tbsp scallions, chopped
¼ tsp oregano
1 tomato, peeled, seeded and finely chopped
3 tbsp parsley, chopped
1 tbsp capers
½ tsp salt
½ tsp dry mustard
¼ c white wine

Mix all ingredients together thoroughly. You may add anything you choose to this sauce: chopped pickle, hard boiled eggs, chopped olives, etc. Spoon the sauce over the frog legs. Serves 4.

FROG LEGS A LA NICOISE

12 pr med frog legs
plain flour
2 tbsp tomato fondue a la nicoise
salt and pepper to taste
2 ozs butter
2 tbsp consommé
3 tbsp parsley, chopped

Season and flour frog legs. Sauté in butter. When cooked, add tomato fondue a la Nicoise. Mix well and cook over high flame for 3 minutes. Heap in a serving dish. Pour consommé over the mixture and sprinkle with parsley. Serve.

TOMATO FONDUE A LA NICOISE

1 med onion, chopped
1 tsp tarragon, chopped
1 tsp chervil, chopped
6 tomatoes, peeled, seeded and chopped
1 clove garlic, chopped
1 tsp parsley, chopped
salt and pepper to taste
2 ozs butter

Gently cook onion in butter. When onion begins to color add tomatoes, garlic, and season to taste with salt and pepper. Cook gently until the liquid of the tomatoes has almost disappeared. Add parsley, tarragon and chervil at the last minute.

FROG LEGS A LA MIREPOIX

12 pr med frog legs
Mirepoix
2 tbsp parsley, chopped
plain flour
salt and pepper to taste
2 tbsp consommé
2 ozs butter

Season and flour the frog legs. Sauté them in butter and when they are cooked, add the Mirepoix. Cook these ingredients over a very high flame mixing well. Heap on a serving dish and pour over the mixture the consommé and sprinkle with parsley. Serve.

MIREPOIX

2 sm carrots
pinch powdered thyme
1 bay leaf
¼ c celery
½ c onions
2 ozs butter
salt

Dice extremely fine the carrots, onions and celery and cook them slowly in butter. Season with salt, thyme and bay leaf. Cook until the vegetables are very tender. Remove bay leaf.

FROG LEGS A LA LYONNAISE

12 pr med frog legs
vinegar
parsley, chopped
salt and pepper to taste
2 tbsp onion, minced
plain flour
2 ozs butter

Season the frog legs and dip them in flour. Sauté in butter. When they are well browned add the onion which has been lightly browned in butter. Sauté the frog legs and onion together. Sprinkle with parsley. Heap on a platter covered with a sauce made by heating a little vinegar in the cooking butter. Serve.

FROG LEGS DILLOISE

2 doz frog legs
salt and pepper to taste
1 oz butter
2 ozs bacon
1 pt tomato sauce
tabasco sauce

Cut frog legs in two and season with salt and pepper and sauté with butter and bacon cut in small squares. Fry for 5 minutes until the bacon is nearly crisp. Then add the tomato sauce and simmer for 10 minutes. Add a few drops of tabasco sauce and season well. Serve.

FROG LEGS PAYSANNE WITH APPLE BRANDY

8 pr frog legs
⅔ tbsp parsley, chopped
plain flour
⅔ tbsp chives, chopped
juice of 1 lemon
⅔ tbsp tarragon, chopped
2 ozs butter
2 tbsp apple brandy
2 cloves garlic, crushed
2 tbsp dry white wine
salt and pepper to taste

Sprinkle legs with lemon juice and keep cold for 1 hour. Wash, dry and dust with flour. Put butter in frying pan with garlic, a little salt, plenty of black pepper and fry briskly 2-3 minutes. Discard the garlic. Heat butter to smoking, add the frog legs and parsley, chives and tarragon. Toss legs until golden all over and then add the apple brandy. Set alight and let burn out while still stirring. Finally add the wine and stir. Serve.

FROG LEGS ITALIAN STYLE

3 doz frog legs, separated
butter
fresh mushrooms, chopped
half glaze
parsley, chopped
salt and pepper to taste
shallots, chopped
½ pt white wine
tomato sauce

Season frog legs with salt and pepper, and fry lightly in butter. Add some shallots, mushrooms, wine, some half glaze, and a little tomato sauce. Let cook for about 5 minutes. When done, season to taste and finish with a piece of fresh butter. Pour into a chafing dish and sprinkle with a little parsley. Serve.

HALF GLAZE

1 pt espagnole sauce*
1 Mirepoix*
2 tbsp fish glaze
salt and pepper to taste
1 pt brown fish stock
butter
½ pt sherry

Combine the espagnole sauce, brown fish stock, Mirepoix and fry in butter. Reduce to half. Add fish glaze, sherry, and season to taste. Strain through a fine sieve.

TOMATO SAUCE

- 2 ozs butter
- carrots, sliced
- celery, sliced
- peppercorns
- 1 bay leaf
- 4 doz lge fresh tomatoes
- salt and pepper to taste
- onions, sliced
- parsley sprigs
- 2 ozs raw ham, lean
- whole cloves
- 1 clove garlic, chopped
- 1 qt chicken or beef stock
- a little powdered sugar

Put butter into a saucepan and add onions, carrots, celery, garlic, ham, peppercorns, cloves, and bay leaf. Fry to a nice color. When done add tomatoes and stock. Season with salt, pepper, and powdered sugar. Slowly simmer for 45 minutes. Strain through a fine sieve. If sauce is too thin, reduce and thicken with a little kneaded butter* and strain again.

BROWN FISH STOCK

- butter
- onions, sliced
- celery, sliced
- bay leaves
- peppercorns
- 1 pt sherry
- 6 lbs fish bones, cut up
- carrots, sliced
- parsley sprigs
- whole cloves
- garlic, sliced
- 8 qts water

Put a large saucepot on the fire with a piece of butter. Add fish bones, vegetables, and other ingredients except sherry and water. Fry together. Drain off butter and add sherry and water and let boil for 1 hour. Strain through a cloth. When strained try to have 4 quarts of broth.

FISH GLAZE

Prepare brown fish stock. Clarify the strained stock and let it reduce to half. Strain again and let it reduce way down to a thick syrup.

LES PROVINCES FROG LEGS

- 18 pr sm frog legs, separated
- plain flour
- 1 c unsalted butter
- 2 tbsp beef stock
- 1 tsp scallion, minced
- milk
- salt and pepper to taste
- ¼ c dry white wine
- 2 cloves garlic, minced
- 1 tbsp parsley, chopped

Soak the legs in milk overnight in a refrigerator. Drain well and roll or shake frog legs in seasoned flour*. Heat about half the butter in a large heavy frying pan over medium heat. Add legs and sauté until golden brown. Transfer to a heated platter and keep warm. Wipe out the pan drippings. Return pan to medium high heat and add the remaining butter and heat until foamy. Add the wine, stock, garlic and scallions and stir well. Pour over frog legs and sprinkle with parsley and serve immediately. Serves 4-6.

GRANT'S GRENOUILLES

frog legs
2 c buttermilk
few dashes tabasco sauce
few shakes garlic salt
self rising flour
salt
3 eggs, beaten
3 dashes worcestershire sauce
lots of black pepper
oil

Salt the frog legs and set aside. Make a batter with all other ingredients except the flour and oil. Dip the legs in the batter and then roll in the flour and fry in a moderately hot deep oil until golden. Serve.

GRENOUILLES FRITES

24 sm frog legs
oil
1 clove garlic, crushed
pinch of freshly ground pepper
bouquet fried parsley
juice of ½ lemon
1 tsp parsley, chopped
pinch of salt
frying batter

Marinate the legs in the lemon juice, a little oil, parsley, garlic, salt and pepper for 1 hour. Drain and dip the legs one at a time into the batter and plunge them into hot oil. Fry until golden all around. Remove and serve them on a hot platter placed on a napkin. Decorate with the bouquet of fried parsley. Serves 2-3

FRYING BATTER

2½ c sifted plain flour
3½ tbsp melted unsalted butter
water
beer (optional)
2 eggs, beaten
salt
4 egg whites, beaten stiff (optional)

Mix the flour with the melted butter and eggs and add salt and then dilute with enough water to obtain a semi liquid batter. Prepare the batter at least 1 hour before it is to be used. Just before using it add the egg whites. You can also make the batter lighter by replacing the water with beer.

FRIED FROG LEGS IN WINE SAUCE

2 lbs frog legs, separated
¼ c milk
1 c oil
½ tsp black pepper
1 egg, beaten
1 c plain flour
½ tsp salt
french bread

Mix the egg and milk in a bowl. Pat dry the frog legs and dip in the egg mixture and then dredge in the flour, shaking off excess. Fry in the oil until they are brown. Drain and pour wine sauce over the legs. Serve with french bread. Serves 4.

WINE SAUCE

2 sticks butter
2 cloves garlic, chopped
4 tbsp worcestershire sauce
½ tsp salt
juice of 1 lemon
4 tbsp parsley, chopped
½ c white wine

Melt the butter in a saucepan and sauté the garlic until soft and then add all the other ingredients, cover, and simmer for 20 minutes.

FROG LEGS SAUTÉ SALONAISE

10 frog legs
½ sm eggplant, cut fine
1 onion, chopped
lemon quarters
salt and pepper to taste
1 c heavy cream
1 tomato, peeled and cut fine
½ tbsp butter
3 tbsp olive oil
¼ c plain flour
1 clove garlic, chopped

Rinse frog legs and wipe dry. Dip each in cream. Sprinkle with seasoning and roll in flour. Fry legs in olive oil in large skillet until golden brown. Remove to a pan and keep hot. Sauté onion, eggplant and tomato together 10 minutes in the skillet in which the frogs were cooked. Put garlic and butter in small pan and cook for 3 minutes. Add to vegetables and mix. Pour over frog legs on warmed serving platter. Serve with lemon wedges. Serves 3-4.

FROG LEGS OSBORN

18 pr med frog legs
½ c butter
1 bell pepper, julienne strips
1 c mushrooms diced
1½ c brown sauce 2
salt and pepper to taste
1 tbsp onion, minced
6 tomatoes, peeled
parsley, chopped

Season frog legs and sauté in butter over high heat until golden brown on all sides. Transfer to a heated casserole. Add to butter remaining in skillet the onion, bell pepper, tomatoes cut into eights, and mushrooms. Cook for 5 minutes, stirring constantly, and then add to frog legs. Pour over them the brown sauce, cover and bake at 400°F for about 15 minutes, or until they are tender. Serve sprinkled with parsley.

BROWN SAUCE 2

2 tbsp oil
1 onion, minced
3 tbsp tomato puree
2 tbsp plain flour
2 c brown stock
salt and pepper to taste

In a saucepan lightly brown onion in the oil. Add the flour and cook, stirring until brown. Add stock and tomato puree and cook, stirring until the sauce thickens. Continue to cook until sauce is reduced to about 1 cup. Season to taste.

BROWN STOCK

3-4 lbs beef bones
2 lbs veal bones
5 qts water
1-2 leeks
1 tbsp salt
1 roast chicken carcass (optional)
2 lge carrots, peeled, sliced
2 lge onions, sliced
5 celery stalks
4 sprigs parsley
6 peppercorns

In a roasting pan place bones cut into small pieces and spread 1 carrot and 1 onion over them. Place in a 400°F oven for 30-40 minutes or until all is a good brown. Transfer to a large kettle and add most of the water. Discard the fat from the roasting pan and add a little water and bring to a boil while scraping brown bits clinging to pan. Pour into the soup kettle. Add the celery along with the leaves, leeks, 1 onion, 1 carrot, parsley, salt and peppercorns. Add any trimmings from the beef or veal, raw or cooked, and if possible a chicken carcass. Slowly bring to a boil, removing any scum that accumulates, and simmer at least 3 hours. There should be 3 or more quarts liquid at the end of cooking. Remove bones and vegetables from kettle and strain through cheesecloth. Pour into jars and cool as quickly as possible. Remove fat from the surface. A fine sediment may settle in the bottom of the jars and in removing the stock do not disturb this. brown stock will keep 4-5 days in the refrigerator and can be used in soups or sauces.

FROG LEGS BORDEAUX

8 pr med frog legs
italian seasoning mix
½ c white onion, chopped
1 clove garlic, minced
1 tbsp unsalted butter
½ c olive oil
½ c plain flour
¼ c scallion tops, minced
1½ c Bordeaux
salt and pepper to taste

Heat the oil in a large skillet over medium heat. Sprinkle the legs with the italian seasoning mix and then dredge them in flour. Sauté in oil about 15 minutes until done. Remove to a heated serving bowl and keep warm. Pour off all but 1 tbsp oil and add the onion, shallots and garlic and sauté covered until the onion is transparent, about 5 minutes. Add the Bordeaux and bring to a boil. Boil briskly until the liquid is reduced to ½ cup. Remove from the heat and stir in the butter and pour the sauce over the legs and serve with cajun rice. Serves 4.

ITALIAN SEASONING MIX

½ tsp salt
½ tsp garlic powder
½ tsp comino
½ tsp marjoram leaves, dried
1 tsp onion powder
½ tsp celery salt
½ tsp sweet basil leaves, dried
½ tsp oregano leaves, dried

Combine and mix thoroughly all ingredients.

CAJUN RICE

2 c raw rice
1½ tbsp onion, minced
1½ tbsp bell pepper, minced
½ tsp salt
¼ tsp white pepper
¼ tsp thyme, dried, crumbled
2½ c chicken stock
1½ tbsp celery, minced
1½ tbsp unsalted butter, melted
¼ tsp garlic powder
¼ tsp cayenne pepper
¼ tsp oregano, dried, crumbled

Preheat oven to 350°F. Combine all ingredients and mix well. Cover and bake until the rice is tender, about 1 hour. Serves 8-10.

FROG LEGS GRUYERE

18 sm frog legs
½ c light cream
½ c butter
2 tsp scallions, minced
1 c dry white wine
3 lge oysters
dash cayenne pepper
½ c lemon juice
salt and white pepper to taste
1 tbsp onion, minced
½ cup Marsala
½ c mushrooms, sliced, sautéed
½ c Gruyere cheese, grated
lemon wedges

Soak legs in very cold water for 3 hours and then drain. Dip in a mixture of cream and lemon juice. Sprinkle with salt and white pepper. Melt the butter in a medium skillet and stir in scallions and onion. Add the frog legs and sauté over low heat for 3 minutes on each side. Add the white wine and ½ the Marsala. Poach gently for 15 minutes or until tender turning at least once. Remove frogs to an oven proof platter and strain pan juices into a small saucepan. Add oysters, mushrooms and rest of Marsala and simmer for 10 minutes. Season with cayenne pepper and salt to taste. Pour sauce over frog legs and sprinkle with the cheese. Bake at 400°F for 3-5 minutes until cheese is bubbly and lightly browned. Garnish with lemon slices. Serves 6.

FROG LEGS JERUSALEM

12 frog legs, boned, cut up
1 tbsp celery, chopped
3 scallions, chopped
3 ozs butter
1 c evaporated milk
salt and pepper to taste

Put in a sauté pan chopped celery, scallions and butter and simmer for about 5 minutes, then add frog legs seasoned with salt and pepper; simmer for 5 minutes. Add milk and cook for 10 minutes. Serve in a chafing dish.

FROG LEGS GREENWAY

12 frog legs
chives, chopped
4 ozs white wine
chervil, chopped
2 ozs butter
1 egg yolk
parsley, chopped
½ pt chicken broth
salt and pepper to taste
4 ozs evaporated milk

Cut frog legs in two and sprinkle with salt and pepper. Melt butter in a saucepan and add the frog legs and simmer for 5 minutes. Then add wine, chicken broth, a little chives, parsley and chervil, and cook for 5 minutes. Before serving, season well. Bind the egg yolk and milk and add. Serve.

FROG LEGS A LA NEWBURG

4 pr large frog legs
3 egg yolks, slightly beaten
1 c cream
½ c madeira
½ c consommé
cayenne pepper
salt

Drop legs in boiling salted water and simmer for 20-25 minutes or until tender. Drain. Heat consommé, add wine, season to taste, and boil 3 minutes. Mix cream and egg yolks. Add and simmer 2 minutes stirring constantly. Pour over frog legs. Serves 4.

FROG LEGS COOKED IN WHITE WINE

2 pr large frog legs
salt and pepper to taste
1 tbsp parsley, chopped
3 ozs fresh cream
pinch of sugar
1 lge white onion
1 tsp plain flour
1 4 oz can mushrooms, drained
1 tbsp butter
3 ozs dry white wine
1 tbsp chives, chopped
1 ripe tomato
juice of ½ lemon

Chop onion finely and sauté in half the butter until cooked (not brown) and add frog legs. Add 2½ ozs wine, cream, salt, pepper, mushrooms and tomato chopped in large pieces. Cook covered, slowly for 15 minutes. Mix flour with remaining butter and put in the preparation while stirring and boiling. Finish with a pinch of sugar, ½ oz wine, finely chopped chives, parsley and lemon juice. Mix well and serve. Serves 1.

CASSEROLED FROG LEGS

12 pr lge frog legs
cayenne pepper
2 tbsp brandy
salt and pepper to taste
marinade*
¼ pt cream
2 tbsp chopped truffles, if available
2 ozs butter
2 tbsp sherry
nutmeg

Marinate the legs for 1 hour. Remove and wipe them. Sauté legs in the butter for 10 minutes turning frequently. Transfer the legs to a shallow casserole and pour the brandy over and set alight. When the brandy has burnt out add the cream seasoned well with salt, pepper, cayenne and nutmeg. Add the sherry and cover with finely chopped truffles. If truffles are not available use fresh mushrooms. Cover the casserole and cook in a moderate oven at 325°F for 25 minutes. Serve the casserole with freshly made buttered toast.

FROG LEGS WITH WHITE WINE AND MELBA TOAST

12 frog legs
6 sprigs chives, chopped
1 glass white wine (8 oz)
Melba toast
3 mushrooms, chopped
1 tbsp hollandaise sauce
juice of ½ lemon
1 oz butter
2 scallions, chopped
salt and pepper to taste

Put the frog legs, scallions, mushrooms, salt, pepper, wine and butter in a saucepan. Cover and cook on a moderate heat for 10 minutes. Remove the frog legs and keep warm on a dish. Cook the wine gravy quickly for 3-4 minutes. Add the hollandaise sauce, lemon juice and chopped chives, but on no account let it reboil or it will curdle. Pour the sauce over the legs and serve with Melba toast.

FROG LEGS PROVENCAL

12 pr medium frog legs
1 tbsp chives, chopped
4 tbsp olive oil
2 cloves garlic, chopped
½ c milk
1 tbsp parsley, chopped
1 lemon
plain flour
½ lb butter
salt and pepper to taste
juice of 1 lemon

Dip frog legs in milk and then roll in seasoned flour*. Heat 2 ozs butter with about ½ the olive oil and let it get hot but not smoky. Add the frog legs and cook on both sides for about 10 minutes, until the legs are nicely browned. Sprinkle with the lemon juice, chives and parsley. Season to taste. Put on a serving dish and keep warm. Slightly brown the remaining butter and olive oil in a clean pot and add the garlic. Pour over the frog legs and serve with lemon wedges.

BRANDY FROG LEGS PROVENCAL

8 pr frog legs
salt and pepper to taste
lemon juice
2 tbsp parsley, chopped
plain flour
2 tbsp chives, chopped
2-3 tsp garlic, crushed
2 tbsp dry white wine
5 tbsp butter
1 tbsp brandy
2 tbsp tarragon, chopped

Wash frog legs well in lemon juice and water. Dry and dust lightly with flour. Heat butter in pan to foaming; add garlic. Cook 1 minute, then put in the frogs and shake the pan until they are golden brown on each side. Add salt, pepper, tarragon, chives, and parsley, and cook for 1 minute. Heat the brandy and wine and set them alight, then pour over the frogs. Serve at once in a hot dish. Serves 4.

MUSHROOM FROG LEGS PROVENCAL

12 frog legs
1 clove garlic
8 fresh mushrooms, sliced
1 sm onion, thinly sliced
4 tomatoes, chopped
1 carrot, finely chopped
¼ lb butter
1 tsp basil

Sauté lightly in butter the onion, mushrooms and frog legs. Remove the legs. Add carrot, tomatoes, garlic and basil. Simmer until it thickens in about 10 minutes. Season to taste. Pour over frog legs in buttered baking dish and bake at 350°F for 15-20 minutes. Serves 4.

LEMON FROG LEGS PROVENCAL

12 pr frog legs
juice of ½ lemon
1 clove garlic, minced
plain flour
6 tbsp butter
4 tbsp parsley, minced
milk
salt and pepper to taste
12 tomatoes, peeled
2 tbsp oil

Cut out tomato cores, seeds, and squeeze to remove as much juice as possible; coarsely chop. Heat 2 tbsp butter in a saucepan, add garlic, and sauté a few seconds. Add tomatoes and 2 tbsp parsley. Simmer uncovered until slightly thickened. Season with salt and pepper. Keep warm while preparing the frog legs. Place frog legs in a shallow dish, cover with milk, and let stand for 10 minutes. Dip in seasoned flour* to coat lightly. Melt 4 tbsp butter in a skillet until bubbly; add oil. Sauté frog legs a few at a time until lightly browned, turning often. Place on a warm platter and top with tomato sauce and parsley. Serves 6.

SPICY FROG LEGS PROVENCAL

8 pr lge frog legs
¼ c extra virgin olive oil
¼ c white wine
lemon juice
¼ c tomatoes, peeled, seeded, minced
¼ tsp thyme
salt and pepper to taste
plain flour
¼ c minced garlic
1 tbsp pernod or herbsaint or
  2 tbsp white vermouth
½ c parsley, minced
tabasco sauce

Rinse the legs and pat dry and separate. Dredge in flour and shake off excess. Heat the oil over high heat and when hot add the legs. Do not crowd. Brown well and remove and keep warm. Pour off about half the oil and add the garlic. Cook and toss for a minute and add the wine and carefully the pernod, herbsaint or vermouth. Reduce to about a tsp and add the tomatoes, parsley and thyme. Cook and toss for 5 minutes and add the lemon juice and tabasco sauce and correct the seasoning. Serve over frog legs. Serves 4.

LES CUISSES DE GRENOUILLE PROVENCALE

12 frog legs
1 egg, beaten
1 c milk
1 tbsp onion, chopped
½ tsp parsley, chopped
1 c white wine
1 tbsp bread crumbs
1½ c plain flour
1½ c water
1 c butter
½ tbsp garlic, chopped
juice of 1 lemon
salt and pepper to taste

Mix flour, salt and pepper. Combine egg, milk, ½ cup water, and beat well. Pass the frogs through the flour, dip in egg wash, then flour again and then back to egg wash twice and flour 3 times. Set on a platter separated from each other. Melt butter in a skillet large enough for half the legs. Brown on all sides and repeat until all legs are fried. Add the onion, garlic, and parsley to the butter and brown over low heat, stirring constantly. Add the lemon juice, wine, and remaining water and bring to a boil. Add the bread crumbs and place the legs in the skillet on a very low fire. Cover and let simmer until ready to serve. Serves 2.

EASY FROG LEGS PROVENCAL

18 pr med frog legs
salt and pepper to taste
lemon juice
2-3 tsp garlic, chopped
cold water
5 tbsp butter
parsley, chopped

Soak the frog legs in cold water for 2 hours; drain and dry thoroughly. Season with salt and pepper. Place 2 tbsp butter in a skillet and sauté legs on all sides until golden. Arrange on a heated platter and sprinkle with lemon juice and parsley. Add remaining butter and garlic to the pan and when the butter is nut brown pour it quickly over the legs. Serve with the garlic butter still foaming.

FROG LEGS PAYSANNE PROVENCALE

18 pr med frog legs
12 sm tomatoes, peeled, seeded
2 sprigs dill, minced
1 sm clove garlic, crushed
salt and pepper to taste
1 tbsp parsley, chopped
cold water
½ c olive oil
1½ tsp parsley, minced
½ tsp sugar
chicken stock

Soak the legs in cold water to cover for 2 hours. Drain and dry thoroughly. Heat the oil to the smoking point with dill, minced parsley, garlic, sugar, salt, and pepper. Add tomatoes and cook over high heat, stirring constantly, until it is very hot. Blanch legs in boiling chicken stock; drain well. Add to the tomato mixture and simmer gently for 15 minutes. Arrange the legs in a ring on a heated serving platter, cover with sauce, and sprinkle with chopped parsley.

WINE FROG LEGS PROVENCALE

4 pr frog legs
½ lemon
2 ozs clarified butter
1 clove garlic, minced
½ tsp tarragon, crushed
3 ozs white wine
½ c plain flour
2 ozs olive oil
2 tsp shallot, minced
2 tsp parsley, minced
salt and pepper to taste

Dredge the legs in the flour. Pour oil and butter into a pan and let it get hot but not burning. Add legs, shallots, garlic, parsley, and tarragon. Place lemon in cheesecloth and squeeze juice over the legs. Add the wine, lower heat, and simmer. Season with salt and pepper. Serves 2.

FROG LEGS PROVENCALE & RICE

8 pr frog legs
¼ c oil
2 tbsp onion, minced
½ tsp dill seeds
1 tsp parsley, chopped
1 11 oz can condensed bisque or tomato soup
seasoned plain flour*
1 clove garlic, chopped
1 tsp chives, chopped
1 tomato, peeled, chopped
½ c dry white wine
cold water
hot cooked rice

Soak legs in cold water for 2 hours. Drain and pat dry. Dust with seasoned flour and shake off excess. In a skillet heat oil; add the garlic, onion, and sauté until lightly browned. Add legs and sauté until brown. Add remaining ingredients and simmer until the legs are tender, about 15-20 minutes. Serve over hot rice.

FROG LEGS A LA MEUNIERE

12 pr medium frog legs
parsley, chopped
salt and pepper to taste
cayenne pepper to taste
parsley, chopped
3 ozs butter
juice of 2 lemons
cornstarch
milk

Parboil legs for 3 minutes, then soak them in milk for 1 hour. Drain and pat dry, season, and pass the legs through cornstarch. Sauté in butter until browned. Heap them on a heated platter. Add the lemon juice to the butter remaining in the skillet and raise heat to high and scrape the sides and bottom to dissolve any particles clinging to it. Cook until butter is lightly browned. Pour the Meuniere sauce over the legs and sprinkle with parsley. Serve at once. Serves 3-4.

FROG LEGS BELLE MEUNIERE

8 pr med frog legs, separated
¼ c scallions, chopped
¼ c white wine
juice of 1 lemon
¼ lb butter
¼ c mushrooms, sliced
3 tsp cognac
salt and pepper to taste

Salt and pepper the legs and sauté slowly in the butter until brown. Add the scallions and mushrooms and simmer for 5 minutes. Add the wine and lemon and simmer for 1 minute. Finish by adding the cognac and parsley.

MOUNTAIN CHICKEN

12 pr frog legs, separated
3 cloves garlic, crushed
1 tsp salt
1 tbsp malt vinegar
1 c plain flour
1 med onion, grated
½ tsp cloves, ground
½ tsp white pepper
1 c oil
lime halves

Mix onion, garlic, cloves, salt, pepper and vinegar in a large bowl. Add frog legs and allow to stand for 1 hour turning occasionally. Heat oil in a heavy frying pan. Pat the legs dry and pass in the flour. Fry ¼ at a time for 5 minutes on each side. Drain and serve with lime on the side. Serves 6.

FROG LEGS WITH WHITE WINE AND ARTICHOKE BOTTOMS

8 pr lge frog legs
salt and pepper to taste
¼ c peanut oil
2 tbsp scallions, minced
4 artichoke bottoms, cooked, cubed
½ tsp dill weed
2 tbsp dry sherry
¼ c chicken stock
4 tbsp butter, at room temperature
black pepper to taste
plain flour
cayenne pepper
2 tbsp butter
1 tsp garlic, minced
½ tsp tarragon
¼ c fresh tomato, minced
¼ c white wine
¼ c heavy cream
salt and white pepper to milk

Soak the legs in milk for 1-2 hours. Drain and dredge with flour seasoned with salt, black and cayenne pepper. Heat oil over moderately high heat. Cook the legs a few at a time until golden brown; remove and keep warm. Pour off the oil and add 2 tbsp butter to the pan. Add scallions, garlic, and artichokes. Add tarragon, tomato and sherry; cook 5 minutes. Add wine and stock and reduce by two-thirds. Add in the cream and reduce by half. Whisk in the 4 tbsp butter a little at a time and correct the seasonings. Serve sauce over the frog legs. Serves 4.

FROG LEGS CONTINENTAL

6 pr frog legs, separated
1 c chicken broth
1 tbsp parsley, chopped
2 cloves garlic, minced
1 tsp sugar
1 c milk
3 tomatoes, peeled
1 tsp basil
4 tbsp olive oil
salt and pepper to taste

Wash legs in cold water, then soak in milk for 1-2 hours. Dry. Poach in chicken broth for 5 minutes. Cut tomatoes in half, seed, and chop into a skillet. Add the parsley, basil, garlic, oil, and sugar. Season sauce with salt and pepper to taste; bring to a boil, then simmer for 5 minutes. Add frog legs and cook another 10 minutes. Serve on rice. Serves 2.

FROG LEGS POULETTE

- 12 pr frog legs
- 1 tbsp onion, minced
- ½ c dry white wine
- 2 tbsp plain flour
- 1 tbsp water
- 1 tbsp parsley, chopped
- salt and pepper to taste
- 2 tbsp butter
- 1 egg yolk

Soak frog legs in cold water for 2 hours. Drain and dry thoroughly. Sauté frog legs in hot butter for 2 minutes on each side. Remove. Sprinkle the flour in the remaining pan juices and blend. Add wine, salt, pepper, parsley, onions and frog legs. Simmer, covered, over moderate heat for 10-15 minutes or until tender. Remove frog legs to hot serving dish and keep hot. Beat together the egg and water. Remove sauce from heat. Add a little of the hot pan liquid to egg and blend thoroughly. Pour egg mixture into remaining sauce and stir over lowest possible heat or in top part of double boiler over hot water until thickened. Do not boil again. Pour sauce over frog legs and serve immediately. Serves 4-6.

CAYENNE FROGS A LA POULETTE

- 6 frog legs
- 2 tbsp butter
- ½ 4 oz can mushrooms, sliced
- 2 c + water
- 2 egg yolks, well beaten
- 2 tbsp plain flour
- 1 lemon
- salt and pepper to taste
- cayenne pepper
- juice of ½ lemon

Bone the frog legs and cut the meat in small pieces about 1 inch in size. Have ready a stew pot and add the meat, salt and pepper, and ½ the lemon cut up fine. Cover with water and let it cook until the meat is tender. Drain and put the meat aside and now make the sauce. Take 1 tbsp butter and the flour and rub together until smooth and put in the remaining butter to melt. Add the 2 cups of water and stir well. When it begins to boil well add the mushrooms and season with salt and cayenne pepper. Then add the frog meat and season again to taste and let this boil for 10 minutes. Take the frogs off the fire and add the egg yolks and stir in thoroughly. Add the juice from the remaining ½ lemon and serve.

GRENOUILLES A LA POULETTE

24 sm frog legs
few drops lemon juice
salt and pepper to taste
1 bouquet garni*
3 egg yolks, well beaten
parsley, chopped
white wine
5½ tbsp unsalted butter
1 onion, julienne strips
¼ lb mushrooms, sliced
½ c cream

Poach the frog legs in a small amount of wine mixed with a few drops of lemon juice and 2 tbsp melted butter. Season with salt and pepper, onion and a bouquet garni. As soon as the liquid comes to a boil add the mushrooms. Cook slowly, covered, over a low flame. Drain the frog legs and mushrooms; keep warm. Reduce the liquid and strain through a double cheesecloth. Thicken with egg yolks and cream. The sauce must be quite thick, so be sure the cooking liquid is sufficiently reduced before adding the thickening ingredients. After the sauce has thickened, add the frog legs and mushrooms and 3½ tbsp unsalted butter. Mix well and correct the seasoning; pour very hot into a dish. Sprinkle with parsley. Serves 2.

MUSHROOM FROG LEGS POULETTE

24 frog legs
salt and pepper to taste
chicken stock
2 egg yolks, well beaten
1 doz mushrooms, sliced
butter
½ pt white wine
kneaded butter
¼ c cream

Fry the frog legs lightly in butter in a sauté pan and season with salt and pepper and moisten with the wine and some chicken stock. Cover and let simmer for 5 minutes. Place the legs in a chafing dish. Reduce the stock and thicken with kneaded butter. Let boil for a few minutes and then thicken with the egg yolks and cream with a piece of fresh butter. Strain the sauce and add the mushrooms and pour the sauce over the legs and serve very hot.

KNEADED BUTTER

1 lb butter
cayenne pepper
½ lb plain flour, sifted
salt
nutmeg, grated

Put butter in a bowl and let it get a little soft and then beat it well with a whip. Add the seasonings and incorporate slowly the flour.

FROG LEGS POULETTE WITH FISH STOCK

18 pr med frog legs
1½ c cold milk
1 sm onion, minced
1½ tbsp plain flour
¼ c fish stock*
salt and pepper to taste
1 tbsp parsley, minced
3 egg yolks, slightly beaten
patty shells
1 c salted water
1 sm carrot, minced
3 tbsp butter
¼ c white wine
1 tbsp lemon juice
pinch cayenne pepper
½ c + 2 tbsp light cream
1 tbsp parsley, chopped

Marinate the legs in the salted water and milk for 2 hours. Drain and dry the legs. Sauté lightly the carrot and onion in 2 tbsp butter and then add the legs and sauté for 5 minutes, stirring constantly. Sprinkle the mixture with the flour and blend it well and add the wine, stock, and lemon juice. Season with salt, pepper, cayenne pepper and chopped parsley. Bring to a boil, stirring constantly, and lower the heat and simmer gently for 15 minutes. Bone the legs and keep the meat hot. Strain the sauce through a fine sieve into another saucepan and add the ½ cup cream and bring to a boil twice and remove the pan from the heat. Stir in the egg yolks mixed with the 2 tbsp cream slowly. Finish the sauce with 1 tbsp butter and minced parsley. Heat the meat in the sauce but do not bring to a boil and serve the mixture in individual ramekins or in patty shells.

FROG LEGS MARINER STYLE

24 frog legs
salt and pepper to taste
few fresh mushrooms, chopped
a little chicken stock
2 tbsp cream
butter
4 scallions, minced
½ pt white wine
kneaded butter*

Fry the legs lightly in butter in a sauté pan and season with salt and pepper. Add scallions and mushrooms, and moisten with the wine and stock. Cover and simmer for 5 minutes. Place legs in a chafing dish. Reduce the stock and thicken with a little kneaded butter and let it boil for a few minutes. Finish the sauce with the cream and a piece of butter. Pour the sauce over the legs and serve very hot.

FROG LEGS A LA BECHAMEL

12 pr medium frog legs
1 tbsp white wine
few tbsp thick bechamel sauce
  mixed with some fresh cream
2 ozs + 1 tsp butter
salt and pepper to taste

Season frog legs with salt and pepper. Put frog legs in a frying pan with butter and white wine. Cover and cook. Add a few tbsp thick bechamel sauce and simmer for a few moments. At the last minute add 1 tsp butter. Serve in a pyramid on a platter.

BECHAMEL SAUCE

¾ lb white roux*
a little nutmeg
salt and pepper to taste
6 ozs lean veal, diced
2½ qts milk
1 sm bouquet garni*
2 ozs onion, sliced
butter

Moisten roux with boiled milk and mix well. Add to the sauce the veal which has been cooked in butter until white with the onion. Season with salt, pepper, nutmeg and the bouquet garni. Simmer for 1 hour and then strain. Keep hot in a double boiler and dabbing the surface of the sauce with butter to prevent a skin from forming. veal can be left out if desired.

FROG LEGS AU GRATIN AND MUSHROOMS

12 pr med frog legs
4 ozs butter
cream sauce
8 ozs sliced mushrooms, drained
plain bread crumbs
salt and pepper to taste

Season the frog legs and sauté quickly in smoking hot butter. Put them on a round dish covered with a cream sauce surrounded by a border of sliced mushrooms. Sprinkle with bread crumbs. Pour on butter and brown in oven.

CREAM SAUCE

1 c bechamel sauce*
3 tbsp butter
½ cup + 4 tbsp cream

Cook down by a third the bechamel sauce to which has been added ½ cup cream. Remove from heat and add butter and remaining cream.

FROG LEGS AND SCALLIONS GRATIN

12-16 pr med frog legs
1 clove garlic, chopped
2½ c bread crumbs
1 lb butter, melted and cooled
2 scallions, minced
2 tbsp parsley, chopped
salt and pepper to taste

Mix the scallions, garlic and parsley with the bread crumbs. Season the frogs with salt and pepper and dip them in the butter. Coat the legs evenly in the crumb mixture. Take a large fairly deep gratin dish and butter it generously. Lay the legs in it and fill the spaces between them with the left over crumb mixture and pour in the rest of the butter. Bake in a 400°F oven for 15-20 minutes or until the top is lightly browned. Serves 4.

FROG LEGS AND PEPPERS

6 pr frog legs
1 sprig thyme
2 tbsp butter
¼ c parsley, chopped
salt and pepper to taste
2 cloves garlic, crushed
4 onions, sliced
4 bell peppers, diced
6 tomatoes, chopped
1 c chicken stock
1 bay leaf

Slowly brown frog legs in butter and add salt, pepper and onions and cook until onions are brown. Add tomatoes, bay leaf, thyme, parsley and garlic. Cover and cook slowly for 30 minutes. Add bell peppers and chicken stock and continue to smother until frog legs are tender. Serve

FROG LEGS AND LEMON WEDGES

6 frog legs
2 tbsp butter
oil
lemon wedges
1 egg
1 tbsp vermouth
almonds
1 clove garlic
plain flour
a little milk

Dip frog legs in a wash of milk and egg. Then flour. Sauté until brown. Skim off oil and add garlic, vermouth, butter, salt and pepper. Transfer to a serving dish. Sprinkle with almonds. Garnish with lemon wedges.

BATTER FRIED FROG LEGS WITH DILL

1½ lbs frog legs, separated, feet removed
½ c ice water
2 tbsp plain flour
2 tbsp cornstarch
½ tsp salt
safflower or peanut oil
4 lge sprigs fresh dill
2 lge eggs, lightly beaten
½ c dry white wine
½ c water
2 sprigs, parsley
6 shoots fresh chives
⅛ tsp ground white pepper

In a glass loaf dish combine wine, water, parsley, chives, pepper, and frog legs. Refrigerate for 2 hours, turning frequently. Drain and dry thoroughly. In a chilled bowl add the eggs, ice water, flour, cornstarch and salt; mix well. Heat oil at a depth of 1½ inches to 375°F. Dip the legs in the batter and deep fry for 2-4 minutes. Drain on paper. Dip dill sprigs in batter and deep fry until golden. Drain on paper. Garnish each serving of fried frog legs with a sprig of dill. Serves 4.

FROG LEGS MARINIERE

24 frog legs
1 tbsp parsley, chopped
1 tsp chives, chopped
salt and pepper to taste
3 ozs butter
1 pt allemande sauce
4 ozs white wine
6 scallions, chopped

Place frog legs in a sauce pan with the butter, season, and simmer for 5 minutes. Then add scallions and simmer for 3 minutes. Add wine and simmer until nearly dry and then add the allemande sauce and simmer for 5 minutes. Serve with chives and parsley sprinkled over the top.

ALLEMANDE SAUCE

3 tbsp butter
1 tbsp lemon juice
2 egg yolks, beaten
1¼ c milk
2 tbsp plain flour
1 chicken bouillon cube
¼ tsp salt

Make a white sauce of the butter, flour, milk, salt and bouillon cube cooking it in the top of a double boiler over hot water until smooth and thick or for about 15 minutes. Add egg yolks slowly to hot sauce, stirring thoroughly. Return to double boiler and continue cooking for 5 minutes, stirring often. Add lemon juice gradually, stirring constantly.

GRENOUILLES A L'INDIENNE

12 pr frog legs
4 ozs butter

Sauté frog legs in butter. Serve them in a border of curried rice. Pour curry sauce a l'indienne over the frogs.

CURRIED RICE

1½ tbsp onion, minced
1 pt consommé
3 ozs butter
½ lb rice
1 tsp curry powder

Cook onion in half the butter until golden brown. Add the curry and rice. Shake the pan on the stove until the rice is well covered with butter and curry. Pour in consommé, cover, and cook in the oven 18 minutes. After cooking, fluff up the rice and mix in the remaining butter.

CURRY SAUCE A L'INDIENNE

1 pt button mushrooms, drained
pinch cayenne pepper
few slices lean ham
pinch pimento
1 oz butter
1 bouquet garni
4 tbsp chicken consommé
1 onion, chopped
a little mace
3 cloves
saffron
½ pt mushrooms, chopped
1 recipe allemande sauce*

Put into a saucepan the ham, onion, bouquet garni, chopped mushrooms, cloves, a good pinch of pimento, cayenne pepper and mace. Add the chicken consommé and simmer over a very low heat. Strain through a napkin and remove all grease. When it is cooked down, mix in the Allemande sauce; when it is cooked down to a desired point add a small infusion of saffron to color it a nice yellow. Then strain again through a cloth. Just before serving put in butter and mushroom buttons.

BOUQUET GARNI

A bouquet garni is any combination of parsley, thyme, bay leaf, sweet basil, celery, chervil, tarragon, rosemary, savory, etc. tied in a cloth bag. Use several items. Discard after use.

FROG LEGS A L'ANGLAISE

12 frog legs
2 eggs
3 ozs butter
plain flour
1 tbsp oil
salt and pepper to taste
2 c plain bread crumbs

Season the frog legs with salt and pepper. Flour and dip in bread crumbs a l'anglaise (eggs, 1 tbsp oil and bread crumbs). Sauté in butter. Serve on a long dish on a bed of maître d'hôtel butter, surrounded with boiled new potatoes.

MAÎTRE D'HÔTEL BUTTER

½ c butter
1½ tbsp parsley, chopped
pinch fine ground black pepper
1 tsp salt
dash lemon juice

Mix all ingredients and stir with a spoon until it forms a smooth paste and is thoroughly mixed.

FROG LEGS A LA MORNAY

12 pr med frog legs
1 egg yolk, beaten
¼ c white wine
4 ozs butter
juice of 1 lemon
parmesan cheese, grated

Cook the frog legs gently in a little white wine, ½ the butter and lemon juice. Drain them (reserve drippings) and dry them. Serve them on a round dish lined with mornay sauce and surrounded with a border of duchess potatoes. Cover with mornay sauce to which the pan drippings have been added. Sprinkle with cheese. Pour on remaining butter. Brush the duchess potatoes with beaten egg. Brown in the oven or broil.

MORNAY SAUCE

⅓ c bechamel sauce*
½ c fresh cream
¼ c Gruyere cheese, grated
3 tbsp butter
¼ c parmesan cheese, grated

Boil down about a the bechamel sauce mixed with the fresh cream. Add the grated cheeses. Mix and incorporate the butter.

DUCHESS POTATOES

1 lb potatoes, peeled
1 egg, beaten
little nutmeg
3 tbsp butter
salt and pepper to taste

Boil cut potatoes briskly in salted water. Drain and put in the oven for a few moments to evaporate excessive moisture and quickly rub through a sieve. Put the puree into a saucepan, dry off a few moments on the fire, turning and add butter, seasonings and bind with beaten egg.

FRIED FROGS IN BEER BATTER

frogs
salt and pepper to taste
lemon wedges
beer batter*
garlic powder
oil
creole tartar sauce*

Heat oil. Season frogs (whole or just legs) and coat in beer batter and fry until golden. Serve with lemon wedges and creole tartar sauce.

Cayenne Fried Frog Legs

12 frog legs
½ c plain flour
½ tsp cayenne pepper
salt and pepper to taste
oil
¼ tsp baking powder
2 eggs, well beaten
¼ c mustard
2 tbsp milk

Separate frog legs and dry. Season with salt and pepper. Rub with mustard and place in a refrigerator for several hours. Make a batter of well beaten eggs, baking powder, flour and milk. Roll frog legs in the batter and then fry in deep oil until golden brown, about 5 minutes. Serves 6.

Buttermilk Fried Frog Legs

frog legs
½ tsp tabasco sauce
salt and pepper to taste
1 clove garlic, crushed
plain flour
1 tsp worcestershire sauce
4 ozs butter, melted
oil
2 tbsp lemon juice
buttermilk

Soak frog legs in buttermilk for 1 hour. Mix flour and seasonings to taste in a paper bag. Shake 4-6 legs at a time in bag. Fry in hot oil until golden brown. Combine butter, tabasco sauce, lemon juice, worcestershire sauce and garlic in a saucepan and heat. Serve frog legs on a large platter with the sauce drizzled liberally over them. Allow 4 legs per person.

Fried Frog Legs with Creole Tartar Sauce

2 doz frog legs
4 eggs
3 tsp salt
creole tartar sauce*
oil
1 c plain flour
lemon wedges
1 c yellow cornmeal
1 tsp cayenne pepper
¼ c milk

Pour oil into a deep fryer to a depth of 2-3 inches and heat to 375°F. Meanwhile season the frog legs on all sides with the cayenne and salt. Combine the cornmeal and flour in a shallow bowl and stir to mix well. Combine the eggs and milk in another shallow bowl and beat them together until the mixture is smooth. Roll frog legs in cornmeal mixture. Immerse in the egg mixture and then turn into the cornmeal mixture again to coat it evenly. Deep fry the frog legs in the hot oil, turning them with a slotted spoon for about 5 minutes or until they are golden brown and crisp on all sides. Keep hot. To serve, mound on a heated platter and arrange the lemon wedges around the edge. Serve at once accompanied by creole tartar sauce.

PEPPER FRIED FROG LEGS

2 lbs frog legs
2 tbsp mayonnaise
1 tbsp lemon juice
¼ tsp salt
⅔ c plain flour
oil
2 eggs, beaten
1 tbsp cornstarch
½ tsp baking powder
⅛ tsp black pepper
⅓ c bread crumbs, seasoned

Arrange frog legs in a shallow container. Mix eggs, mayonnaise, cornstarch, lemon juice, baking powder, salt and pepper and stir until smooth. Pour over frog legs and cover and chill for at least 30 minutes. Combine the flour and bread crumbs in a paper bag. Remove legs from marinade, shaking off excess, and place 2-3 legs at a time in the flour mixture. Shake until legs are well coated. Pour oil to a depth of 2-3 inches in a dutch oven and heat to 375°F. Fry legs 1-2 minutes or until dark golden brown, then drain on paper towels. Serves 6.

BISQUICK FRIED FROG LEGS

1 doz frog legs
1 c Bisquick
3 lge eggs, beaten
4 ozs butter or olive oil
½ c non-dairy creamer
½ tsp garlic powder
¼ tsp Lawry's Seasoned pepper
salt to taste

Mix eggs and non-dairy creamer together. Dredge frogs in this solution and then the Bisquick that has the seasonings added. Fry in butter or oil.

GRENOUILLES EN BEIGNETS

frog legs
Frying Paste
oil
salt and pepper to taste
vinegar
parsley, chopped

Marinate frog legs in salt, pepper, parsley and a little vinegar for 2 hours turning occasionally. Dry them just before cooking and dip them in frying paste. Fry for 5 minutes in deep hot oil. Serve with fried parsley.

FRYING PASTE

½ c plain flour
½ tsp salt
2 tbsp olive oil
6 ozs warm water
2 egg whites, beaten

In a bowl mix flour, salt, olive oil and warm water. Make a smooth paste and let stand for 1 hour. Just before using, add the beaten egg whites.

ONION FRIED FROG LEGS

12 med frog legs
½ tsp cloves, ground
1 tbsp malt vinegar
lime wedges
½ c onion, finely grated
1 tsp salt
freshly ground black pepper
3 cloves garlic, crushed
1 c plain flour
1 c oil

In a large glass bowl combine the onions, garlic, vinegar, cloves, salt, a few grindings of pepper, and stir. Wash the frog legs under cold running water and pat them completely dry. Then place them in the onion mixture and turn them about to coat evenly. Marinate at room temperature for at least 1 hour turning the legs from time to time. In a skillet heat the oil over moderate heat until it is very hot but not smoking. Dip the frog legs in flour and shake vigorously to remove the excess. Fry the frog legs 7-8 at a time for about 4 minutes on each side, turning so that they color quickly and evenly without burning. To serve, mound the legs attractively on a heated platter and arrange the lime wedges around them.

DRUNK FROG LEGS AND MUSTARD

12 med frog legs
plain flour
¼ c vermouth
½ tsp cayenne pepper
oil
¾ fine plain bread crumbs
½ c milk
2 eggs, beaten
1 clove garlic, minced
2 tbsp orange zest, grated
salt and pepper to taste

Put the frog legs in a shallow glass dish and pour the milk over them and turn occasionally for 1 hour. Drain the legs and pour the milk into the beaten eggs and add the vermouth, garlic, cayenne pepper, salt and pepper. Add the orange zest to the bread crumbs. Dredge the legs in the flour, shake off excess, then the egg mixture and then roll in the crumbs. Pat lightly to adhere crumbs. Refrigerate until ready to use. Heat the oil to about 365°F and fry legs until golden brown. Serve with cold mustard sauce. Serves 4-6.

COLD MUSTARD SAUCE

2 egg yolks
2 tbsp lemon juice
1 tbsp bell pepper, minced
1 clove garlic, minced
1 tbsp parsley, minced
1 tsp basil
2 tbsp creole or dijon mustard
1 c peanut oil
2 dill gherkins, minced
1 shallot, minced
2 tbsp tomato sauce
salt and pepper to taste

Allow the yolks to come to room temperature and put in a bowl. Beat with a whisk with the mustard and lemon juice for 3 minutes. Slowly drizzle in the oil forming an emulsion. Stir in the remaining ingredients and correct seasoning.

DRUNK FRIED FROG LEGS

6 frog legs, more if desired
milk
oil
1 tbsp vermouth
salt and pepper to taste
lemon wedges
egg
plain flour
1 clove garlic, minced
2 tbsp butter
almonds as desired

Dip the frogs in a mixture of egg and milk and then in flour. Sauté in a small amount of oil until brown. Skim off excess oil and add the garlic, vermouth, butter, salt and pepper. Transfer to a serving dish and sprinkle with almonds. Garnish with lemon wedges.

DEEP FRIED FIVE SPICE FROG LEGS OR ALLIGATOR

frog legs or alligator
¼ tsp msg
1 tbsp soy sauce
¼ tsp five spice powder*
2 slices ginger
2 egg yolks, beaten
1 tsp salt
1 tsp sugar
½ tsp sesame oil
1 tbsp sherry
2 shallots
oil

Combine all ingredients except eggs, cornstarch and oil and let stand 30-60 minutes. If using alligator, let stand overnight. Dip each leg in egg and then in cornstarch and deep fry in oil over a medium heat for 3-4 minutes and then turn up the heat and fry for an additional minute.

Paella

- 3 lbs frog legs
- 1 bell pepper, chopped
- 1 tsp instant chicken broth
- 1 8 oz can mushrooms, sliced
- 1 clove garlic, minced
- 1 pimento, cut into strips
- 1 8 oz bottle clam juice
- 2 tbsp plain flour
- ¼ c olive oil
- 8 ozs sweet peas, fresh or frozen
- ¼ tsp pepper
- 1½ lbs shrimp, cleaned
- 1½ c water
- ½ tsp salt
- 1 c onions, chopped
- 4 tomatoes, peeled, sliced
- 2 8 oz cans minced clams
- 1½ c long grain rice
- several strands of saffron

Coat frog legs (which have been separated at the joints) with flour and brown in oil in a large skillet. Place browned frog legs in a paella pan or 12 cup shallow casserole. Sauté rice, onion, garlic, mushrooms, bell pepper, and pimento in oil in same pan, stirring often, or until rice is golden. Spoon over and around frog legs in dish; top with peas and tomatoes. Combine clam juice, water, instant chicken broth, salt, pepper, and saffron in skillet and bring to a boil. Then pour over rice mixture in dish and cover. Bake at 350°F for 30 minutes. Add clams with liquid and cleaned shrimp. Cover and bake 30 minutes, or until frogs and rice are tender. If fresh steamer clams are available buy 12-18 and use them in place of canned minced clams. To cook, scrub the shells well, place in large saucepan with 1 cup water, cover and bring to boiling. Simmer 3-5 minutes or until shells open. Lift out with tongs. Strain broth through cheesecloth to remove sand and measure (you should have 1 cup). Substitute for bottled clam juice and use clams in shells in place of canned clams.

Rice Valencia with Frog Legs

- 3 lbs frog legs
- 1½ c long grain rice
- 2 tsp salt
- 10 ozs sweet peas, fresh or frozen
- 1 tsp thyme, crumbled
- ¼ tsp pepper
- roasted peppers
- 4 c chicken broth
- 3 tbsp olive oil
- 1½ lbs small white onions cut in half

Sprinkle rice over heated oil in a large skillet, stirring constantly, until rice turns a rich brown. Remove to a 12-cup shallow casserole. Brown the cut edge of the onions in same pan, adding more oil if needed. Place on rice in casserole. Separate frog legs into sections at the joints and brown in same skillet. Remove and place over rice in casserole. Stir chicken broth, salt, thyme and pepper into pan drippings, heat to boiling, stirring constantly. Pour into casserole, stir and cover. Bake in 350°F oven for 45 minutes. Stir sweet peas into casserole, cover and bake 15 minutes longer, or until liquid is absorbed and the rice is tender. Place roasted peppers around casserole as a garnish just before serving.

Roasted Peppers

- 2 each lge red and green bell peppers, quartered, seeded
- olive oil

Brush peppers with olive oil. Arrange in a single layer on a large cookie sheet. Broil 4 inches from heat 10 minutes, turn and brush with oil. Broil 10 minutes longer.

STEWED FROGS

frogs
salt and pepper to taste
3 cloves garlic, crushed
2 lge onions, chopped
1 c brown roux*
1 8 oz can mushrooms, sliced
tabasco sauce
water
butter

Clean frogs and use only the backs, legs, or both. To the roux add cold water to bring to the desired thickness, (frogs will give off water and thin the roux). Sauté onions, garlic, and drained mushrooms in a small amount of butter until soft and golden. Add to roux, add the frogs, and cook over medium heat until frogs are cooked. Season with salt, pepper, and tabasco sauce to taste. Serve over rice.

LOUISIANA STEWED FROGS

12 frog legs
1 tbsp plain flour
1 c oyster water
1 sprig parsley
1 sprig marjoram
1 clove
2 doz oysters
salt and pepper to taste
1 tbsp butter
1 c water
1 sprig thyme
1 bay leaf
10 allspice berries
croutons
1 egg yolk, beaten
cayenne pepper to taste

Put the butter in a frying pan. When it begins to melt add flour and stir constantly. When brown add water and oyster water. When it begins to boil throw in the frog legs and add salt, peppers, thyme, bay leaf, marjoram, allspice berries, clove, and oysters. Let simmer 15 minutes and take off the heat. Add egg yolk blending well, and serve immediately with garnishes of croutons fried in a little butter, and place oysters on them.

FROGS CHASTANT

12 frogs, dressed
1 med bell pepper, minced
plain flour
water
salt and pepper to taste
3 lge onions, minced
1 10 oz can Ro-Tel tomatoes drained
corn oil
scallion tops, chopped
parsley, chopped

Separate frog legs from bodies. Roll frogs in seasoned flour* and partially fry in hot oil. Remove frogs and keep warm. Discard the oil, except for enough to sauté onions and bell pepper, retaining flour in the oil. Sauté the onions and bell pepper until well wilted and then add tomatoes. When sautéed, replace frogs and add a little water and cover. Cook, adding more water if needed, until the frogs are done. Add scallion tops and parsley before removing from the heat. Serve.

FROG BACKS A LA THIBODEAUX

20 frog backs
1 lb fresh mushrooms
1 bell pepper, chopped
1 clove garlic, chopped
salt and pepper to taste
16 ozs sour cream or plum yogurt
2 lge onions, chopped
¼ c dry white wine
butter
1 tsp worcestershire sauce

Wash and clean backs. Dry and season with salt and pepper. Brown frogs in butter and put aside. Add onions, garlic, bell pepper, and sauté until soft. Set aside. Sauté mushrooms in butter and add worcestershire sauce before removing mushrooms. Stir to mix. Add sour cream or yogurt to mushroom mixture and stir on low heat until well mixed. Add frogs and vegetables. Cook for 10 minutes. Add wine and cook for 5 minutes. Serve over rice or toast.

FROG BACK SPAGHETTI

20-30 frog backs
26 ozs ketchup
12 ozs chili sauce
1 8 oz cup sour cream
1 tsp oregano
1 lb can whole tomatoes
¼ c capers
1 4½ oz can small ripe olives, chopped
2 lge onions, chopped
1 lge bell pepper, chopped
butter
2 cloves garlic, chopped
salt and pepper to taste

Salt and pepper the frogs and brown in butter in heavy dutch oven. Remove and keep warm. Sauté onions, garlic and bell pepper until soft in the same pot. Add the chili sauce, ketchup, tomatoes and oregano and cook on low heat 60 minutes. Add olives, capers and frog backs and cook 30 minutes more. Add sour cream, mix well, and serve over spaghetti or noodles.

FROG LEGS BEURRE NOISETTE

12 pr small frog legs
⅛ tsp mace
⅛ tsp cayenne pepper
½ tsp black pepper, freshly ground
2-3 c cold milk
oil (peanut and corn equally)
1½ cups plain flour
1½ tsp salt

Separate the frog legs and rinse under running water, drain on paper towels, and dry. Place in a dish and pour cold milk to cover; soak at room temperature 1-1½ hours. Combine the flour, salt, pepper, cayenne and mace and mix thoroughly. Remove frog legs and roll in the flour. Coat thoroughly and then place on platter to dry for 2 minutes. Fry in hot oil until golden brown, about 5-7 minutes, turning often to brown evenly. Remove, drain, and keep warm. To serve place 3-4 pairs in individual dishes with raised edges. Spoon about ¼ cup beurre noisette sauce over each portion, being sure to include some of the browned solid matter that has settled to the bottom of the pan. Serve immediately.

BEURRE NOISETTE SAUCE

1 c butter
5 tsp lemon juice
2 tsp parsley, minced

Melt butter over medium heat in a small saucepan and cook until butter begins to turn light brown. Remove from heat and add the lemon juice and parsley. The sauce will foam up. When the foaming ceases, return the pan to heat and cook for 1 minute. Stir thoroughly and remove from heat.

MOCK MOUSSE DE CREVETTES DE GRENOUILLES

frog legs
1 tbsp lemon juice
1 6 oz package Gervais cheese
1 tbsp truffles or ripe olives, finely chopped
1½ c crawfish meat, cooked
1 tbsp dill, chopped
1 tsp dijon mustard
¼ c Armagnac brandy
salt and pepper to taste

Sauté seasoned frog legs in butter until done. Remove legs, bone and cut into chunks. Reserve. Combine all remaining ingredients in blender and whirl at high speed until smooth. Sides of blender will have to be scraped 2-3 times during blending. Chill. Serve over chopped frog legs.

FROG LEGS WITH CRAWFISH TAILS

4 pr frog legs, separated
2 cloves garlic, minced
¼ c white wine
½ tsp dill weed
½ lb crawfish tails
2 tbsp butter
2 tbsp scallion tops, minced
½ cup heavy cream
2 tbsp parsley, minced
salt and pepper to taste

Heat the butter over moderately high heat and add the legs (at room temperature). Cook and toss for 4 minutes or until just done through. Remove legs; add the garlic and scallions to the butter remaining and cook for 1 minute. Add cream, dill, and parsley. While that is reducing by half remove the meat from the legs. Add the meat and crawfish and allow to heat through. Serves 4.

OYSTER SAUCE FROG LEGS

10 ozs frog legs
3 slices ginger
2 tbsp water
3 tbsp sherry
3 tsp oyster sauce*
3 cloves garlic, mashed
½ c meat stock
2 tsp cornstarch
1 tsp sugar
3 scallions
3 tbsp oil
½ tsp sesame oil
½ tsp salt

Tie scallions in 3 small bundles. Mix sherry, sugar, salt and oyster sauce; add frog legs. Mix cornstarch with water. Heat oil and brown the garlic. Add frog leg mixture, scallion bundles, ginger, and stir fry for 2 minutes, adding liquid if needed. Add the meat stock and simmer on low heat 15-20 minutes. Remove the garlic, scallions, and ginger. Thicken sauce with cornstarch and add sesame oil. Serve.

SPICY FROG LEGS

1 lb frog legs
½ tsp salt
3 tbsp oil
2 tbsp hoi sin sauce*
2 tsp cornstarch
2 hot red peppers
⅓ c soup stock
2 lge bell peppers
1 tbsp sherry
1 tsp light soy sauce
2 tbsp water
1 clove garlic, mashed
5 slices ginger

Wash and dry frog legs. Slice the bell and hot peppers discarding the stems and seeds. Mix the hoi sin sauce, soy sauce, sherry, salt, ginger and soup stock. Heat the oil in pan; when smoking add peppers and garlic and stir fry a few seconds; add the frog legs and stir fry 1 minute. Add the Hoi Sin sauce mixture, cover, and simmer 4 minutes. Thicken with cornstarch and water. Serve.

STIR-FRIED GINGER FROG LEGS

1 lb frog legs
2 tsp vinegar
⅓ c peanut oil
1 tbsp light soy sauce
¼ c sherry
2 tbsp cornstarch
½ tsp sesame oil
6 oz-can fried vegetable steak*
2 tsp sugar
1 scallion, minced
1 tsp salt
6 slices ginger, minced

Cut vegetable steak into ½ inch chunks. Clean frog legs and cut each into 2-3 small pieces. Mix sesame oil, sherry, salt, sugar, vinegar and soy sauce thoroughly. Marinate the frog legs in the mixture for ½ hour. Place the cornstarch in a paper bag and add the marinated frog legs and shake to coat. Heat the peanut oil in a skillet until hot. Brown everything in the oil. Add the vegetable steak, scallions and ginger and cook 3 minutes more with a small amount of water.

NYMPHES A L'AURORE

10 pr frog legs, separated
tarragon leaves, fresh
chaud froid sauce
chervil, minced
court bouillon
champagne aspic

Poach the frog legs simmering in the court bouillon for 15 minutes. Cool and dry the legs. Dip the legs in the refrigerated chaud froid sauce to coat with a thin uniform glaze. Arrange them on a bed of champagne aspic and garnish with the chervil and tarragon, or you can cover them with diced champagne aspic

COURT BOUILLON

1 qt white wine
1 qt water
salt
1 c onions, minced
1 c chopped herbs, parsley, tarragon, etc
peppercorns

Mix all ingredients well.

CHAUD FROID SAUCE

6 tbsp. butter, melted
1 qt stock, chicken or fish
white pepper, to taste
1 pkg gelatin
1 c heavy cream
8 tbsp. plain flour
salt to taste
nutmeg, to taste
1 c stock
paprika

Blend the butter and flour over low heat and stirring for 15 minutes to make a white roux*. Slowly add the quart of stock and seasonings and bring to a boil. Cook over low heat for 30 minutes to reduce it by ¼, skimming off the foam. Soften the gelatin in 3 tbsp liquid from the cup of stock. Add to sauce and continue simmering and stirring. Slowly stir in the cream and cook until the sauce has a thick consistency. Remove from the fire and stir in the paprika. Refrigerate.

CHAMPAGNE ASPIC

3 c chicken stock
3 pkg gelatin
2 egg whites
2 c dry champagne, chilled

Mix the gelatin in a ½ cup of cold stock. Beat the egg whites and mix with the gelatin and stock. Bring to a light boil, stirring constantly. Let stand 5 minutes while the egg mixture clarifies and then pass through a sieve. Set a bowl with the aspic inside another bowl filled with ice and refrigerate. When cooled add the champagne.

HAWAIIAN FROG LEGS

3 lbs frog legs
1 c coconut flakes
parsley
1 tbsp soy sauce
½ tsp curry powder
3 tbsp rum
½ tsp cayenne pepper
1 lge onion, diced
2 tsp salt
1 12-16 oz can crushed pineapple
4 ozs butter
1 8 oz can coconut milk
1 c plain flour

Marinate frog legs in coconut milk, rum and soy sauce for 2 hours. Discard all but ¼ cup. Mix salt, flour and cayenne pepper in a paper bag and shake frog legs to coat. Brown frog legs in butter. Sauté onion until clear. Add frog legs, onion, coconut flakes, pineapple and curry into a covered casserole dish with the ¼ cup of marinade. Cook for 40 minutes at 375°F. Garnish with parsley.

CAJUN JAMBALAYA

3 lbs frogs
2 tbsp plain flour
3 c water
2 c rice
2 tbsp oil
1½ tsp salt
½ bu scallions, minced
1 lb ham
4 tomatoes, peeled and diced
½ tsp black pepper
½ tsp tabasco sauce
salt and pepper to taste
2 med onions, chopped
½ tsp thyme
1 pt oysters
1 clove garlic, chopped
½ bu parsley, minced
1 lb shrimp, cleaned
1 lge bell pepper, chopped

Separate frog legs from bodies and season with salt and pepper. Heat oil in a heavy pot and fry until browned. Remove from pot and set aside. Make a roux by adding the flour to the drippings and stir until light brown. Add the ham, frogs, shrimp, onions, bell pepper, tomatoes and garlic. Cook, stirring constantly, for about 10 minutes. Add oysters, water (use water from oysters, if any), salt, thyme, tabasco sauce and black pepper. When the water boils pour in the rice. Let the mixture come to a boil again and stir thoroughly to combine all the ingredients. Cover and simmer 30-45 minutes until the rice is cooked. Add parsley and scallions and stir lightly to mix. Cover for 5 minutes more. Serve hot. Serves 8.

FROG CROQUETTES

2 c cooked frog meat, chopped
2 tbsp plain flour
1 c light cream
1 tsp worcestershire sauce
⅛ tsp paprika
1 tbsp parsley
cracker crumbs
3 tbsp butter
1 tsp salt
1 tsp catsup
2 eggs, beaten
⅛ tsp black pepper
2 tbsp cold milk
oil

Melt butter and add flour and cold cream slowly stirring until smooth and creamy. Add seasonings and parsley and boil 4 minutes. Add frog meat. Mix well and pour out on a platter to cool. When cool enough to handle take a large spoonful of the mixture in floured hands and shape into croquettes. Put into a cold place until firm. When firm roll in finely crushed cracker crumbs and then in eggs beaten with 2 tbsp milk and then again in the cracker crumbs. Fry in deep hot oil until golden brown.

FROG MEAT A LA KING

2 c cooked frog meat, cubed
1 c mushrooms
¼ c celery, minced, sautéed
1 pimento, chopped
lemon wedges
1 c white sauce
¼ c cooked peas
1 bell pepper, minced, sautéed
few drops worcestershire sauce

Add frog meat and all other ingredients to the white sauce and cook until thoroughly heated. Serve on hot toast or hot biscuits and garnish with lemon wedges. Serves 2.

FROG MEAT AU GRATIN

1 c frog meat, cooked, chopped
1 c frog stock
1 stalk celery, minced
2 tbsp butter
1 tsp salt
water
¾ c cream
1 sm slice onion, minced
2 tbsp cornstarch
2 eggs, well beaten
1-3 tsp paprika
2 tbsp water

Boil frog legs in a little water until cooked. Remove, bone and chop. Combine the cream and stock from boiling frogs. Heat onion and celery in the creamy stock. Melt butter and add the cornstarch to the 2 tbsp water. Then add butter and cornstarch mixture to creamy stock, cook, and stir until creamy. Then add the eggs. Then add the seasonings and frog meat. Blend and pour into a well-buttered baking dish and bake at 350°F for 25 minutes.

ESCALLOPED FROG

2 c frog meat, cooked, diced
1½ tsp butter
1 tbsp cold water
1 c celery, cooked
1 tbsp cornstarch
3 cups mashed potatoes, seasoned

Blend the cornstarch and water and add to the butter and frog meat. Use the potatoes to line a casserole or baking dish. Pour the thickened frog meat and celery into the potato-lined dish and bake at 375°F or 400°F just long enough to thoroughly heat and to slightly brown the surface of the potatoes. Serves 3-4.

FROG LUNCHEON WITH CORN

1 c frog meat, cooked, diced
½ c heavy cream
1½ c canned whole corn
1 tsp salt
butter
1 tbsp butter, melted
¼ c bell pepper, chopped
1 tbsp onion, chopped
2 eggs, well beaten

Combine all ingredients mixing well and pour into a well buttered baking dish and bake 25 minutes at 350°F and serve hot. Serves 2-3.

FROG LOAF

2 lbs frog meat, diced fine
1 egg, beaten
½ bell pepper, chopped
salt and pepper to taste
whole cooked onions
tomato sauce
2 c bread crumbs
1 onion, sliced
parsley, chopped
whole cooked carrots
whole sm potatoes, cooked

Combine meat, egg, bell pepper, salt, pepper, bread crumbs, sliced onion and parsley and mix well. Form into a loaf and bake in a moderate oven. Serve with tomato sauce, whole cooked carrots, onions, and potatoes. Serves 2-3.

ROAST FROG

1 bullfrog, dressed
salt and pepper to taste
butter
stuffing (your favorite)
plain flour

Fill body cavity of frog with stuffing and truss securely in a compact shape; lay on its back in a roasting pan. Dust with flour, salt, pepper and bits of butter and place in a hot oven. Baste every few minutes until meat is tender. Serves 1.

FROG POT PIE

2 lbs frog meat, cubed
1½ c water
½ onion, minced
cornstarch
1 carrot, diced
1 c peas
salt and pepper to taste
pie crust

Place meat into a shallow saucepan with water, carrots, peas, onion, salt and pepper and cook slowly until tender. Thicken with a little cornstarch and pour into a casserole. Cover with the pie crust and put several small slits in it. Bake in a moderate oven until the crust is a golden brown.

FRIED BULLFROG TADPOLES

lge bullfrog tadpoles
oil
vinegar

Place tadpoles in vinegar overnight in a refrigerator. Drain and deep-fat fry.

TOAD LEGS

The toad leg recipes below come to us from our friends down under in Australia. The Cane Toad (*Bufo marinus*) is an introduction that has become an absolute pest in the country. Only the legs are being eaten. The toad has to be thoroughly skinned and washed very thoroughly as the skin is poisonous and poisonous secretions are given off from the large parotid glands on the sides of the head. These toads and others have been eaten by native peoples and references point it out for the American Indians. Skinning thoroughly and eating the legs only render them harmless except for the Colorado River Toad (*Bufo alvarius*) which has poisonous glands on the legs as well. Survival books also point out that toads can be eaten with the proper precautions.

TOAD LEGS IN WHITE WINE

toad legs, dressed
1 oz butter
parsley, chopped
1 chicken stock cube
1 egg yolk
parsley (for garnish)
1 c white wine
1 shallot, minced
1 tsp cornstarch
1 tbsp thick cream
salt to taste
lemon (for garnish)

Wash and cut off feet. Salt and set aside for 30 minutes. Boil wine with butter, shallot, and parsley. Add salted legs and cook until tender. Remove legs and keep warm. Thicken the wine sauce with cornstarch. Add stock cube and cream. Mix well and simmer for 1 minute. Remove from heat. Add the egg yolk and beat sauce with a fork. Return toad legs to sauce and serve hot garnished with parsley and lemon.

CRUMBED TOAD LEGS

toad legs, feet removed
plain flour
bread crumbs
oil or butter
milk
eggs, beaten
salt and pepper to taste

Thoroughly wash. Mix the eggs and milk, adding salt and pepper. Then dip the legs in and roll in flour and then into the eggs and then roll in bread crumbs. Deep fry legs in oil or shallow fry in half butter and half oil.

MARINATED TOAD LEGS

toad legs, feet removed
pepper
oil
salt
soy sauce

Wash thoroughly and dry. Salt and pepper the legs and marinate in enough soy sauce to cover for 5 hours. Barbecue or shallow fry in oil.

TOAD LEGS IN BEER BATTER

toad legs, feet removed
salt
¾-1 c beer
1 pinch bicarbonate of soda
lemon pepper seasoning
oil
1 c plain flour

Thoroughly wash legs, dry and sprinkle with salt. Set aside for 30 minutes. Make a batter of the flour, soda and beer. The batter will be runny. Pass salted legs in flour and lemon pepper seasoning. Dip in batter and deep fry until golden brown. Serve immediately.

KILLER TOAD IN GARLIC

3 pr toad legs, feet removed
1 clove garlic, crushed
3 tbsp lemon juice
seasonings
mango
1½ glasses Tolland wine
1 sm onion, chopped
½ pt cream
butter
kiwi

Thoroughly wash and sauté the legs in frying pan with butter. Remove and keep pan hot and add the onion, garlic, wine and lemon juice until it bubbles. Add the legs and cream making sure it doesn't curdle. Remove from heat and arrange the legs on plates and add sauce. Garnish with mango and kiwi slices.

Alligator Meat Cuts

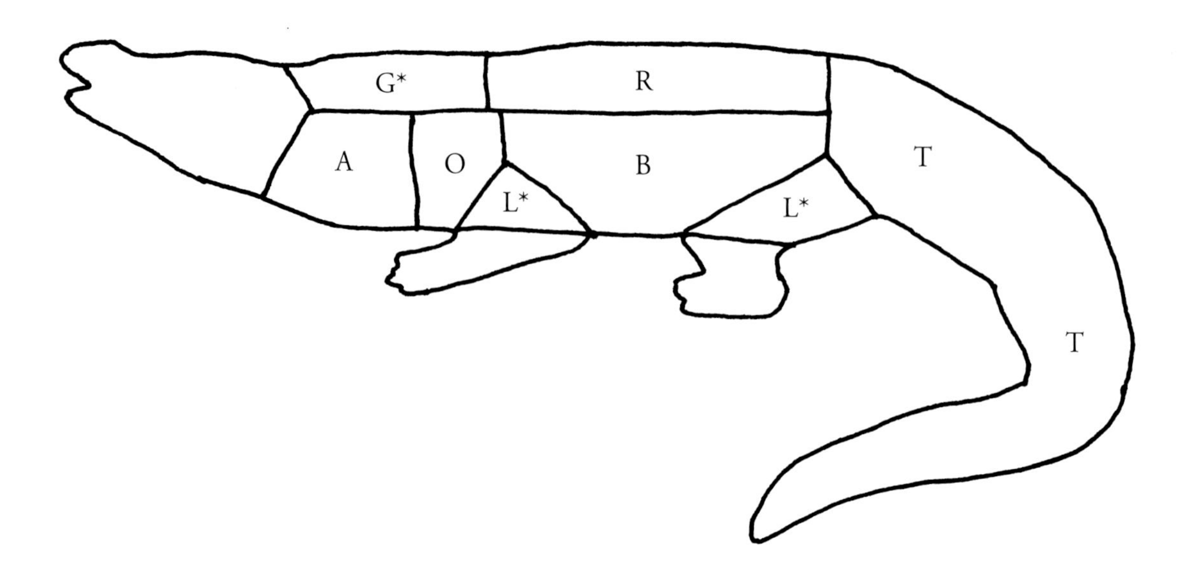

G* Top of neck, cube or use mallet.
A Jaw, very tender. *Appearance*— White to light pink with no fat deposits.
T Tailmeat, very tender. *Appearance*— Light pink to white with internal bands of hard white fat that appear circular in cross section and run lengthwise near tail vertebrae.
O* Neck meat, cube or use mallet.
R Back strap, tender. *Appearance*— Similar to meat from tail without fat bands.
B Body meat. *Appearance*— Similar to meat from tail without fat bands.
L* Leg meat, must cube or use mallet. *Appearance*— Darker in color with small fat deposits along tendons.
* Less tender cuts

Adapted from: Anonymous (n.d.) *The Alligator Cookbook.* Division of marketing, Florida Department of Agriculture and Consumer Services, Talahassee, Florida, 50 pp. used with permission.

CROCODILIANS

Preparation—Chances are slim that you will be processing an alligator (or crocodile) from a live specimen so I will give a few pointers, based on my experience, on processing the carcass after it has been skinned for its hide. Be sure to have plenty of free time. Gut and discard the entrails and wash alligator thoroughly. The easiest way to section the alligator is to remove the tail by cutting it off just behind the rear legs. Set aside. Cut off head if still attached. Remove the legs at the joint where it connects to the body. Remove the shoulder blades. Remove the two jowl strips of meat, one on each side. I consider this to be the choicest part of the alligator. Discard the rest of the head unless it is fit to make a skull. Many hunters usually shoot them or use an axe which ruins the head for skull purposes. You can now process the body by either removing all the meat or cutting it up in pieces with the bone in. Rib racks are excellent when barbecued. No matter which method you use keep the types of meat separate and remove all fat. The fat will make your dish taste rancid and foul. Now to process the tail. The tail meat can either be separated from the spinal column in 2 to 4 sections. I find 4 sections easier. Cut along the seam on the side of the bone and the length of the tail. You should encounter a lateral "wing" of the vertebrae. Stay on one side of this. Curve your knife gently at the base next to the vertebrae and then work slowly outward. On the tip and bottom of the tail is a crease. Work toward this. After the first strip is out the others become easier. After the 4 sections are removed discard the remainder of the tail. In two of the sections is a strip of meat easily removed that resembles a pork loin. I consider this the second choicest part. The remainder of the tail is the third best. This loin is fat free but has a thin membrane over it. Remove this membrane. The loin was encased in a layer of fat on the tail fillets. By laying these fillets flat with the fat up and with a sharp knife you can cut this fat off fairly easily. After the major part of the fat is removed you can fine-tune the fat removal. The meat can now be frozen until ready to use after giving it a final washing. You can see from the diagram (page 92) the types of meat and their appearance and locations.

ZIPPY ALLIGATOR DIP

½ lb alligator meat
½ tsp liquid crab boil*
1 tsp salt
½ lemon
2 tbsp scallions, minced
2 tbsp celery, minced
2 tbsp bell pepper, minced
2 tbsp onions, minced
2 tbsp Teriyaki or Soy sauce
2 tbsp parsley, minced
1 tbsp sweet pickle relish
2 tbsp mayonnaise
1 tsp mustard

Boil alligator with first 8 ingredients for 10 minutes. Chop meat and vegetables used in boiling in a food processor or blender. Add remaining ingredients and continue to mix well. Serve chilled with crackers.

ALLIGATOR DIP

1 lb alligator meat
¼ c scallions, minced
2 tbsp sweet relish
1 tsp liquid crab boil*
¼ c celery, minced
¼ c mayonnaise
2 tsp salt
2 tbsp worcestershire sauce
2 tsp mustard
2 tbsp lemon juice
¼ c parsley, minced
1 c water

Cut alligator body meat into small pieces and boil in the water, crab boil and lemon juice for 10 minutes or until tender. Drain and chop well in a blender and place in a mixing bowl with the other ingredients. Mix well and chill until ready to serve.

ALLIGATOR BLUE CHEESE DIP

½ c alligator, ground, cooked
1 c sour cream
4 ozs blue cheese, crumbled
⅓ c evaporated milk
1 tsp worcestershire sauce
2 tsp parsley flakes

Combine all ingredients thoroughly and chill for at least 4 hours or more. Serve with dipping ingredients.

VEGETABLE GATOR DIP

1 c gator meat, cooked, flaked
1 avocado, peeled, mashed
1 8 oz cream cheese, softened
2 tbsp lemon juice
½ tsp curry powder
1 scallion, minced
½ tsp salt
1 jalapeno pepper, minced
¼ tsp worcestershire sauce

Mix thoroughly the meat, avocado, cream cheese and lemon juice. Add remaining ingredients mixing well. Chill for several hours before using. Makes 2½-3 cups dip.

GOOD AND EASY ALLIGATOR DIP

1 c gator meat, cooked*, ground
1 3 oz cream cheese
salt and pepper to taste
small amount evaporated milk
1.4 oz Hidden Valley Salad Dressing Mix
1 pt sour cream
½ c mayonnaise

Combine well all ingredients using milk to attain desired thickness and season to taste and chill well before serving. Makes about 3 cups.

ALLIGATOR CHIP DIP

2 c gator meat, cooked*, flaked
1 8 oz sour cream
1 8 oz cream cheese, softened
½ c mayonnaise
1 sm jar pimento, chopped
1 scallion top, minced
1 sm onion, grated
few tbsp milk, if needed

Combine all ingredients except milk and mix well. Add the milk a little at a time to reach desired consistency. Chill for several hours or overnight. Makes about 3 cups.

GATOR EGGPLANT DIP

1 c gator meat, cooked*, ground
1 lge eggplant
1 med onion, chopped
1 sm bell pepper, chopped
½ tsp salt or to taste
¼ c olive oil
1 10 oz can Ro-Tel Tomatoes
¼ c sauterne wine
1 clove garlic, minced
dash tabasco sauce

Peel, cube and boil eggplant in salted water until tender. Drain well. Heat the oil and saute the onion, bell pepper and garlic until tender. Add the eggplant, tomatoes, meat, salt and pepper. Cook for 30 minutes on low until thickened and then add the wine and stir in well. Remove from the heat and cool to room temperature and put into glass containers and refrigerate and serve cold with crackers or rye bread toast. Makes 2-3 cups.

ALLIGATOR ONION DIP

1 c gator meat, ground, boiled
1 pt sour cream
1 envelope dried onion soup mix
¼ c mayonnaise
1 tsp worcestershire sauce
dash tabasco sauce

Combine all ingredients, cover, and refrigerate for several hours. Makes 2-3 cups.

GATOR AND CARROT DIP

2 c alligator meat, cooked*
1 tsp horseradish, prepared
½ tsp salt
dash tabasco sauce
¼ c sweet pickle relish
1 sm jar pimento, chopped, ground or flaked
1 carton sm curd cottage cheese
1 pt sour cream
1 lge carrot, shredded

Mix well all ingredients and chill overnight. Let stand for 30 minutes before serving. Makes about 4½ cups.

BLUE DEVIL GATOR DIP

1 c gator meat, cooked*, ground
1 c sour cream
½ tsp dry mustard
1 tsp lemon juice
tabasco sauce to taste
1 3 oz cream cheese, softened
1 envelop blue cheese salad dressing mix
salt to taste

Combine the cream cheese, sour cream, salad dressing mix, mustard and lemon juice and blend until there are no lumps. Season with salt and tabasco sauce to taste and add the gator meat blending well. Refrigerate for several hours. Makes about 3 cups.

ALLIGATOR POINTE-AUX-CHENES

2 lbs alligator tail, cubed
peanut oil
cornmeal fish fry, seasoned*

Marinade

½ c honey
½ tsp cayenne pepper
1 lge onion, cut in chunks
4 lge cloves garlic, mashed
⅓ c yellow mustard
black pepper to taste
1 med lemon, sliced thin seeded
white wine (if needed)

Combine all marinade ingredients except wine in a medium bowl and add the alligator and stir thoroughly. If meat is not covered with the marinade add enough wine to cover. Chill in a refrigerator for 6 hours, stirring every hour. Remove and allow to come to room temperature and then drain the alligator. Dredge in seasoned fish fry until well coated. Shake off excess. Fry in oil over a medium heat until golden brown. Do not overcook. Drain and serve hot as an hors d'oeuvres.

ALLIGATOR APPETIZER

2-2½ lbs alligator tail meat, cut in finger size strips
juice of 3 lemons
¼ c parsley, chopped
¼ c bell pepper, chopped
oil
sm bottle tabasco sauce
½ c + 1-2 tbsp water
½ c celery, chopped
¼ c scallions, chopped
salt and pepper to taste
plain flour (optional)

Marinate meat in tabasco sauce, ½ cup water and lemon juice in the refrigerator for a few hours or overnight. Remove meat from marinade and dry. Season with salt and pepper to taste. Meat can be lightly floured if desired. Using a small amount of hot oil quick fry the meat, vegetables and parsley. Add 1-2 tbsp water and cover. Simmer 3-5 minutes. Serve hot. Serves 6-8.

NUT-CRUSTED ALLIGATOR

1 lb alligator fillets, 1 inch slices
2 tbsp wheat germ, toasted
3 tbsp mayonnaise
½ tsp soy sauce
¼ c cashews or walnuts, chopped
⅛ tsp paprika
1 tsp lemon juice
parsley sprigs

In a shallow dish combine nuts, wheat germ and paprika mixing well. In another small bowl stir together the mayonnaise, lemon juice and soy sauce. Dip the slices into the mayonnaise mixture and then roll in the nut mixture to coat. Arrange pieces in a greased shallow baking pan and bake at 400°F until alligator flakes easily. Transfer to a serving platter and garnish with the parsley. Serves 4.

BUTTERED GATOR FINGERS

2 lbs white alligator meat, in finger sized strips
1 4 oz can mushrooms, sliced
2 tbsp parsley, chopped
salt to taste
1 stick butter
½ c scallions, chopped
½ c white wine
¼ c lemon juice
tabasco sauce to taste

Make a marinade of the wine and lemon juice and soak the meat overnight in the refrigerator. Drain and reserve the marinade. Melt the butter in a large skillet and add the mushrooms with the liquid and cook until most of the liquid is gone. Add the meat fingers in a single layer and cover tightly and let steam for 15 minutes or until tender, stirring carefully. Add the scallions and simmer 5 minutes longer. Season and add the parsley just before serving. Serves 4-8.

ALLIGATOR COCKTAIL

1 lb white meat, cubed, steamed
1 c celery, chopped
2 hard boiled eggs, chopped
3 tbsp dill pickle relish
½ c mayonnaise
lettuce
1 tbsp lemon juice
2 tbsp ketchup
½ tsp worcestershire sauce
½ tsp salt
½ tsp cayenne pepper
lemon slices, thin, seeded

Mix all ingredients except meat and lettuce and chill well. Chill meat. To serve mix gently the sauce and meat together and serve in cocktail cups or on lettuce leaves. Garnish with lemon slices. Serves 4.

COCKTAIL ALLIGATOR

1 c alligator meat, cooked*, diced
1 tsp worcestershire sauce
½ tsp salt
dash tabasco sauce
1 tbsp onion, minced
lettuce
1 tsp lemon juice
1 c tomato juice
2 tbsp prepared horseradish
1 tbsp bell pepper, minced
lemon wedges

Combine all ingredients except lemon and lettuce and mix well and chill for several hours. Serve on lettuce and garnish with the lemon wedges. Serves 4.

MEAT BALL SURPRISE A LA GATOR

2 c ground alligator, boiled
2 eggs, beaten
¼ c milk
¼ tsp garlic powder
½ tsp tabasco sauce
1 tsp salt
stuffed olives, whole
plain flour
bread crumbs, if needed
oil

Combine the meat, seasonings, eggs and milk mixing well. Use finely ground bread crumbs to make the mixture thick enough to hold its shape. If too stiff add a little more milk. Using wet hands take a small amount of the mixture and shape around an olive so that the olive is in the center. Dredge the balls in the flour and fry in hot oil until browned. Drain and keep warm and serve with toothpicks and your favorite cocktail sauce. Makes several dozen.

GATOR COCKTAIL BALLS

1 lb alligator meat
1 lge onion
1 lge bell pepper
potato flakes
parsley flakes
oil
3 med potatoes, peeled
1 bay leaf
boiling water
salt and pepper to taste
plain flour
water

Place meat, onion, pepper, potatoes and bay leaf in enough boiling water to cover and boil until meat is tender. Drain and reserve 1 cup of the liquid. Grind the meat and mash or grind the vegetables removing the bay leaf and add the reserved liquid. Add enough potato flakes so that the mixture is stiff. Season to taste and add the parsley flakes. Shape into 1 inch balls and dredge in flour. Cook in the oil until brown. Drain and keep hot. Serve with chili or seafood sauce. Makes about 50 balls.

POLYNESIAN ALLIGATOR SALAD

2 c alligator tail meat, cooked, chopped
1½ c pineapple, diced or
white seedless grapes
½ c sour cream
1 tsp lemon juice
½ c carrots, shredded
1 c celery, chopped
½ c almond halves, toasted
½ c mayonnaise
1 tsp curry powder
salt and pepper to taste

Combine all ingredients well and chill. Arrange on salad greens to serve.

ALLIGATOR SEVICHE

1 lb alligator white meat in bite size pieces
1½ c lime juice
2 tbsp red pepper flakes
2 cloves garlic, minced
1½ c lemon juice
2 qts boiling water
1 lge onion, minced
2 tbsp salt

Place the meat in a colander and pour the boiling water over it seeing that all the meat is scalded well and set aside. Mix remaining ingredients well and add the meat being sure the meat is covered. If not add more juices and adjust for the seasoning. Refrigerate covered for 3-5 days. Serve with beer and crackers.

ALLIGATOR STUFFED LETTUCE

1 c alligator meat, cooked, flaked
1 sm jar pimento, chopped
5-10 stuffed olives, chopped
½ tsp celery seeds
½ tsp tabasco sauce
pickles
½ tsp salt
1 med head lettuce
1 c cheddar cheese, grated
½ c sour cream
sliced olives

Core the lettuce by removing the center from the bottom up leaving about a 1 inch shell. Place bottom side down to drain after washing. Combine all ingredients except the sliced olives and pickles and mix thoroughly. Spoon into the lettuce shell and pack firmly. Chill in the refrigerator for 6-12 hours wrapped tightly. To serve cut into 4-6 wedges and garnish with the sliced olives and pickles. Serves 4-6.

TANGY GATOR SALAD

1 lb gator meat, cubed, boneless
1 qt boiling water
1 carrot, whole
1 bay leaf
1 c celery, sliced
1 red onion, sliced
1 carrot, sliced
½ c mayonnaise
½ c lemon juice
1 tbsp parsley, chopped
½ c bell pepper, chopped
½ tsp salt
4 lge tomatoes or 2 avocados, halved
½ tsp tabasco sauce

Add the meat to the water along with the carrot and bay leaf and cook until the meat is tender. Drain and reserve the meat. Toss the meat which has been cooled, celery, onion and carrot in a large bowl and mix well. Mix the mayonnaise, lemon juice, parsley, pepper and seasonings and mix well. Pour over the meat mixture and again mix well. Cover and chill for several hours or overnight. To serve drain slightly and serve on tomatoes which have been cut into eight and spread open on lettuce leaves or halve the avocados and spoon the mixture over. Serves 4.

ALLIGATOR SWAMPGARDEN MOLD

1 lb alligator meat, boneless
2 c water
2 tbsp lemon juice
1 sm onion, cut in wedges
½ tsp salt
1 bay leaf
1½ c rice, cooked
1 carrot, shredded
1 rib celery, minced
1 sm bell pepper, minced
1 scallion, chopped
1 tbsp sweet pickle relish
½ c mayonnaise
1 tsp soy sauce
½ tsp curry powder

In a saucepan bring the water to a boil and add the meat, lemon juice, onion, salt and bay leaf and bring to a boil again and then reduce the heat and simmer for 30-45 minutes or until the meat flakes easily. Drain, cool and flake the meat discarding the liquid or save the stock for some other recipe. Add all the remaining ingredients in a large bowl and add the meat and mix well. Oil a bowl or mold large enough to hold the mixture and pack tightly into the mold. Chill for 12 hours or longer. To serve, unmold on a bed of lettuce and garnish with items of your choice. Serves 4-6.

FANCY MOLDED ALLIGATOR

1 c alligator meat, ground, cooked
1 tbsp mayonnaise
1 tbsp lemon or lime juice
tabasco sauce to taste
½ tsp curry powder
2 8 oz cream cheese
½ lb sharp cheddar cheese, grated
1 tsp garlic salt

Bring all ingredients to room temperature and combine in a large bowl and mix thoroughly. Pack into a greased 1 quart mold and chill overnight or longer. To serve unmold on a bed of lettuce leaves and add garnishes if you wish.

ONION GATOR SALAD

2 c gator meat, ground, cooked
2 tbsp lemon juice
1 c mayonnaise
6 scallions, thinly sliced
salt to taste
fresh salad greens (optional)
4 c potatoes, diced, cooked
½ c celery, chopped
¼ c dill pickle relish
1 lge carrot, thinly sliced
black pepper to taste
paprika (optional)

Combine all ingredients except salad greens and paprika and toss lightly to mix well. Season to taste and refrigerate for at least 24 hours before serving. Spoon into beds of fresh greens and sprinkle with paprika if desired. Serves 6-8.

MEXICAN ALLIGATOR FIESTA SALAD

1 lb gator meat, ground, cooked
1 sm head lettuce, shredded
1 lge onion, chopped
4 med tomatoes, chopped
¼ lb cheddar cheese, grated
1 8 oz Catalina Salad Dressing
1 sm bottle taco sauce
1 lge avocado, peeled, cubed
1 lb can kidney beans, drained
1 5 to 7 oz bag corn chips
¼-½ tsp salt
butter

Cook the meat in beef broth if possible. Sprinkle the avocados with a little lemon juice. Heat the meat with a very small amount of butter and keep warm. Line a large salad bowl with lettuce and top with the tomatoes and onion and sprinkle with half the cheese and corn chips. Add the salad dressing and taco sauce (mild or hot) to the meat and toss well adding the beans. Pour about half of this over the chips and then add the remaining chips and top with the remaining meat mixture. Then add the avocados and remaining cheese on top. Serve like this but toss with a wooden spoon before serving. Serve with hot buttered tortillas and cool lemonade or cold beer. Serves 6-8.

PARADISE SALAD A LA GATOR

2 c gator meat, cubed
2 onions
2 carrots
dash tabasco sauce
1 chicken bouillon cube
1 c celery, chopped
1 c raisins, white
paprika
water
2 c coconut, grated
1 tbsp curry powder
1 c mayonnaise
¼ c reserved stock
1 tbsp lemon or lime juice
salt and white pepper to taste
parsley, chopped
1 13 oz can pineapple chunks, drained

Boil the meat with the onions, carrots, tabasco sauce and bouillon cube in enough water to cover until the meat is tender, about 30-45 minutes. Drain and reserve the stock. Soak the raisins in the drained pineapple juice for 20 minutes and then drain. Place the meat in a large bowl and add the celery, raisins, pineapple and coconut and mix well. Set aside in the refrigerator. Add the curry powder to the mayonnaise and add the ½ cup of reserved stock, pineapple and lemon juice and mix well and season with the salt and pepper. Add to the meat mixture and toss well. Chill for a few hours. Garnish with the parsley and paprika before serving. Serves 4-6.

BRANDIED ALLIGATOR

2 c alligator meat, cubed,
cooked, from tail tenderloin
½ c mayonnaise
1 tbsp ketchup
2 tsp lemon juice
1 tsp parsley, minced
¼ c brandy
tabasco sauce to taste
salt to taste
pinch oregano, ground

Combine all ingredients except the meat and mix thoroughly and then add the meat and toss to coat all pieces. Chill for several hours. Serve on lettuce leaves or avocado halves and use lemon slices for garnish and sprinkling. Serves 4.

GATOR STUFFED CANTALOUPE

2 c gator meat, cooked, ground
4 sm to med cantaloupes
3 stuffed olives, chopped
2 tbsp sweet pickle relish
1 tbsp creole mustard
1 c sour cream
taste
1 tbsp butter, softened
pinch onion powder
pinch celery seed or salt
1 lge apple, cored, diced
4 hard boiled eggs
salt and white pepper to paprika

Halve the cantaloupes lengthwise and remove the seeds and drain with cut side down. Chop the egg whites and grate the yolks. Mix the olives, mustard, relish, butter and meat together mixing well. Add the onion powder, celery seed or salt and apple and toss well. Add the sour cream and the egg whites and stir in gently. Season to taste. Fill halves with the salad mixture and chill well before serving. Sprinkle with the egg yolks and paprika before serving. Serves 8,

ALLIGATOR MOUNDS

1 c gator meat, ground, cooked
1 envelope gelatin, unflavored
1 10 oz can cream of asparagus
 or cream of celery soup
¼ c mayonnaise or salad dressing
¼ c radish, chopped
½ c celery, chopped
¼ tsp salt
dash tabasco sauce
½ c cold water

Place the gelatin in the water and set aside. Heat the soup in a saucepan and when very hot add the gelatin and lower the heat and cook, stirring constantly, until the gelatin is dissolved. Set aside to cool. When barely warm and thickened add the remaining ingredients. Spoon into 4 individual cups or a mold that has been lightly oiled and chill until firm. Unmold and serve on lettuce leaves. Garnish with mayonnaise and radish slices or rosettes if desired. Serves 4.

ALLIGATOR SALAD MOLD SUPREME

1 c gator meat, cooked, ground
1 envelope gelatin, unflavored
½ c cold water
1 10 oz can tomato soup
2 3 oz pkgs cream cheese
1 c mayonnaise
½ c celery, minced
1 tsp lemon juice
½ c stuffed olives, chopped
½ c pecans, chopped
4 hard boiled eggs, chopped
1 tsp parsley flakes
½ tsp paprika
½ c onion, minced

Soak the gelatin in the cold water. Heat the soup and stir in the cheese and simmer over low heat until very smooth. Add the gelatin and stir until well blended. Let cool and then add the remaining ingredients and mix well. Place in an oiled mold and chill until firm. Unmold on a bed of salad greens and garnish if you desire. Serves 8-10.

HOT ALLIGATOR SALAD

1-2 lbs alligator meat, boneless
1 bay leaf
1 carrot, whole
1 tbsp tabasco sauce
1 tsp celery seed
1½ c celery, diced
1 c potato chip crumbs
1 med onion, diced
1 med bell pepper, diced
1 c mayonnaise
1 tsp worcestershire sauce
½-1 tsp creole seasoning*
water

Boil over medium heat the meat, bay leaf, carrot, tabasco sauce and celery seed in enough water to cover until the meat is tender. Drain, cool and remove the meat and discard the seasonings. To the meat add all remaining ingredients except crumbs and mix well. Place in a buttered casserole and cover with the crumbs. Bake in a moderate oven at 350°F until the center is hot and bubbly. Serves 6.

SUPPER GATOR SALAD SUPREME

1 lb gator meat, cooked, ground
2 envelopes gelatin, unflavored
¼ c cold water
3 tbsp lemon juice
1 tbsp onion, minced
¼ tsp cayenne pepper
¾ c mayonnaise
1 c celery, diced
¼ c bell pepper, diced
1 sm jar pimento, diced, drained
½ tsp salt

Soften the gelatin in the cold water and lemon juice and then dissolve completely in the hot water and set aside to chill until thickened and syrupy. Fold in the remaining ingredients and put into a 2-quart oiled mold and chill until very firm. To serve unmold on a bed of lettuce. Serves 4-6.

ORANGE GATOR MOLD

1 c gator meat, cooked, flaked
1 3 oz pkg orange gelatin
1½ c boiling water
1 c sour cream
extra sour cream (optional)
½ c carrot, grated
1 8 oz can crushed pineapple, well drained
½-¾ c pecans or walnuts, chopped
carrot curls (optional)

In a quart bowl combine the gelatin with the boiling water and stir to dissolve and then add the sour cream and blend well. Refrigerate until partially set. Add the meat, carrots, nuts and pineapple and stir to mix well. Pour into an oiled mold and chill at least 6 hours. Unmold on a bed of lettuce and garnish with mounds of sour cream and a few carrot curls if desired. Serves 4-6.

ALLIGATOR RAVIGOTTE

2 c alligator meat, cooked, finely diced
1 tbsp prepared horseradish
2 tbsp pimento, chopped
1 hard boiled egg, chopped fine
½ tsp lemon juice
1 tbsp dry mustard
1 c mayonnaise
2 tbsp parsley, minced
2 tbsp capers, minced

Thoroughly mix all the ingredients except the meat. Add the meat and toss lightly and chill for several hours. Serve as h'oeuvres with crackers or toast points. As a salad place into individual ramekins and garnish with tomato, pimento or olive slices. Serves 4.

FLAMING GATOR

2-3 lbs alligator tail fillets,
1 stick butter
¼ c cognac or whiskey
1 c milk + water to cover meat
1 tsp white pepper
½ tsp salt
2 cloves garlic, minced

Cut the meat ½ inch thick across the grain and marinate in enough milk and water to cover for several hours before cooking. Drain and dry. Melt the butter and add the pepper and pour over the meat and cook for 20 minutes covered tightly on a very low fire. Uncover and check the meat and continue to cook uncovered until the meat is tender. Season with the salt and drain off the excess drippings and serve it in a sauce boat. Place the meat in a serving platter and warm the cognac or whiskey in a small saucepan and ignite and pour over the meat to serve. Serves 4-6.

RED GATOR, RED GATOR

½ lb alligator meat, ground
¼ c butter
¼ c plain flour
1½ c milk
1 6 oz can tomato paste
1 tsp salt
½ tsp garlic salt
2-4 drops tabasco sauce
¼ tsp curry powder
2 tsp brandy or sherry
½ lemon, sliced, seeded
water

In unseasoned water boil the meat for about 30 minutes until tender. Drain well. Melt the butter and add the flour stirring constantly and then add the milk and cook until smooth and thickened stirring constantly. Add the tomato paste and a can of water and all the seasonings and stir in the meat and heat well. Keep warm in a chafing dish and serve with crackers.

SPICY AND SWEET ALLIGATOR

1 lb alligator meat, cubed
1 bay leaf
1 tbsp tabasco sauce
water to cover
1 carrot, peeled and stemmed

Sauce

¼ c A-1 sauce
¼ c ketchup
¼ c heavy cane syrup
¼ c apple cider vinegar

Boil the meat with the bay leaf, tabasco sauce and carrot until very tender but not falling apart. Drain and rinse and set aside. Combine all the sauce ingredients in a heavy saucepan and stirring constantly heat to boiling and then simmer for 5 minutes. Add the meat and heat thoroughly. Serve over rice or in hollowed out french buns or in a chafing dish and serve with crackers for dunking. Makes 4 cups.

GATOR ROUNDS

1 c cooked* ground alligator
1 3 oz cream cheese, softened
¼ c mayonnaise
2 tbsp white onion, minced
1 clove garlic, minced
24-30 bread rounds about 2 in. diameter
butter
1 c swiss or romano cheese, grated

Mix well all the ingredients in the left hand column in a glass bowl and chill until needed. Butter the bread rounds on one side and place in a single layer on a cookie sheet under the broiler for a few minutes to brown lightly. Turn and brown the unbuttered side and repeat until all rounds are browned. Cool the rounds slightly and set aside until needed. To serve spread small amounts of meat mixture on the unbuttered side of the rounds and sprinkle with the cheese. Place on the cookie sheet and broil about 4 inches from the heat for 3-5 minutes until cheese is lightly browned. Serve hot. Makes 24-30 rounds.

ALLIGATOR TONGUE ACADIAN

8 alligator tongues, boiled, skinned, diced to 1 inch size
¾ c celery, minced
1 pkg Knorr Oxtail Soup Mix
½ c dry vermouth
cayenne pepper to taste
2 c onions, minced
1 c bell pepper, minced
2 sticks butter
1 pkg Lipton Onion Soup Mix
4 c fresh mushrooms, sliced

Saute the onions, bell pepper and celery in the butter until the vegetables are well-wilted. Add the soup mixes and the vermouth and cook until the gravy thickens. Add the alligator and simmer for 15 minutes and then add the mushrooms and a little cayenne. Serve over rice. (This can also be done with beef tongue.)

PICKLED ALLIGATOR

1 lb alligator tail fillets
water to cover
1 onion
1 carrot
dash tabasco sauce
2 tbsp oil
½ c water
½ c vinegar
2 tbsp onions, minced
1 tsp pickling spice
1 bay leaf
½ bell pepper, diced
1 tsp salt
¼ tsp white or cayenne pepper

Slice the meat into finger strips and boil gently in water to cover with the onion, carrot and tabasco sauce. When tender drain and reserve the stock for other uses. Discard the vegetables. Combine the remaining ingredients mixing well and add the meat and gently turn. Cover and let stand in the refrigerator at least 24 hours turning occasionally. Serve cold as a relish or a main dish. Makes about 2½ cups pickle. Will last in a refrigerator about 2 weeks.

ALLIGATOR REMOULADE

1 lb alligator meat, cubed, boiled
1 tbsp lemon juice
¾ tsp paprika
salt and pepper to taste
½ c ketchup
3 scallions, minced
2 tbsp brown mustard

Add all ingredients in a bowl and mix well. Cover and chill until ready to serve. Serve on lettuce leaves and garnish as desired. Serves 4-6.

QUICK SANDWICH GATOR FILLING

1 c gator meat, cooked, chopped
¼ c sweet pickle relish
1 tsp soy sauce
¼ c mayonnaise
buttered white or rye bread
1 tbsp prepared mustard
¼ tsp celery salt
1 hard boiled egg, chopped
8-10 olives, chopped
salt and pepper to taste

Combine well all ingredients and season to taste and chill for several hours or overnight. Spread on buttered bread for a sandwich. Makes 4-8 sandwiches.

OPEN FACE GATOR SANDWICHES

1 c gator meat, ground, cooked
½ c ripe olives, chopped
4 hard boiled eggs, chopped
⅔ c mayonnaise
black pepper to taste
salt and pepper to taste
sliced bread, crust removed
butter
garlic salt to taste
1 slice swiss cheese per sandwich

Combine the meat, olives, eggs and mayonnaise and season to taste. Butter the crustless bread slices and spread with the meat mixture. Place a slice of cheese on each and place on a cookie sheet. Melt a little butter in a heavy skillet and add the crusts and season with garlic salt and pepper and toss for a few seconds. Then place the bread crusts on the cookie sheet with the sandwiches and bake in a 400°F oven for 8-10 minutes or until the sticks are crisp. Serve hot. Makes 4-8 sandwiches.

ALLIGATOR POCKET SANDWICHES

2 c gator meat, cooked, ground
1 med tomato, chopped
¼ c dill pickle relish
1 sm onion, sliced
2 tsp milk
½ tsp creole seasoning*
½ c mayonnaise
6-8 slices bacon or 1 c ham, cubed
pinch lemon pepper seasoning
2 pita breads, cut in half

Combine the gator meat and tomato in a large bowl. Cook the bacon or ham in a skillet and then drain and reserve the pan drippings. Add the onion and seasonings to the drippings and saute until the onions are soft but not browned. Add to the meat-tomato mixture and mix well. Crumble the bacon, if using, and reserve a little for topping and add remainder to meat-tomato mixture. Fill the pockets with the filling. In a small saucepan heat the mayonnaise, relish and milk over a low fire stirring constantly. Spoon over each pocket sandwich and sprinkle with the reserved bacon or ham. Serve immediately or wrap in foil and keep in a low oven until needed. Serves 4.

NUTTY GATOR SPREAD

1 c gator meat, cooked, ground
¼-½ c mayonnaise
1 3 oz cream cheese
1 c pecans, chopped
1 4¼ oz can olives, chopped
½ tsp lemon juice
salt and pepper to taste
pinch nutmeg

Mix all the ingredients using only ¼ cup of the mayonnaise to start and add more if needed to obtain a good spread. Use as a dip or spread on crackers or rye bread.

GATOR ITALIAN MEAT PIES

1 lb alligator meat, ground
½ c milk
2 tbsp parmesan cheese, grated
1 tsp italian seasoning
1 8 oz can pizza sauce
1 4 oz can mushrooms, sliced
1 egg, beaten
¼ c bread crumbs
1 tbsp minced dried onion
1 2 oz pkg sliced smoked beef, cut up
½ c mozzarella, cheddar or swiss cheese, shredded

In a mixing bowl combine the egg and milk and stir in bread crumbs, parmesan, onion and italian seasoning and mix well. Add the alligator. Press the meat mixture into the bottom and up the sides of a greased 9 inch pie plate. Bake meat mixture at 375°F for 15 minutes. Top with the drained mushrooms and smoked beef. Spoon the pizza sauce over all. Bake meat pie 10 more minutes. Sprinkle shredded cheese on top. Bake about 5 minutes more or until the cheese is melted. Let stand for 5 minutes before cutting the meat pie into wedges. Serves 6.

POO-YI PIES GATOR-STYLE

1 lb gator meat, cooked, ground
1 tbsp oil
2 tbsp plain flour
1 onion, chopped
3 scallions, chopped
¼ c ketchup
oil (for frying)
2 c plain flour, sifted
1 tsp baking powder
¼ c butter, melted
1 egg
2-4 tbsp milk
salt and pepper to taste

Make a roux* of the oil and 2 tbsp flour until thick and very dark. Add the onions ketchup and meat and cook until thick and pasty and add seasonings to taste. Set aside. Sift the flour and baking powder and add the butter and egg stirring well. Add enough milk to make a stiff dough. Roll out very thin and using a biscuit cutter cut out circles of the dough. Reroll the dough and repeat until all is used. Put a spoonful of the meat mixture on a circle and place another circle over it and seal the edges with a fork or your fingers until completely sealed. Repeat until all pies are made. Fry in deep, hot oil until a golden brown. Serve hot. Makes about 20 small pies.

CAJUN ALLIGATOR TAMALE PIE

2 lbs gator meat, ground, boiled
¼ stick butter
3 red onions, diced
¼ tsp cayenne pepper
2 bay leaves, crushed
1 tsp sugar
1 c boiling water
1 14½ oz can chicken broth
1 14½ oz can water
1 8 oz can tomato sauce
½ tsp cumin powder
¼ tsp dry mustard
2 c yellow cornmeal
2 tsp salt

Cook the cornmeal in the boiling water with the salt to make a mush. Cool until it can be handled and then pour it into a 2 quart shallow casserole and shape the mush up the sides. Combine all the remaining ingredients in a heavy pot and bring to a boil and then reduce the heat and simmer for 30 minutes stirring frequently. Spoon the meat mixture carefully into the lined casserole. Bake in a 400°F oven for 20-30 minutes or until crust is browned. Cool a little before serving and cut into slices or squares. Serves 8.

TORTILLA CASSEROLE A LA GATOR

1 lb alligator meat, cooked, diced
1 10¾ oz can cream of mushroom soup
¼ c oil
1 4 oz can green chilies, seeded, minced
1 jalapeno pepper, minced
1 lge onion, chopped
1 c cheddar cheese, grated
lge bag tortilla corn chips

Heat the oil and saute the onion until onion is soft. Add the chilies and jalapeno and stir in the soup and half the cheese. Cook on a low fire until the cheese is melted stirring constantly. Oil a 3-quart casserole and place enough chips on the bottom to cover. Sprinkle over half the meat and then top with half the soup mixture. Repeat and top with the remaining cheese and some chips. Bake uncovered at 325°F for 45-60 minutes until casserole is bubbly and heated through. Serves 8.

GATOR TACO CASSEROLE

1 lb gator meat, cooked, ground
1 10 oz can Ro-Tel Tomatoes
1 envelope taco seasoning mix
tomatoes, chopped
1 onion, chopped
4-6 taco shells
shredded lettuce
cheddar cheese, grated

Heat the meat, Ro-Tel Tomatoes and onion over a medium heat until very warm. In a casserole break up the taco shells and top with the meat mixture and toss lightly. Bake at 350°F until hot. Sprinkle cheese on the top and place back in the oven until the cheese melts. Remove from the oven and top with the lettuce and chopped tomatoes and serve while hot. Serves 5-8.

GATOR ENCHILADAS

1 lb gator meat, cooked, ground
½ stick butter, melted
1 16 oz can refried beans
1 bean can water
salt and pepper to taste
1 lge onion, chopped
1 jalapeno pepper, minced
½-1 lb monterey jack cheese, grated
8-12 flour tortillas, warmed

Add the beans and water to the butter and bring to a boil and then add the onions and cook for 15 minutes stirring often. Add the meat and the chopped jalapeno and continue to cook for 45-60 minutes or until thick. Season to taste and remove from the heat. Add about a cup of the cheese and stir until it is melted. Put a spoonful of the mixture in the center of each tortilla and roll up. Place the seam side down in a baking dish and top with the remaining cheese. Bake at 300°F for 20-30 minutes or until the cheese is melted and bubbling. Serves 6-8.

ALLIGATOR TACOS

2 c gator meat, ground, cooked
1 tsp lemon juice
1 tsp salt
1 sm onion, diced
2 tbsp butter
½ head lettuce, shredded fine
1 c cheddar cheese, shredded
1 lge tomato, diced
taco sauce, hot or mild
8 taco shells

In a large bowl combine the meat, lemon juice and salt and set aside. In a heavy small saucepan, saute the onion in the butter and then add this to the meat mixture and mix well. Heat the taco shells according to directions. Let each person assemble his own taco by spooning in the meat mixture and topping with lettuce, tomato, cheese and sauce. Other toppings may include sour cream, chopped onion or jalapeno pepper if desired. Serves about 6.

ALLIGATOR FRITTERS

1½ lbs alligator, ground
1 onion, minced
½ tsp baking soda
1 tsp salt
oil
½ c bread crumbs
⅓ c milk
1 egg, beaten
¼ tsp black pepper

Mix all ingredients except oil. Shape into ½ inch meatballs and refrigerate.

⅔ c milk
2 eggs, beaten
1⅓ c plain flour
1½ tsp salt
mustard sauce
½ tsp baking soda
4 tsp oil
1 tsp double-acting baking powder
¼ tsp black pepper

Mix all ingredients into a bowl to make a batter. In a 3-quart saucepan heat about 3 inches of oil over a medium flame. Coat meat balls with batter and fry about 3 minutes until golden. Drain well and keep warm. Top with mustard sauce 2. Serves 6.

MUSTARD SAUCE 2

1 c mayonnaise
½ tsp sugar
¼ tsp paprika
½ c prepared mustard
¼ tsp salt
⅛ tsp cayenne pepper

Mix well all ingredients.

FRITTERS A LA GATOR

1 lb gator meat, cooked, ground
1 c plain flour
1 tsp baking powder
oil
½ tsp salt
dash tabasco sauce
2 eggs, beaten with 2 tbsp milk

Sift the flour and baking powder and then add the remaining ingredients except oil mixing well. If it is too stiff add a little more milk to make a very thick batter. Drop a tsp at a time into hot oil and cook until browned. Drain well. Makes about 3 dozen.

ALLIGATOR FRITTERS 2

2 lbs alligator meat, boiled, diced
3 lge onions, chopped
1 c plain flour
salt and pepper to taste
3 bell peppers, chopped
1 can beer
oil

Mix the beer and flour and let stand at room temperature for 3 hours. Mix the meat, onions and pepper and stir into the beer batter. Season to taste. Drop by spoonfuls into hot oil and fry until brown.

ALLIGATOR CREPES

1 c alligator tail meat, diced, cooked
¼ tsp salt
¼ tsp white pepper
1 egg yolk, beaten
4 crepes
parsley, minced
¼ c white wine
¼ stick butter
¼ c scallions, minced
¼ c plain flour
¾ c alligator stock

Over medium heat in a large skillet melt the butter and saute the onions until tender and then blend in the flour and stir constantly and cook slowly for about 5 minutes. Do not brown the flour. Remove from the heat and add the stock, wine, salt and pepper and mix well. Add the egg yolk and stir until completely mixed and smooth. Add the meat and return to a medium heat and when almost boiling lower to very low and cook for 15 minutes stirring constantly. Roll the crepes with 3-4 tbsp of the sauce inside. Serve with remaining sauce on top. Garnish with the parsley. Serves 4.

CREPES

¾ c sifted flour
pinch of salt
milk
2 eggs
1 tsp sugar
oil

Mix the eggs with the flour, sugar and salt and then add milk stirring well until the batter is the consistency of evaporated milk. Beat until very smooth. Lightly oil a 6 inch skillet and pour about 2 tbsp of the batter into the pan tilting quickly to distribute the batter evenly. Cook for 1 minute or until edged brown and turn to brown the other side. Keep the cooked crepes in a towel to keep moist and warm. Oil pan before each crepe. Makes 4 crepes.

LOUISIANA ALLIGATOR MEAT ROLLS

1 c alligator meat, ground,
1 tbsp butter
¼ c bell pepper, minced
1 scallion, chopped
4-6 crepes*
black pepper to taste
pinch of salt
½ c white wine
1 10 oz can cream of celery soup
½ lb american cheese, grated
1 12 oz can evaporated milk

Combine the soup and milk and mix thoroughly. Saute the pepper and onion in the butter until tender. Add the salt and wine and simmer until hot. Add the meat and 1 cup of the soup mixture. Add more salt and pepper if needed. Place in each crepe 2 tbsp of filling in the center and roll. Place on individual baking dishes or in a casserole. When all the crepes are filled top with the remaining soup mixture. Sprinkle with the grated cheese and cook in a 375°F oven until the cheese is melted and the sauce is bubbling. Makes 4-6 crepes.

GATOR AND RICE CROQUETTES

1 c gator meat, cooked, ground
1 c + italian bread crumbs
1 tbsp parsley flakes
1½ c rice, cooked
2 eggs, slightly beaten
dash tabasco sauce
½ tsp salt
oil

Combine the meat, cup of bread crumbs, rice, parsley, eggs and seasonings and mix well. Shape into balls and roll in additional bread crumbs. Chill the balls for 2-4 hours and then deep-fry in oil until brown. Serve hot with ketchup or cocktail sauce. Makes 8-10 large croquettes.

ALLIGATOR DELIGHTS

2 c alligator tail meat, cooked, ground
¾ c monterey jack cheese
⅓ c mayonnaise
2 eggs, beaten
1 tbsp lemon juice
salt to taste
freshly ground black pepper
tabasco sauce to taste
plain bread crumbs
3 slices white bread
½ stick + butter
1 bell pepper, chopped
1 med onion, chopped
1 clove garlic, mashed
3 scallions, chopped
2 ribs celery, chopped
3 tbsp parsley, chopped

Saute the vegetables in the ½ stick butter until soft. Soak the bread in water and squeeze dry. Grate the cheese and mix with the remaining ingredients, except the bread crumbs and remaining butter, and add the vegetables, mixing well. Fill 8-10 crab shells or ramekins with the mixture and top with bread crumbs and dot with butter. Bake at 300°F for 30-40 minutes. A casserole dish can be used instead of individual servings. Serves 8.

ALLIGATOR GRILLADES

1 lb alligator tail
plain flour
¼ tsp white pepper
½ med onion, minced
1 stalk celery, minced
1 16 oz can italian plum tomatoes
4 sprigs parsley, chopped
¼ c olive oil
½ tsp salt
⅛ tsp cayenne pepper
1 red bell pepper, minced
1 tsp garlic, minced
¼ c dry white wine
3 leaves basil, chopped

If possible get a young alligator and cut the cutlets against the grain about ¼ inch thick. Tenderize if necessary. Lightly dust the cutlets with flour and a little salt and white pepper. Heat the oil in a skillet until almost smoking and saute the cutlets for about 1 minute on each side. Remove and keep warm. In the remaining oil saute the onion, bell pepper, celery, garlic and parsley until tender. Chop the tomatoes and add them to the skillet along with the tomato juice, basil, salt, pepper, cayenne and wine. Bring to a simmer and return the cutlets to the pan and cook them in the sauce for 5 minutes. Serve.

TRAPPER'S OMELET A LA GATOR

1 lb alligator meat, cooked, diced fine
6 eggs, beaten slightly
¼ c cream
salt and pepper to taste
½ c onion, chopped
2 tbsp butter
3 slices bacon, fried, crumbled
½ c bell pepper, chopped

Saute the bell pepper and onion in the butter on low heat until tender. Add the meat and simmer for a few minutes. Add the bacon and cream to the eggs and mix well. Add to the meat mixture and cook on a low heat until the eggs are set, occasionally turning. Season to taste and serve at once. Serves 4-6.

ALLIGATOR SCRAMBLED EGGS

1 c alligator meat, cooked,
  ground or finely diced
¼ stick butter
1 scallion, tops, chopped
8 eggs, beaten
salt and pepper to taste

Place meat and scallions in the butter and saute until hot and bubbly. Add the eggs and cook over a low heat until eggs are set, stirring occasionally. Season. Serves 6.

GATOR PLATTER

2-3 lbs alligator meat, cubed
4 med onions, peeled
1 lge can or jar spiced peaches
tabasco sauce to taste
water
¼ c butter
½ c plain flour
salt to taste
hot rice or mashed potatoes

Boil the meat in water for 20 minutes and then drain and rinse. Again boil the meat in fresh water with the onions until the onions are tender. Drain. Drain the peaches and set aside. Melt the butter and add the flour and cook until flour is lightly browned, stirring often. Add the peach juice and bring to a boil and then add the meat and cook over a low fire until the meat is very tender. Add the peaches and the cooked onions and cook until they are hot. Season to taste. Serve over the rice or potatoes. Serves 6-8.

CREAMED EGGS AND ALLIGATOR

2 c alligator meat, cooked, finely diced
1 tsp prepared horseradish
4 hard boiled eggs, chopped
2 tbsp parsley, chopped
½ c bread or cracker crumbs
½ tsp tabasco sauce
½ c butter
½ c plain flour
2 c milk
1 tsp salt

Melt butter and add the flour and mix well and cook for 3-5 minutes on a low heat stirring constantly. Continue stirring and add the milk slowly and cook until thickened. Add the salt, tabasco sauce, horseradish and meat and cook on low for 15 minutes stirring often. Add the parsley and eggs and stir in gently. Preheat oven to 350°F. Pour the mixture into a baking dish and top with crumbs and bake for 20-30 minutes or until very hot. Serve at once over hot biscuits or english muffin halves. Serves 4-6.

STUFFED ALLIGATOR STEAKS

4 alligator steaks (1 in thick)
½ c oil
salt and pepper to taste
½ c scallions, chopped
1 lb alligator meat, ground
1 c water
¼ c parsley, chopped
2 cloves garlic, chopped

Mix ground meat with scallions and parsley and season to taste with salt and pepper. Cut a slit lengthwise in the middle of each steak to form a pocket. Stuff about 4 ounces of the seasoned meat into each pocket. Rub each steak with salt and pepper. Heat oil and fry the steaks until brown on each side. Add water a little at a time. Cover pot tightly and cook on low heat for about 1 hour. Serves 4.

STUFFED ALLIGATOR ROLL

1 lge alligator tail tenderloin
2 tbsp brown sugar
2 tbsp mint, chopped
2 med apples, cored, diced
1 c bread cubes
¼ c milk
butter, melted
¼ tsp ginger, ground
water
½ c celery, chopped
2 tbsp butter
1 tsp salt
lemon juice

Soak the alligator in water with some crushed mint for several hours and then drain and dry. Make a lengthwise pocket in the tenderloin about 1 inch from the ends. Rub the meat inside and out with melted butter. Cook the celery in the 2 tbsp of butter until soft. Add all the remaining ingredients except lemon juice and mix well. Fill pocket with the stuffing and tie. Place in a buttered baking dish and use leftover stuffing and place around the meat. Cover and bake at 325°F for 45 minutes and then baste with melted butter and lemon juice as needed and continue to bake for another 30-45 minutes or until tender.

ALLIGATOR SPANISH RICE

½ lb alligator, ground
¼ c onion, chopped
½ tsp salt
⅛ tsp paprika
1 14½ oz can stewed tomatoes
2 slices bacon, chopped
steamed rice for 4
⅛ tsp black pepper
¼ tsp chili powder

Saute bacon pieces. Add onion and alligator, breaking meat into small pieces. Add the seasonings and cook until almost done. Add tomatoes and simmer 25 minutes. Put in the rice and simmer an additional 5 minutes. Serves 4.

BAKED GATOR CABBAGE ROLLS

1 lb gator meat, cooked, ground
1 med cabbage
2 tbsp butter
1 lge onion, chopped
dash tabasco sauce
romano or parmesan cheese, grated
1 6 oz can tomato paste
2 paste cans water
1-1½ c cooked rice
1 c V-8 Juice or tomato juice
salt and pepper to taste

Core and steam the cabbage until the large outer leaves peel easily and set aside. You should have 8-12 leaves. Chop the remaining cabbage. Melt the butter and add the onion and saute until the onion is clear on a low heat. Add the meat, paste, water and chopped cabbage and bring to a boil and then simmer until the cabbage is tender in about 15 minutes. Stir in the rice and season to taste. Spoon about a ½ cup of mixture solids on each leaf and roll up and secure with toothpicks. Place in a shallow baking pan and when all are done pour the remaining juices with tabasco added on top of the rolls. Sprinkle with the cheese and bake at 350°F for 1 hour. Serve.

ALLIGATOR CABBAGE ROLLS

1 lb alligator meat, ground
½-1 c water
2 onions, chopped
2 tbsp butter
1 med cabbage
1 8 oz can tomato sauce
bread crumbs
1 14½ oz can stewed tomatoes
¼ c long grain rice, raw
salt and pepper to taste
pinch oregano, ground
½ tsp sugar
parmesan or romano cheese, grated

Core and steam the cabbage until the outer leaves come off easily. Chop the rest of the cabbage coarsely. Cook the meat, onions, butter, tomatoes and ½ c water for 30 minutes stirring often. Add the rice and seasonings and reduce the heat and cook for 30 minutes or longer until the rice is tender. Add water if necessary. Adjust seasonings if necessary. Fill the cabbage leaves with the filling mixture and roll up and secure with toothpicks. Place in a baking dish and combine any remaining filling with the chopped cabbage and spread around the rolls. Top with the can of tomato sauce and sprinkle with the cheese and bread crumbs. Bake at 350°F for 1 hour. Serves 8-10.

ALLIGATOR CHOWDER

alligator fillets
salt and pepper to taste
white onions, thinly sliced
cream
water
butter
soda crackers
bacon
irish potatoes, peeled

In an iron dutch oven or deep iron skillet place a layer of sliced bacon. On top of this, layer alligator fillets, then layer thinly sliced, raw irish potatoes, a layer of onions and a layer of soda crackers. Dot with butter and salt and pepper to taste. Repeat layers in the same order until the pot is filled. Add water halfway to height of vessel, cover. Cook slowly until alligator, onions and potatoes are done. Liquid must boil away so bottom layer of bacon and alligator is brown. Add cream to cover contents, heat to boiling and serve immediately.

ALLIGATOR CHOWDER DELUXE

2½ lbs alligator, ground*
8½ lbs potatoes, peeled, diced
1½ gal water
1½ lbs onions, chopped
3 12 oz cans tomatoes
¼ tsp thyme
4 ozs salt pork, cubed
2 tbsp bacon drippings
2 lge cloves garlic, chopped
½ tsp sweet basil
¼ tsp oregano
salt and pepper to taste

Place the potatoes in a large dutch oven and add water and salt and bring to a boil. Cover and simmer until the potatoes are soft. In a skillet brown the salt pork in the bacon drippings. Drain and add onions and saute until tender. Put tomatoes in a blender for a second to puree. Add tomatoes and chopped garlic into the skillet mixture. Bring to a boil and simmer until the liquid is absorbed or until thick. Add this mixture to the potatoes. Next add the ground alligator. Add spices and bring to a boil. Cover and reduce heat to simmering for 45 minutes.

*Leg or body meat is preferred.

KID'S ALLIGATOR CASSEROLE

3 c alligator meat, cooked, finely diced
1 4 oz can mushrooms, sliced
½ c parmesan cheese, grated
1 6 oz pkg egg noodles
1 10¾-can cream soup, any kind
1 can vegetables, drained, your choice

Boil the noodles for 15 minutes according to package directions. Drain. Mix all remaining ingredients, except cheese, with the noodles and place in a greased 9-10 inch pie plate. Top with the cheese. (Bread crumbs or crushed potato chips can be substituted or added.) Bake at 325°F for 25-35 minutes until very hot but being careful that it doesn't get too brown. Serves 4-6.

TOMORROW'S GATOR BROCCOLI CASSEROLE

1 c gator meat, cooked, diced
1 pkg broccoli, frozen
1 c sharp cheddar cheese, grated
1 tsp onion powder or onion salt
butter
2 tbsp lemon juice
dash tabasco sauce
1 c sour cream
bread crumbs

Cook the broccoli and then drain. Combine all the ingredients except the bread crumbs and butter and place in a buttered casserole. Cover and refrigerate for 24 hours. Preheat oven to 300°F while letting the casserole stand at room temperature for 15 minutes. Sprinkle with the bread crumbs and dot with butter. Bake for 30-45 minutes or until very hot and lightly browned. Serves 4-6.

HASENPFEFFER A LA ALLIGATOR

2 lge or 4 sm alligator legs
5 c water
2 c vinegar
½ c sugar
1 lge onion, sliced
2 12 oz cans beer
Kitchen Bouquet, if needed
2 tsp salt
1 tsp tabasco sauce
1 tsp pickling spice
¼ c oil
¼ c plain flour
3-5 c reserved marinade
rice, cooked

Cut the legs at the joints and remove any visible fat. Marinate the meat in 2 cups water, vinegar, sugar, onion, beer, salt, tabasco sauce and spice for 48 hours or more turning often. Drain and reserve the marinade. In a large skillet heat the oil and add the flour and cook slowly, stirring, until the flour is browned. Add 1 cup each of water and reserved marinade and bring to a boil. Add the meat and then enough water and marinade to cover the meat with an inch or so of liquid. Heat on a medium fire and then turn down to low and simmer for several hours. Remove the bones as the meat falls off. Season to taste and add Kitchen Bouquet if needed. Serve over rice. Serves 8-10.

ALLIGATOR MARINIERE

1 lb gator meat, boiled, diced
½ c butter
¼ tsp cayenne pepper
⅓ c white wine
1 egg yolk, beaten
½ tsp salt
1 c scallions, minced
2 tbsp plain flour
2 c milk
paprika

In a heavy skillet melt the butter and saute the scallions until tender. Add the flour, mix well and cook slowly for 3-5 minutes stirring constantly. Add the milk and mix until smooth. Add the seasonings (except paprika) and wine and cook for 15-20 minutes on low. Remove from the heat and add the egg yolk stirring as quickly as possible as it is added. Stir in the meat and return to heat and cook on medium until very hot. Spoon into ramekins and sprinkle tops with paprika. Place under broiler to heat before serving. Serves 3-5.

CRAWGATOR STEW

12 lbs crawfish, live, purged**
6 lbs alligator meat, cut up
¼ c garlic, chopped
4 c tomato paste
3 c onions, chopped
2½ c potatoes, diced
1 c parsley, chopped
1 tsp Cajun Magic Seafood Seasoning*
½ lb butter
½ lb plain flour
½ c olive oil
4 c canned whole tomatoes
2½ c celery, chopped
1 carrot, chopped
1 gal chicken stock
4 c water

Melt butter in a large pot and add the flour and cook, stirring, 8-10 minutes until browned. Slowly add the stock while stirring. In another pot heat the oil and saute the onions until clear. Add the garlic and saute until golden. Add the tomato paste and water and bring to a boil. Simmer 10 minutes. Add whole tomatoes and once the tomatoes are heated add the alligator meat and cook slowly 1 hour or more. Add the celery and carrots and cook until al dente. Add potatoes and cook until soft. Add parsley and seasoning and simmer briefly. Cool and let set overnight. One to 2 hours before serving bring to a boil and add the crawfish alive and return to a boil and simmer 45 minutes. Rice is served on the side.

**To purge crawfish place into well salted water for a few minutes. Drain and wash thoroughly.

ALLIGATOR GOULASH

alligator meat, diced
paprika
1 bay leaf
carrots, sliced
new potatoes, diced
oil
alligator stock
onions, chopped
thyme
celery, chopped
plain flour
salt and pepper to taste

In a hot skillet sear the alligator in the oil and season with salt, pepper, paprika, thyme, bay leaf and vegetables. Add the flour and cook for 10 minutes. Then add the stock and stir until thickened and smooth. Simmer for 45-60 minutes or until the meat is tender. Serve hot with sourdough bread.

ALLIGATOR AND HERBS

2 lbs alligator tail
3 tbsp butter, softened
2 tsp salt
½ tsp cayenne pepper
1 lge lemon, sliced, seeded
¼ c hot water
2 med onions, sliced
1 c ketchup
1 tbsp worcestershire sauce
1 tsp chives, dried
1 tsp parsley, dried
1 tsp thyme, dried

Slice the meat in ½ inch pieces. Place in a single layer in a shallow baking dish. Rub with butter and sprinkle with the seasonings. Cover with the onions and lemon slices. Combine the ketchup, worcestershire sauce, herbs and water and pour over the meat and onions. Cover tightly and bake at 350°F for 1 hour or until meat is tender. Serves 5-6.

ALLIGATOR MANDARIN

1 lb alligator meat, boneless, cubed
½ tsp black pepper
juice from mandarin slices
paprika
oil
½ tsp salt
1 lb can mandarin orange slices, drained
dash tabasco sauce
parsley, minced

Oil a shallow casserole and place the meat in a single layer and sprinkle with the salt and pepper. Combine the mandarin juice and tabasco sauce and pour over the meat. Cover and bake at 350°F for 30 minutes. Remove the cover and arrange the mandarin slices over the meat. Place under a broiler for 6-10 minutes. Remove and dust with paprika and sprinkle parsley on top. Serve. Serves 4.

Gator J'aime

2 lbs gator meat, cubed, boiled
½ c butter
½ c scallions, chopped
¼ c plain flour
⅓ c ketchup
2 tbsp worcestershire sauce
parsley, chopped
⅓ c bottled chili sauce
1 14½ oz can chicken broth
1 c sherry
1 8 oz can mushrooms buttons
¼ tsp salt
dash tabasco sauce

In a large heavy skillet saute the onions in melted butter until soft and then blend in the flour and stir until very brown. Add the ketchup, worcestershire and chili sauces. Stir in the broth, sherry, mushrooms and seasonings. Heat and simmer for 30 minutes on low. Add the meat and cook 20 minutes longer. Remove from the heat and let cool slightly. Place in a serving dish and garnish with chopped parsley and serve piping hot. Serves 4.

Alligator in Brown Gravy

2 lbs alligator, 1 in cubes
1 bell pepper, minced
garlic to taste
salt to taste
oil
2 lge onions, minced
2 ribs celery, minced
cayenne pepper to taste
parsley, chopped, to taste
water

Brown the alligator in a little oil and remove and set aside. To the remaining oil add the onions, bell pepper, celery and garlic. Cook the vegetables in the oil for 20 minutes or longer until they become very well caramelized. Add water to thin and stir until well-blended. Add the browned meat, cayenne, salt and parsley and simmer on a low heat until the meat is tender.

Baked Alligator in Milk

1 lb alligator tail fillets
¾ c evaporated milk
¾ c water
butter
2 tbsp plain flour
½ tsp salt
½ tsp white or cayenne pepper

Combine all ingredients except butter and let stand in a refrigerator for several hours. Transfer to a buttered baking dish, covered, for 25-45 minutes at 350°F or until meat is tender. Uncover after the first 20 minutes. Serves 4-6.

GATOR BURGUNDY

1 c gator meat, cooked, diced
2 tbsp butter
1 sm onion, chopped
1 scallion, chopped
3 cloves garlic, crushed
¼ c plain flour
1 14½ oz can stewed tomatoes
1 sm carrot, diced
¼ c burgundy
salt and pepper to taste

Saute the onions in the butter in a heavy pan until soft. Add the garlic and saute a few minutes longer. Stir in the flour and continue cooking until the flour is brown stirring often. Then add the tomatoes, carrot and seasonings and cook for 20 minutes or until the carrots are tender. Add the meat and cook for 15 minutes and then add the burgundy and continue to cook for 15 minutes. Serve as a soup or over hot rice or noodles. Serves 2-4.

ALLIGATOR IMPERIAL

1 lb alligator tail meat, cooked, flaked
1 16 oz can mushrooms, buttons, drained, reserve juice
½ tsp salt
½ tsp tabasco sauce
½ c sherry
¼ tsp ginger, powdered
½ stick butter
¼ c plain flour
1 pt sour cream
1 sm onion, minced
1 tbsp parsley flakes

Melt the butter in a heavy skillet over a low heat and add the flour and stir until smooth. Stir in the sour cream and remaining ingredients mixing well and cook over a low heat for 30 minutes. Add juice from mushrooms if needed during cooking. Serve over rice or egg noodles. Serves 6-8.

ALLIGATOR SAUSAGE IN CREOLE SAUCE

5 lbs alligator sausage
¼ c bell pepper, chopped
1 toe garlic, minced
1 8 oz can tomato sauce
1 sm can whole tomatoes
Tony Chachere's Creole Seasoning*
¼ c onion, chopped
⅛ c celery, chopped
⅛ c peanut oil
½ c cold water
1 10 oz can Ro-Tel Tomatoes

Saute in oil in a heavy pot the onion, bell pepper, celery and garlic over medium heat, uncovered, until the onions are wilted. Add the tomato sauce, tomatoes, water and seasonings to taste. Continue cooking over a medium heat. Place the sausage in a skillet with ½ inch of water. Cover and steam for 15 minutes turning every few minutes. Remove from the skillet and set aside. Cut into 1 inch pieces and add to the sauce and continue cooking for about 40-45 minutes. Serve over rice.

ALLIGATOR SAUSAGE AND CRAWFISH CASSEROLE

1 lb alligator sausage, fresh
1 lb smoked alligator sausage
1 10 oz can Ro-Tel Tomatoes
1 10 oz can beef consomme
1 lb mushrooms, sliced
2 bay leaves
1 tbsp black pepper
1 lb crawfish tails
2 c Uncle Ben's Converted Long Grain Rice
1 10 oz can onion soup
1 bu shallots, chopped
1 clove garlic, minced
1 tbsp Tony Chachere's creole Sesoning*
2 stick butter

Remove the fresh alligator sausage from its casing and crumble and brown over a medium heat. Drain off excess grease. Cut the smoked sausage into 1 inch pieces and mix with the remaining ingredients in a covered casserole and cook in a preheated 350°F oven for 1 hour. After 30 minutes, stir well and continue cooking. Let cool, uncovered, for a few minutes and then serve. Serves 10-20.

ALLIGATOR SUPREME

2 lbs alligator meat, ground
2 tbsp salt
½ c butter
4 tbsp celery, minced
4 c whole tomatoes
2-4 tbsp chili powder
½ c plain flour
½ c onions, chopped
½ tsp cinnamon

Saute the onions and celery in butter until soft. Stir in the flour, chili powder, salt, cinnamon and tomatoes and blend thoroughly. Simmer about 10 minutes, stirring often. Add alligator and cook until meat is done. Serves 8-12.

ALLIGATOR SUPREME 2

6 alligator tail fillets
butter, about ½ stick
4 tsp marjoram
½ c sherry
2 c half and half cream
3 egg yolks, slightly beaten
1 tbsp parsley, chopped
1 8 oz can mushrooms, sliced
½ c swiss or monterey jack
cheese, grated

Melt the butter in a heavy skillet on low heat and saute the fillets until tender being careful not to break when turning. Cover lightly while cooking. Remove the fillets and add the seasoning and sherry to the pan juices and cook until almost evaporated. Add the cream and egg yolks while stirring constantly and continue to cook and stir until thickened. Place the fillets in a shallow baking dish and arrange the mushroom slices which have been drained over them. Pour the sauce evenly on top and sprinkle with parsley and then top with the cheese. Bake at 425°F for 5-10 minutes or until the cheese is melted and slightly browned. Serves 6.

LO-CAL ALLIGATOR CASSEROLE

1 lb alligator meat, cubed
⅔ c V-8 Juice
2 tbsp apple cider vinegar
¼ c red bell pepper, chopped
1 1.4 oz envelope low calorie italian salad dressing mix
¼ c scallions, chopped
oil

Combine all ingredients in a glass bowl mixing well and let stand in the refrigerator for several hours turning often. Oil a casserole baking dish and place the meat in a single layer and pour the marinade over it. Cover tightly and bake at 350°F for 45 minutes. Remove cover and baste and bake for 20-30 minutes more. Serves 4-5.

FILLETS OF ALLIGATOR IN WINE

6 alligator tail fillets
1 lemon, sliced, seeded
3 tbsp butter
6 mushrooms, sliced
water
salt and pepper to taste
2 scallions, chopped
¾ c white wine
2 tbsp plain flour
dill, fresh, chopped (if not available, use parsley)

Place the fillets in a shallow pan and sprinkle with water and place the lemon slices over them and let stand in a refrigerator for several hours or overnight. Turn several times. When ready to use remove the lemon slices and pat dry. Melt 1 tbsp of the butter in a heavy skillet and add the scallions and mushrooms. Add the fillets and half the wine and cover tightly and cook on a low heat for 20-30 minutes or until the meat is tender. Add remaining wine as needed during cooking. In a small skillet heat the remaining butter and add the flour stirring constantly until thick and smooth. Remove the fillets to a hot serving platter and add the roux* to juices remaining. Add the seasonings and pour over the fillets and sprinkle with the dill or parsley. Serves 6.

WINE AND DINE GATOR

3 lbs gator leg meat, boneless
2 c dry red wine
3 cloves garlic, minced
½ tsp basil, dried
1 tsp salt
½ tsp thyme
1 med onion, sliced
½ c oil
½ c plain flour
pepper to taste

Combine the meat, wine, garlic, basil, thyme and onion and marinate overnight or for at least several hours in the refrigerator. Drain and reserve the marinade. Heat the oil and add the flour and stir and cook until a caramel color. Add the meat and stir while cooking for 5 minutes. Add the reserved marinade slowly with the heat on high. When boiling reduce the heat and simmer for 45-60 minutes or until the meat is tender. Add more wine and water mixed half and half if it gets too dry while cooking. Serve with rice or creamed potatoes or with french bread for dunking. Serves 6.

ALLIGATOR WITH CREOLE SAUCE

4 alligator loin slices
plain flour
4 sprigs italian parsley, chopped
creole sauce
milk
2 tbsp extra-virgin olive oil
salt and white pepper

Get alligator loin slices about 6-ounces each and tenderize until about twice their original size. Flour. Heat the oil in a skillet and saute the alligator until lightly browned. Add the creole sauce and parsley and bring to a boil. Adjust the seasonings and serve. Serves 4.

CREOLE SAUCE 2

¼ c olive oil
1 med onion, sliced
1 tsp garlic, chopped
1 tbsp tomato puree
2 bay leaves
½ tsp salt
pinch white pepper
1 bell pepper, coarsely chopped
2 ribs celery, coarsely chopped
1 c canned italian plum tomatoes, chopped
1 tbsp tomato paste
1½ tsp fresh thyme leaves
¼ tsp black pepper

Heat the oil in a large skillet and saute the onion, bell pepper and celery until lightly browned. Add all the other ingredients and bring to a boil and then reduce the heat to a simmer and cook for 10 minutes, or until thickened.

ALLIGATOR V CHANZEE

1 lb alligator meat, strips or steaks
2 ribs celery, chopped
1 c mushroom slices, fresh
2 c white sauce*
parsley, chopped
1 onion, chopped
½ stick butter
½ c wine
1 bell pepper, chopped

Saute the mushrooms in the butter and remove and reserve. Add the wine, pepper, onion and celery and simmer until soft. Put the meat in a buttered baking dish and top with the butter/wine and vegetable mixture. Cover tightly and bake at 350°F until the meat is tender, up to 1½ hours. Heat the mushrooms in the white sauce and pour over the meat and put back in the oven to heat thoroughly and then sprinkle with parsley and serve.

ALLIGATOR WITH CREAM

1 lb alligator tail, ½ in pieces
¼ c peanut oil
1 clove garlic, minced
4 mushrooms, sliced
½ c cream
1 tbsp parsley, minced
salt and pepper to taste
plain flour
½ c vermouth
2 scallions, minced
1 tbsp wine vinegar
1 tsp tarragon
3 tbsp butter, room temperature

Lightly pound the meat with the side of a knife and dredge in flour and shake off the excess. Heat the oil in a pot over moderate heat. Brown the meat on both sides and set aside. Keep warm on a plate loosely covered with foil. Discard the oil and add the vermouth, garlic, onion, mushrooms and vinegar. Allow to boil scraping the brown bits clinging to the pot. Add the cream, tarragon, scallions and parsley and allow to reduce by half. Off the heat swirl in the butter and correct the seasonings and serve with the alligator at once. Serves 4.

CREAMED ALLIGATOR

1 c gator meat, boiled, diced
1 tbsp plain flour
1 tbsp + butter
¼ tsp lemon juice
¾ c whole milk
pinch tarragon
pinch basil
salt to taste
1 4 oz can black olives, sliced
bread or cracker crumbs

Combine the flour, tbsp butter and milk in a heavy saucepan and cook over a low heat until blended and slightly thickened. Season with the spices and cook for 5 minutes. Mix in the alligator and olives and cook on a low heat until thick enough to mound slightly when spooned. If too thick add a small amount of milk or cream. Spoon into ramekins and sprinkle with the crumbs and dot with butter. Bake at 400°F for 10-15 minutes until slightly browned and very hot. Serves 6.

CREAMED ALLIGATOR 2

2 lbs alligator tail meat
½ c water
¼ c lemon juice
2 med onions, sliced
½ c biscuit mix
1 tsp salt
oil
butter
¼ tsp black pepper
¼ tsp + paprika
1 egg, slightly beaten
½ c half and half cream
dash tabasco sauce
¼ c parsley, chopped (optional)
egg noodles, cooked

Cut the meat into finger size pieces and place in the water with the lemon juice and onions and refrigerate for several hours turning often. Remove the alligator and drain the juice to remove the onions and reserve. Mix the biscuit mix, salt, ½ tsp paprika and black pepper and mix well. In a bowl with the reserved juice add the egg and mix well. Dip the alligator into the egg mixture and then in the biscuit mix and repeat so each is coated twice. Fry in hot oil until browned. Place the meat in a shallow casserole and top with the onions. Combine the cream, tabasco sauce, parsley and paprika and pour over the top and then bake at 325°F until cream is absorbed, about 45 minutes. Serve with hot buttered noodles. Serves 6.

CREAMY GATOR CASSEROLE

1 lb gator meat, boiled, ground
½ stick butter
1 10½ oz can cream of celery soup
¾ c raw rice
2 c water
½ tsp tabasco sauce
1 10½ oz can cream of chicken soup
2 med onions, diced
1 rib celery, diced
¼ c milk
1 tsp salt

Add the meat to the butter and stir well and cook covered until heated. Add the onions, celery, rice, soups, milk and water and bring to a boil and add the seasonings. Pour into a large casserole and cover and bake at 350°F for 45 minutes. Lower the oven to 300°F and bake for 45-60 minutes longer or until the rice is tender and the liquid absorbed. Remove the cover during the last 20 minutes of cooking. Serves 6-8.

ALLIGATOR A LA KING

2 c gator meat, cooked, chopped
2 tbsp butter
½ stick butter
1 c half and half cream
1 sm onion, chopped
½ sm bell pepper, chopped
water, if needed
1 4 oz can mushrooms, sliced
1 tsp paprika
¼ c parsley flakes
½ tsp cayenne pepper
1 tsp salt
½ c white wine
toast points, english muffins or biscuits

Melt the butter and add the flour and cook for several minutes. Add cream slowly while stirring and then add the onion, pepper and undrained mushrooms. Simmer for 2-4 minutes or until hot. Add the meat, seasonings and wine and cook for 20-30 minutes over a low heat stirring often. Add a little water if necessary to maintain a consistency. Adjust seasonings and serve over toast points, english muffins or biscuits. Serves 4-5.

ALLIGATOR SCALLOPINE

2 lbs alligator pink meat, cubed
seasoned plain flour*
oil
2 med onions, sliced
1 bay leaf
1 tsp sugar
1 4 oz can mushrooms, sliced
2 8 oz cans tomato sauce
½ c hot water

Roll the meat in the seasoned flour and place on a chopping board, and pound with a meat mallet until flattened. Fry until golden in the oil and when all meat is fried soften the onions in the pan drippings. Place the meat in an oiled casserole and top with the onions. Mix the remaining ingredients well and pour over the top of the meat. Bake at 350°F for about 2 hours reducing the heat after 1 hour. Remove the bay leaf before serving. Serves 6-10.

ORANGE GATOR

4-6 alligator tail steaks, 1 in thick
¼ c plain flour
2 tbsp sugar
½-1 tsp salt
¼ tsp dry mustard
¼ c butter
1½ c orange juice
pinch of ginger
¼ tsp cinnamon
3-4 c hot cooked rice

Make a marinade of the orange juice, ginger, cinnamon and mustard and add the meat and let stand in the refrigerator for 12-24 hours. Drain and reserve the marinade. Melt the butter and add the flour and slowly cook until lightly browned. Lay the steaks in the skillet over the flour mixture and cook on each side for 5 minutes turning carefully. Cover the pot and simmer until just beginning to get tender, about 20 minutes. Remove the meat and add the sugar, salt and marinade and bring to a boil and simmer until slightly thickened. Place the meat back carefully into the sauce and heat through. Serve over rice. Serves 4-6.

SMOTHER FRIED ALLIGATOR

3 lbs alligator
2 c hot water
juice of 2 lemons
oil

Brown alligator in a small amount of oil. Add hot water and lemon juice. Cover and simmer until tender.

SMOTHERED ALLIGATOR

4 sm alligator legs, jointed
½ stick butter
1 med onion, chopped
1 scallion, chopped
water, as needed
1 tsp salt
½ tsp black pepper
2 lge tomatoes, peeled, diced
2 tbsp parsley flakes

If alligator legs not available use 2-3 cups of boneless meat instead. Melt the butter and add the meat and cover and cook on low for 15-20 minutes. Add the onions and a little water and cook 15-20 minutes longer adding more water if too dry. Add all remaining ingredients and cook for 30-45 minutes adding water as needed. If the meat is boneless you may want to mix 2 tsp cornstarch with ½ c water and add to thicken. If bones are present remove as meat falls away. Season to taste and serve over rice or with fresh french bread. Serves 4-6.

SMOTHERED GATOR

1 lb gator white meat, diced
½ stick butter
2 tsp plain flour
¼ c dry sherry
tabasco sauce to taste
hot cooked rice
water
3 cloves garlic, minced
2 ribs celery with tops, chopped
2 bell peppers, chopped
2 scallions, chopped
salt to taste
3 c hot water

Boil the meat in plain water for 10 minutes and then drain and rinse discarding water. Melt the butter and add the flour and cook until brown on a low heat stirring constantly. Add the peppers, scallions, garlic and celery and cook until soft. Add 3 cups hot water and bring to a boil. Add the meat and sherry and again bring to a boil and lower the heat and simmer for 20-40 minutes or until meat is tender. Season highly. Add a tbsp more sherry after removing from the heat if desired. Serve over rice. Serves 4-6.

ALLIGATOR STEAKS TANDOORI

6-8 alligator tail steaks 1 in thick
½ tsp ginger
½ tsp cumin
3 tsp salt
1 tbsp lime juice
2 tbsp lemon juice
½ c peanut oil
1 tsp dry mustard
2 c yogurt, plain
3 cloves garlic, minced
1 tsp cayenne pepper
½ c cider vinegar
1 tsp red chili powder
butter

Put all ingredients except steaks in a blender and blend until smooth. Place the steaks in a shallow glass casserole and pour the blended ingredients over them. Cover and place in a refrigerator for 12-24 hours. Remove the steaks and place in a clean buttered baking pan reserving the marinade. Bake covered at 350°F for 30 minutes. Remove the cover and continue baking at 325°F until the meat is tender. Baste often with the reserved marinade. Serves 6-8.

ALLIGATOR PIZZA

1 lb gator meat, cooked, ground
1 sm onion, chopped
1 tsp salt, divided
½ tsp oregano, ground or flakes
½ tsp sage
1 c mozzarella cheese, grated
1½ c biscuit mix
½ c + milk
½ tsp garlic powder
½ tsp sugar
2 8 oz cans tomato sauce

Cook the meat, onion, half the salt, oregano, sugar, sage and tomato sauce in a heavy pan and simmer on low until the crust is ready. Combine the biscuit mix, milk, garlic powder and remaining salt and form a stiff dough. Add more milk if necessary. Knead for several minutes on a floured surface and roll out ⅛ to ¼ inch thick and fit into a 9 inch pizza pan. Pour the meat sauce over the crust and sprinkle with the cheese. (Bell peppers, olives or other pizza toppings may also be added to your taste.) Bake in a 400-425°F oven on the middle rack for 20-25 minutes. Serves 6.

ALLIGATOR BALLS 1

3 lbs alligator, coarsely ground
salt and pepper to taste
1 c oil
1½ c instant mashed potatoes
welch sauce
3 eggs
½ c water
1 c scallions, chopped
1 c onions, minced

Combine alligator, eggs, potatoes, onions, scallions, salt and pepper and mix thoroughly. Make 10-12 balls. Brown in oil. Place browned balls in pot and add water or welch sauce. Cover tightly and let cook slowly for about 45 minutes. Serve with hot cooked rice. Serves 5.

WELCH SAUCE

1 10 oz jar Welch Grape Jelly
1 12 oz bottle chili sauce

Heat the jelly and chili sauces in a heavy saucepan.

ALLIGATOR BALLS 2

2 lbs alligator, ground
juice of 1 lemon
2 tsp salt
½ c scallions, chopped
1 tsp cayenne pepper
2 tbsp parsley, chopped
1 egg, beaten
1 c plain flour
¼ c milk
1 c yellow cornmeal
½ c bread crumbs, plain
oil

Mix together well with alligator the salt, cayenne pepper, egg, milk, bread crumbs, lemon juice, scallions and parsley. Shape into small balls. Roll alligator balls in a mixture of flour and cornmeal and fry in deep oil at 350°F until brown.

ALLIGATOR MEAT BALLS

6 lbs alligator, ground
1 tbsp Kitchen Bouquet
1 c bell peppers, chopped
salt and pepper to taste
1 c parsley, chopped
oil
1 tbsp worcestershire sauce
2 c bread crumbs, plain
3 cloves garlic, chopped
2 c scallions, minced
2 eggs

Mix all the above ingredients and let set 24 hours in a refrigerator. Make balls and drop in hot oil 2 inches deep in a pot. Fry covered. Makes approximately 75 balls.

MR. GATOR MEAT BALLS

5 lbs alligator meat
1 tsp pepper
7 sprigs parsley
oil
1 lge bell pepper
1 tbsp salt
¼ tsp cayenne pepper
2 cloves garlic
2 med onions
1 c bread crumbs, plain

Grind alligator, onions, bell pepper, parsley and garlic together and mix thoroughly with pepper, bread crumbs and salt. Shape into balls about the size of golf balls. Fry in deep hot oil until light brown. About 60 balls. Make a gravy of the following ingredients:

8 c boiling water
1 sm cayenne
2 c water
1 c scallion tops, chopped
2 c plain flour
1 c bell pepper, chopped
¼ c oil
2 cloves garlic, chopped
1½ c onions, chopped
salt and pepper to taste
1 8 oz can tomato sauce
3 tbsp parsley, chopped

Brown flour in oil, stirring constantly to prevent burning. Add onions and cook slightly. Add tomato sauce and mix well. Gradually add 2 cups water and mix. Pour mixture into remaining boiling water in a large pot and mix thoroughly. To this boiling mixture add cayenne, scallions, parsley, bell pepper, garlic, salt and pepper to taste. Let boil for awhile. Add alligator balls. Lower heat and simmer for 1½ hours. Serve over hot rice. Serves 20.

MEAT BALLS A LA ALLIGATOR

1 lb alligator meat, chopped
1 egg
1 tbsp onions, minced
2 tbsp celery, minced
1 tbsp parsley, minced
2 tbsp shallots, minced
2 tbsp lemon pepper seasoning
½ tsp salt
¼ c plain bread crumbs
1 c oil
plain flour

Combine ingredients and form into 1 inch balls. Allow to set for 1 hour. Dredge in flour and fry until brown. Serve hot.

GATOR MEAT BALLS

5¼ lbs alligator meat
1 c oil
1 oz black pepper
¼ bu celery
2 ozs salt
2 bu scallions
2½ lbs cracker meal
1½ lbs dry onions
2¼ lbs mashed potatoes
1 bulb garlic
3 ozs Nugget Savory Seasoning
2 bu parsley
8 eggs

Grind alligator meat, onions, scallions, parsley, garlic and celery together. Cook in oil for 30 minutes. Add potatoes, eggs, Nugget Savory Seasoning, salt and pepper. Mix well, chill, and shape into 1 oz balls. Roll in cracker meal and fry in deep oil. Makes 208 1 oz balls.

BAYOU ALLIGATOR MEAT BALLS

2 lbs alligator meat, ground
¼ c celery, chopped
½ tsp mustard
¼ c plain bread crumbs
2 tbsp lemon juice
plain flour
½ c onion, chopped
¼ c parsley, chopped
2 eggs
¼ c bell pepper, chopped
salt and pepper to taste
oil

Mix all ingredients well, except oil and flour, and form into small balls. Roll in flour and fry until golden brown. Use in your favorite spaghetti sauce and simmer 35 minutes.

ALLIGATOR MEAT BALLS IN PINEAPPLE SAUCE

1 lb alligator meat, ground
⅓ c milk
¾ tsp salt
⅛ tsp black pepper
¼ c water
¼ tsp ginger, ground
1 egg, beaten
½ c corn bread stuffing mix
½ tsp sage, ground
½ c pineapple preserves
2 tbsp vinegar

In a mixing bowl combine the egg and milk and stir in the stuffing mix, salt, sage and pepper. Add the alligator and mix well. Shape the meat mixture into 48 meat balls and place on a rack in a shallow baking pan. Bake at 375°F for 20-25 minutes or until done. Remove from the pan and drain the balls. Stir together the pineapple preserves, water, vinegar and ginger. Place the pineapple mixture in a large saucepan and cook and stir over low heat until heated through. Add the meat balls and cover and cook over medium heat about 10 minutes or until the meat balls are heated through. To serve spear with cocktail picks. Makes 48 meat balls.

FRIED ALLIGATOR

- alligator meat
- cayenne pepper
- oil
- salt and pepper to taste
- vinegar
- plain flour
- corn meal

Tenderloin fresh alligator tail. Cut flesh not exceeding 2 x 1 inch thickness. Place cut pieces in pan or dish. Pour on this a small amount of vinegar, add salt and pepper (black and red) to taste. Let alligator soak in this approximately 30 minutes or longer. While soaking, pour in paper bag or other container 4 parts corn meal to 1 part flour. Put about 1 inch of oil in a skillet and heat to approximately 400°F. Drain and roll or shake alligator portions in the meal mixture. Place just enough pieces to cover bottom of skillet. Fry golden brown and serve hot.

MARINATED FRIED ALLIGATOR

- 1 lb alligator meat
- 1 c sherry
- 1 tbsp lemon pepper seasoning
- 1 tsp Season-All
- ¼ c lemon juice
- ½ c italian salad dressing
- plain flour
- oil

Cut meat into bite size pieces and marinate in the first five ingredients for at least 2 hours. Drain and dredge in flour. Fry pieces for about 15 minutes turning often until brown. Drain and serve hot.

ITALIAN FRIED ALLIGATOR

- 1 lb alligator meat
- ½ c parmesan cheese
- ½ c italian bread crumbs
- oil

Cut alligator meat into thin finger strips. Thoroughly mix parmesan cheese and bread crumbs into a paper sack. Add a portion of the meat and shake until meat is well coated. Fry in hot oil for 2 minutes.

FRIED ALLIGATOR IN BEER BATTER

- thin alligator fillets
- oil
- salt and pepper to taste
- cocktail sauce (hot)

Season with salt and pepper. Coat with beer batter and fry in oil until golden. Serve with the cocktail sauce (hot).

BEER BATTER

1 12 oz can beer
¼-½ c plain flour
salt and pepper to taste

Mix together and let stand for 2 hours.

COCKTAIL SAUCE (HOT)

½ c ketchup
2-4 tbsp tabasco sauce
1 tbsp lemon juice
1 tbsp dill or sweet pickle relish

Mix together well and refrigerate. Keeps well for several weeks in a covered container.

ROCKERFELLER FRIED ALLIGATOR

5 lbs alligator meat, bite size
1 tsp garlic powder
1 lge italian salad dressing
1 tsp onion powder
1 tsp salt
1 tsp cayenne pepper
1 tsp black pepper

Mix all ingredients well with meat and chill 3-6 hours or overnight. Drain well.

1½ c plain flour
1½ c yellow corn meal
2 c italian bread crumbs
1 tsp salt
oil
1 tsp cayenne pepper
1 tsp black pepper
½ tsp garlic powder
½ tsp onion powder

Combine above ingredients and then lightly coat each piece. Deep fry until golden brown. Drain well.

ALLIGATOR FRIES

alligator tail meat, cut up
beer
cornmeal
oil
mustard
egg
fish fry coating

Marinate alligator meat in beer. Mix some mustard in beaten eggs and dip meat in and then dredge in equal amounts of cornmeal and fish fry. Deep fat fry until golden.

BRICE PALMAR'S FRIED ALLIGATOR

3 lbs alligator meat, 1 in cubes
1 c mustard
2 c fish fry
oil
3 c evaporated milk
2 tbsp seasonings
2 c pancake mix

Soak meat in milk in refrigerator for 2-3 hours. Drain and season. Add mustard and smear well. Mix fish fry and pancake mix in a paper bag. Drop meat in bag and shake well. Deep fry at 375°F for 5-6 minutes or until golden brown. Serve hot with french fries and seafood sauce.

FRIED ALLIGATOR A LA MARCELLE

1 lb alligator meat, 2 in strips
1 tsp salt
½ tsp cayenne pepper
½ tsp garlic powder
plain flour
1 c sherry
1 tsp black pepper
¼ c lemon juice
¼ c olive oil
oil

Marinate for 2 hours the alligator strips in sherry, salt, peppers, lemon juice, garlic powder and olive oil. Drain and dredge in flour. Deep fry in oil until pieces pop up and are golden brown. Great for dipping in tartar or cocktail sauce.

PANNE'D DEEP FRIED ALLIGATOR

2-3 lbs alligator meat
2 c milk
3 c seasoned bread crumbs* or panne*
salt and black pepper
white pepper to taste
3 eggs, beaten
1 tbsp yellow mustard
¾ gal peanut oil
cayenne pepper to taste

Cut alligator meat into 1½ inch pieces. Season with the peppers to taste. Mix the eggs, milk and mustard together and whisk about 30 seconds. Have the panne or bread crumbs ready and drop pieces of the meat into the egg mixture. Remove and allow the excess to drip off and place into the panne or bread crumbs. Move around until well coated and drop into the hot peanut oil and fry until golden brown and floating which usually takes about 3-5 minutes. Drain. Serve hot.

FRIED ALLIGATOR TENDERLOINS

fresh alligator tenderloins
salt and black pepper to taste
2 c cornmeal
cucumber sauce
vinegar
cayenne pepper to taste
½ c plain flour
oil

Cut the tenderloins into 1 inch thick medallions. Place into a glass or enamel pan and pour over a small amount of vinegar. Add the salt and peppers to taste and let marinate for 30 minutes while turning the medallions occasionally. Place the cornmeal and flour in a paper bag and mix thoroughly. Shake the alligator pieces in the cornmeal mixture and shake off the excess. Put about 1 inch of oil in a skillet and heat to approximately 400°F. Place enough pieces in the skillet to just cover the bottom and fry them golden brown on both sides and serve them hot with the cucumber sauce.

CUCUMBER SAUCE

1 c sour cream
1 tsp prepared mustard
1 tsp lemon juice
1 cucumber, unpeeled, seeded, chopped
1 sm onion, grated
salt and pepper to taste

Combine well all ingredients and chill.

SPICY FRIED ALLIGATOR

5 lbs alligator meat, cubed
juice of 2 lemons
½ tsp salt
cornmeal
1 sm bottle tabasco sauce
½ c water
plain flour
peanut oil

Marinate the alligator in the tabasco sauce, salt, lemon juice and just enough water to cover for 48 hours in the refrigerator in a glass covered bowl. Drain and roll pieces in a combination of equal parts flour and cornmeal and fry until golden in the peanut oil.

CAJUN JOE'S FRIED ALLIGATOR

For Frying

1 lb alligator meat
1 tbsp baking soda
4 tbsp italian dressing
fish fry
2 tbsp lemon juice
1 tsp tabasco sauce
creole seasoning*
oil

For Sauce

½ c oil
2 onions, chopped
1 tsp salt
2 stalks, celery, chopped
1 10 oz can Ro-Tel Tomatoes
¼ c plain flour
2 tsp Season All
1 tbsp tabasco sauce
1 8 oz can tomato sauce
hot water

Cut the meat ½ to 1 inch wide by 3 to 5 inches long. Marinate the meat with the lemon juice, baking soda, tabasco sauce and italian dressing for 1 hour in the refrigerator. Drain and then season with the creole seasoning and dredge in fish fry and deep-fry in hot oil until golden. Now make a sauce by heating the oil and add the flour and stir until brown. Add the onions and cook until slightly brown and then add the tomato sauce and some hot water to thin a little. Cook for ½ hour thinning as needed. Add the tabasco sauce, celery and tomatoes and cook for 15 minutes more and then add the alligator, salt and Season All and cook about 1 hour on a medium fire, half the time with the cover off. Skim excess oil on the top before serving.

FRIED ALLIGATOR TAIL

3 lbs alligator tail steaks
1 tsp black pepper
1 tsp baking soda
cornmeal or fish fry
oil
1 tsp salt
1½ tsp mustard
¼ c vinegar
lemon rind, grated

Marinate in mustard, baking soda and vinegar for 3 hours. Season cornmeal or fish fry with salt, pepper and lemon rind mixing well. Coat meat well in the seasoned cornmeal or fish fry and fry in oil until golden brown. Serve immediately.

FRIED GATOR FILLETS

8 ½ in thick gator fillets
1 c buttermilk
2 eggs, beaten
cold water
salt to taste
1 c self rising flour
1 tbsp oil
peanut oil

Soak fillets in buttermilk for 30 minutes. Drain and salt. Prepare a batter by mixing the flour, eggs, 1 tbsp oil and enough cold water to attain the consistency of a thin pancake batter. Batter each fillet and fry in medium hot peanut oil for 5 minutes.

GLENDA'S FRIED GATOR

1 lb alligator meat
1 sm can evaporated milk
½ bottle Louisiana Hot Sauce
egg sauce
1 egg, beaten
1 tsp yellow mustard
flour

Cut meat into finger size strips. Make a batter of egg, milk, mustard and hot sauce. Dip the meat in flour and then the batter and then flour again and deep-fry for 5 minutes until golden. Serve with the egg sauce.

EGG SAUCE

½ c mayonnaise
1 hard boiled egg, chopped
2 tbsp onion, grated
black pepper to taste
1 tbsp ketchup
1 tsp lemon juice
½ tsp salt

Mix well all ingredients and chill. Keeps for several days in a refrigerator.

BEER FRIED ALLIGATOR

alligator meat
salt
tabasco sauce
parsley flakes
oil
beer
lemon pepper seasoning
fish fry
cayenne pepper

Cut meat into thin strips and marinate in beer, salt and lemon pepper seasoning for ½ hour. Dip into tabasco sauce. Combine fish fry, salt, cayenne pepper and parsley. Dredge meat in this and then deep-fry until golden brown.

FRIED GATOR A LA ITALIENNE

8 ¼ inch thick alligator fillets
dash salt
½ tsp black pepper
1 c italian bread crumbs
1 c buttermilk
2 eggs, well beaten
3 drops tabasco sauce
4 tbsp oil

Pound the fillets with the flat of a cleaver and then soak them in buttermilk and salt for at least 30 minutes. Drain and pat dry. Mix the eggs, pepper and tabasco. Dip each fillet in the egg mixture and then in the bread crumbs. In a skillet heat the oil to 360°-375°F and brown the fillets on both sides. Drain and serve immediately. Serves 4.

FRIED ALLIGATOR HEBERT

3-5 lbs alligator
1 1.2 oz pkg meat marinade mix
1 tbsp Season-All
2 c italian dressing
2 6 oz pkg Oak Grove Fish Fry
1 tsp cayenne pepper

Cut meat into small serving-size bites and marinate for 24 hours in a mixture of italian dressing and meat marinade mix. (If mixture does not completely cover the alligator add a little water.) Mix together the fish fry, Season-All and cayenne pepper. Roll alligator in this mixture and drop in hot oil in a deep fryer. Cook until golden brown and serve immediately.

FRIED ALLIGATOR APPETIZERS

2-2½ lbs alligator tails
½ c + 1-2 tbsp water
½ c celery, chopped
¼ c shallots, chopped
salt and pepper to taste
oil
½ sm bottle tabasco sauce
juice of 3 lemons
¼ c celery, chopped
¼ c bell pepper, chopped
plain flour (optional)

Marinate the meat that has been cut into slivers in tabasco sauce, ½ c water and lemon juice overnight in a refrigerator. Drain and dry and season with salt and pepper. Meat can be lightly floured if desired. Use a small amount of hot oil and quick-fry meat, vegetables and parsley. Add the 1-2 tbsp water and cover. Simmer 3-5 minutes. Serve hot. Serves 6-8.

PETE DIBENEDETTO'S FRIED ALLIGATOR

5 lbs alligator meat, cubed
2 tbsp liquid crab boil
1 bottle beer
plain flour
powdered mustard to taste
black pepper to taste
garlic salt to taste
4 tbsp crab boil seasoning*
4 tbsp oil
juice of 12 lemons
hot sauce to cover meat
yellow cornmeal
salt to taste
cayenne pepper to taste
pinch sweet basil
peanut oil

Cover the meat with enough hot sauce and then add the oil, crab boil, lemon juice and beer and pour over all. Let stand in the refrigerator for 12-15 hours. Drain the meat but do not let it dry out. Mix equal parts flour and cornmeal and add the remaining seasonings and mix well. Roll the meat in the mixture twice. Fry in the peanut oil at 350°F. After 1½ minutes it will float to the top. Let it brown and then remove and drain. Serves 8-10.

FRIED ALLIGATOR A LA TECHE STYLE

2 lbs alligator meat, white, cubed
oil
¼ c ketchup
1 c yellow cornmeal
1 c biscuit mix
Sally sauce
dash tabasco sauce
1 egg, beaten
2 tbsp water
2 tbsp lemon juice
salt and pepper to taste

Marinate the meat in a mixture of the egg, water, lemon juice, tabasco sauce and ketchup for several hours or overnight. Combine the cornmeal and biscuit mix and season with salt and pepper. Drain the meat reserving the liquids and roll in the cornmeal mixture until coated and shake off the excess and dip into the reserved egg mixture and back into the cornmeal mixture. Drop into very hot oil and fry to a golden brown. Drain well and serve with Sally sauce.

SALLY SAUCE

½ c ketchup
¼ c mayonnaise
1 tbsp creole style mustard
1 tsp lemon juice
1 jalapeno pepper, minced
1 sm onion, minced
½ tsp celery seed
tabasco sauce to taste

Combine all ingredients and mix well and refrigerate. Keeps well.

FRIED ALLIGATOR A LA FRANKLIN

1 lb alligator tail meat, cut in cubes or strips
dash tabasco sauce
¾ c french dressing, bottled
seasoned flour*
sauce louie
½ tsp salt
1 c sherry
1 tbsp lemon juice
oil

Make a marinade of the sherry, lemon juice, salt, tabasco sauce and french dressing and add the meat mixing well and refrigerate for up to 24 hours. Drain and coat with the seasoned flour and fry in oil until golden brown. Drain and serve with sauce louie.

SAUCE LOUIE

1 c mayonnaise
¼ c ketchup
1 tsp prepared horseradish
salt to taste
½ tsp prepared mustard
1 tsp lemon juice
tabasco sauce to taste

Mix together well and chill before serving.

BROILED ALLIGATOR A LA WILSON

alligator fillets
lemon butter sauce 2
salt and pepper to taste

Place alligator in a single layer in a broiling pan which has been lightly buttered. (If you want you can marinate the meat in your favorite marinade.) Broil at a high heat with the door slightly ajar for approximately five minutes. Check often and turn frequently for even cooking and to prevent scorching. Serve hot with the lemon butter sauce 2.

LEMON BUTTER SAUCE 2

½ c butter
1½ tbsp lemon juice
¼ tsp white pepper

Melt the butter over a low heat in a small saucepan. Stir in the pepper and lemon juice blending well and heat 30 seconds longer.

OVEN BAKED ALLIGATOR WITH WINE SAUCE

4 alligator fillets, about 1 lb
1 tbsp butter
1 c water
1 tbsp dry white wine
1 2½ oz jar sliced mushrooms, drained
4 scallions, sliced
1 ½ oz pkg chicken gravy mix
½ c sour cream
salt and pepper to taste
paprika

Place fillets on a piece of heavy-duty foil and season with salt and pepper and wrap securely. Bake in a 450°F oven about 30 minutes or until the alligator flakes easily. In a small saucepan cook the scallions in butter over a medium heat until tender but not brown. Stir in the chicken gravy mix and add the water. Cook stirring until the mixture is thickened and bubbly and cook and stir mixture for 1-2 minutes more. Combine the sour cream and wine and add to the gravy mixture. Stir in the mushrooms. Cook until heated through but do not boil. Unwrap the fillets and with a slotted spatula remove the fillets from the foil to a platter or individual serving plates and sprinkle with the paprika. Spoon sour cream mixture over the fillets. Serves 4.

BAKED ALLIGATOR WITH MUSHROOMS

1½-2 lbs alligator tail meat, boneless
½ c water
2 tbsp vinegar
¼ c dry white wine
1 c sharp cheddar cheese, grated
1 8 oz can mushrooms, sliced, drained
½ 10½ oz can cream of celery soup
½ tsp salt
¼ tsp cayenne pepper

Place meat in a shallow greased casserole and add the water and vinegar and cover tightly and bake at 350°F for about 30 minutes. Drain off the liquid. Mix the seasonings, mushrooms, soup and wine and pour over the meat and bake at 325°F for 20 minutes. Sprinkle with the cheese and return to the oven and bake for about 20 minutes. Garnish as desired. Serves 4-6. (cream of asparagus, broccoli or any other cream soup can be used for variation.)

RUSSIAN ALLIGATOR

1 lb alligator fillets, ½ in thick
¾ c russian style salad dressing
2 tbsp lemon juice
bell pepper rings (optional)
½ tsp tabasco sauce
½ tsp salt
1 c mozzarella cheese, grated

Combine the meat, salt, lemon juice, tabasco sauce and russian dressing in a bowl and cover and refrigerate for several hours or overnight turning occasionally. To cook place the meat in a single layer in a suitable shallow baking dish that has been greased. Bake at 325°F covered for 30 minutes. Remove the cover and bake 20 minutes longer. Sprinkle with the cheese and put back in the oven for 15-20 minutes until the cheese melts. Garnish with the pepper rings if desired. Serves 4-6. (monterey jack cheese can be substituted or use a combination of both.)

ALLIGATOR DIJON

8 alligator tail fillets, hand sized, cut ½-1 inch thick
1-2 c milk, with ½ tsp salt/cup
white pepper to taste
1 c sour cream
1 tbsp paprika
2 tbsp worcestershire sauce
3 tsp lemon juice
4 cloves garlic, minced
¾-1 c saltine cracker crumbs
¾ c dijon mustard
1 tsp celery seeds

Soak the meat in enough milk to cover for several hours. Mix the sour cream, mustard, paprika, celery seed, lemon juice, worcestershire sauce and garlic in a bowl. Drain the meat and pat dry and then dip the meat in the sour cream mixture to coat heavily. Sprinkle with the crumbs and place into a shallow baking pan and place on a baking rack. Place in a preheated 400°F oven for 20 minutes and then lower the heat to 300°F and continue to bake for 30-45 minutes or until meat is tender. Serve with the extra dijon sauce on top or in a side sauce boat. Serves 6-8.

ALLIGATOR-SAUTERNE STEAKS

6-8 alligator tail steaks
1 c Sauterne
1 4 oz can mushrooms, sliced
dash tabasco sauce
water
1 5 oz can water chestnuts, sliced
1 stick butter
1 10½ oz can cream of chicken soup
1 tsp salt

Marinate the meat in the Sauterne for several hours and then drain and reserve wine. Place the steaks in a buttered shallow casserole and add about ½ inch of water. Cover and bake at 350°F for 20 minutes. Remove the cover and drain off the liquid. Combine the reserved wine, butter, soup, water chestnuts and drained mushrooms in a saucepan. Heat to almost boiling or until the butter melts and stir in the salt and pour over the steaks and return to the oven uncovered. Bake for 30-45 minutes or until the meat is tender and the sauce is reduced and thick. Serve with noodles or creamed potatoes. serves 6-8.

ALLIGATOR AMADINE

4-6 gator tail fillets, 1 in thick
2 tbsp lemon juice
4-5 drops tabasco sauce
1 tbsp parsley, chopped
½ c almonds, sliced
¼ c plain flour
½ tsp paprika
¼ stick butter, melted

Cut fillets into serving size portions and place in a shallow baking dish. Cover tightly and bake at 325°F for 30 minutes. Let them stay in the oven with the door open while preparing the sauce. Combine the flour, salt and paprika mixing well. Pour the butter over the fillets and then sprinkle with the flour mixture. Place under broiler for 6-10 minutes or until slightly browned. Saute almonds in a small amount of butter and when brown remove from the heat and add lemon juice, tabasco sauce and parsley. Pour over the fillets and serve hot. Serves 4-6.

BAKED ALLIGATOR CREOLE

1 lb alligator meat, cubed
½ tsp salt
½ tsp black pepper
1 clove garlic, mashed
1 c mozzarella cheese, grated
1 lb can stewed tomatoes, mashed

Place the meat in a 2 quart casserole with a cover. Sprinkle with salt and pepper and mix well. Add the tomatoes and garlic and mix well. Bake, covered, at 350°F for 25-45 minutes. When the meat is tender sprinkle the cheese over the top and bake 15 minutes longer without the cover. Serve. Serves 4-6. (monterey jack cheese can be used if desired.)

BARBECUED ALLIGATOR OVEN STYLE

6 lbs alligator steaks
½ tsp salt
2 tbsp parsley, chopped
1 c plain flour or bread crumbs
½ c soy sauce
1 tbsp garlic salt
8 drops tabasco sauce
1½ c oil
½ c lemon juice
¼ tsp pepper

Place alligator steaks in pan. Combine lemon juice, soy sauce, parsley, oil, tabasco sauce, salt, pepper and garlic salt. Pour over steaks and let stand 4 hours turning occasionally. Drain steaks well and roll in flour or bread crumbs and shake off the excess. Arrange steaks in a shallow pan and bake at 350°F for 1 hour, or until tender. Serves 8.

INDOOR BARBECUED ALLIGATOR

2-3 lbs alligator tail steaks
½ tsp liquid smoke
¼ c lemon juice
¼ c lime juice
2 c water
2 tsp worcestershire sauce
1 c ketchup
1 tsp salt
1 tsp celery seed
1 sm onion, minced
¼ c brown sugar or molasses
1 stick butter, melted

Mix the smoke, juices, water and worcestershire and add to the meat and let stand in the refrigerator for 12-48 hours, turning often. Remove meat from the marinade and place in a shallow oiled baking pan and place in a preheated 300°F oven for 2 hours basting with the butter to which some of the marinade has been added. Lower the heat to 275°F and continue baking until tender. Mix the ketchup, salt, celery seed, onion and brown sugar in a small saucepan and add any remaining basting sauce. Use this during the last 45 minutes of baking and serve in separate container for those who like extra sauce. Serves 6-8.

GRILLED ALLIGATOR

alligator fillets
lemon juice
olive oil
salt and pepper to taste
bacon drippings
worcestershire sauce

Marinate the meat in lemon juice for several hours. Wipe dry and rub with olive oil and then salt and pepper to taste. Place fillets on a grill over coals and baste with bacon drippings and worcestershire sauce. Turn often until done.

EASY GRILLED GATOR

4 alligator fillets, ½ in thick
1 lemon
1 tbsp olive oil
2 tbsp tabasco sauce
salt to taste
1 16 oz bottle italian salad dressing
4 tbsp liquid smoke

Place the alligator fillets in a bowl and cover with the salad dressing and marinate for 3 hours in the refrigerator. When ready to grill remove the fillets and dry well with paper towels and sprinkle with salt. Make a basting solution with the olive oil, liquid smoke and tabasco sauce. Spread the basting solution over the fillets with a barbecue brush. Cook very slowly over a low heat. Continue to baste while cooking (make more if you have to) and turn the fillets gently. When the fillets begin to flake the meat is done. Just before removing from the grill slice the lemon and spread a light coat of juice over each fillet. Serve hot.

ROLLED, STUFFED, GRILLED ALLIGATOR ROAST

2 1-lb alligator tail fillets
cayenne pepper
15 cloves garlic, minced
salt and pepper to taste
3 lemons, sliced thin, seeded
paprika
3 tbsp parsley, minced
cotton string

Marinade

2 c red wine
½ c lemon juice
6 cloves garlic, mashed
½ c extra virgin olive oil
1 tsp cayenne pepper

Tenderize the fillets thoroughly on both sides until approximately ½ inch thick or thinner. Mix all the marinade ingredients and cover the alligator in a flat pan for 3 hours, turning every 30 minutes. Remove the meat from the marinade but do not drain. Cover one side generously with cayenne pepper and paprika but don't overdo. Layer the same side of fillets with a layer of the lemon slices and sprinkle with chopped garlic and parsley and salt to taste. Roll each roast and tie tightly with the string about 1 inch apart. Tie in both directions to keep lemon from coming out the ends. Sprinkle each rolled roast with black pepper. Grill about 12 inches above hot coals in a covered barbecue pit turning frequently for about 45 minutes. Remove the string and carve into ½ inch slices and serve.

STOVE-TOP GRILLED ALLIGATOR

4-6 alligator tail fillets, cut
  ½ inch thick across the grain
3 cloves garlic, mashed
1 lemon, sliced, seeded
white pepper to taste
1 clove garlic
about ½ c water
dash tabasco sauce
1 stick butter, melted
salt to taste
1 lemon, thinly sliced
olive or peanut oil

Place fillets in a container and add the seeded lemon, garlic, tabasco sauce and enough water so that meat can be marinated. Let stand in the refrigerator for several hours or up to 36 hours turning frequently. Rub a skillet very well with the olive or peanut oil and the clove of garlic and heat. Drain the meat and strain the marinade liquid. Set it aside. Grill the meat on both sides until tender, covering to let steam if necessary. When cooked put on a hot platter. Mix the butter with the reserved marinade and heat. Pour over the steaks just before serving. Sprinkle with parsley, scallions or paprika if desired. Serves 4-6.

BARBECUED ALLIGATOR

alligator fillets
cayenne pepper
salt
onions, sliced
sliced bacon
butter
lemon juice
lemons, sliced

Wrap bacon slices around edges of fillets and hold in place with toothpicks. Squeeze lemon juice over fillets and then salt and pepper to taste. Place lemon and onion slices over the top. Then put butter slices over this and wrap and seal in aluminum foil. Place on grill. When cooked open the top of the foil and let smoke slightly. Serve.

FOILED ALLIGATOR

4-8 alligator tail slices
salt and pepper to taste
butter
sm mushrooms, sliced
oregano, ground or flakes

For each serving place a slice of meat about the size of a hamburger patty on a square of heavy duty aluminum foil and sprinkle with the seasonings and place mushrooms slices on top. Dot with butter and sprinkle with oregano. Fold ends over and seal to make airtight and place on a cookie sheet and bake at 325°F or 1½-2 hours. Open and let stand a few minutes before serving. These can also be done on an outdoor grill over very low coals for about 3 hours. Serves 4-8.

LINER'S BARBECUED ALLIGATOR

2 lbs alligator fillet
salt and pepper to taste
vinegar
4-5 tsp prepared mustard
barbecue sauce

Sprinkle alligator fillet with vinegar and then roll in the mustard, salt and pepper and let stand for 1 hour. Grill on low fire without basting for about 1 hour. Butterfly fillet by slicing in half and placing the outside down. Baste inside with your favorite barbecue sauce and grill until done.

ITALIAN BARBECUED ALLIGATOR FILLETS

alligator fillets
italian salad dressing

Marinate fillets in italian salad dressing for several hours. Place fillets on aluminum foil over coals in barbecue pit. Brush with marinade while grilling. When near done you can brush with your favorite barbecue sauce or continue with the marinade. (This can also be used with *Amphiuma*, frogs and snakes.)

ALLIGATOR RIBS

½ lb rack of alligator ribs
italian salad dressing
buttermilk
barbecue sauce

Marinate the ribs in buttermilk for 24 hours in a refrigerator and then change the marinade to italian dressing and soak again for another 24 hours. Smoke ribs, then chill to stop cooking process. If you do not have a smoker, broil in oven broiler use liquid smoke or other wood chips placed in steaming water. Lightly brush ribs with barbecue sauce and grill.

BARBECUED GATOR

alligator fillets
oil
barbecue sauce
Season-All
lemon juice

Sprinkle liberally with the Season-All. Marinate the fillets in an adequate amount of oil and lemon juice for 3 hours. Arrange the fillets on a well-oiled grill above a low charcoal fire that has soaked hickory chips added. Grill slow until the fillets are golden brown, occasionally basting with the lemon and oil marinade. When they are done brush with barbecue sauce and cook another 15 minutes.

FRENCH BARBECUED ALLIGATOR

alligator tail steaks
bottle french dressing
¼ c brown sugar
1 tbsp lemon juice
water
1 c barbecue sauce (your favorite)
several dash's tabasco sauce
melted butter
1 tsp soy sauce

Soak the steaks in the french dressing with a dash of tabasco sauce for several hours or up to 24 hours before grilling. Place on a grill over low coals and slowly cook for 1-2 hours basting with a mixture of butter, water and soy sauce to prevent drying. Mix the barbecue sauce, sugar, lemon juice and a few dashes of tabasco sauce and any remaining butter mixture and use it to baste during the last 45 minutes. Serve immediately with any left over sauce.

CHARCOAL BROILED ALLIGATOR FILLETS

alligator fillets, 1-2/person
½ c soy sauce
¼ c dry sherry
juice of 2 lemons
½ tsp onion salt
1 stick butter, melted

Above amounts are for 3-4 fillets. Make a marinade of all ingredients except butter and let the fillets marinate from a few hours to 3 days before cooking. Drain and reserve the marinade. Grill over very low coals turning carefully so that meat doesn't break up. Mix the reserved marinade sauce to the melted butter and baste the meat during cooking. If desired the meat can be wrapped in foil dotted with butter and a little of the marinade before being sealed tight. Serve at once.

CHARCOALED GATOR STRIPS

1-2 lbs gator tail meat, finger sized strips
plain flour
salt to taste
1 stick butter, melted
juice of 1 lemon
2 cloves garlic, crushed
corn meal
milk
black pepper to taste
dash tabasco sauce
1 tsp worcestershire sauce
oil

Soak the meat in milk for several hours or overnight. Drain. Mix equal amounts of corn meal and flour and season with the salt and pepper. Place a heavy pot over the coals and add 3-5 inches of oil and bring to about 375°F. Make a sauce of the butter, tabasco sauce, lemon juice, worcestershire and garlic and set aside. Dredge the meat in the corn meal mixture coating well and drop into the hot oil and fry for 2-3 minutes. Drain. Serve with the sauce poured over. Serves 6-10.

SMOKED ALLIGATOR

alligator fillets
salt and pepper to taste (optional)
italian salad dressing

Marinate fillets in italian salad dressing for several hours if desired. Remove from marinade and place in smoker with a pan of water between meat and coals. Use hickory chips soaked in water to smoke with. Season with salt and pepper. When meat is smoked it can be used as is or added to a jambalaya or a spaghetti.

DRY-SMOKED ALLIGATOR

2 lbs alligator meat, trimmed

Soak meat overnight in a salt water brine. Drain and keep meat in as large a piece as possible. This meat will shrink 20%. Smoke over low heat until meat becomes "dry". Cut and serve.

ALLIGATOR TAIL IN SMOKER COOKER

1 alligator tail
1 med onion, whole
1 bell pepper, whole
2 tbsp parsley, dried
1 tbsp liquid smoke
cayenne pepper
1 c dry white wine
1 lemon, halved
2 tbsp worcestershire sauce
6 drops Peychau's Bitters
salt

Combine the above ingredients except the salt and cayenne pepper in water in the smoker pan. Thoroughly dress the alligator tail by removing all fat. Sprinkle and pat the outside of the alligator with the salt and cayenne pepper. Cook 8-10 hours depending on the size of the tail.

POACHED ALLIGATOR TAIL

alligator fillets
1 bouquet garni*
butter
olive oil
lemon juice
water
plain flour
salt and pepper to taste

Marinate the meat in lemon juice for several hours. Wipe dry and rub with olive oil. Dust with salt, pepper and flour and sear meat in hot butter. Gradually add enough water to cover and then a bouquet garni. Simmer until tender.

BOILED ALLIGATOR

These recipes will vary somewhat from the procedures employed in this book. These recipes are to be used when cooked meat is called for. Basically the meat should be scalded before cooking by placing in scalding water and simmering for about 15 minutes. Drain and wash the meat. Now it is ready to use in the next three methods.

METHOD 1: Boil the meat in seasoned water with bay leaf, hot sauce or cayenne pepper, a carrot or potato and cook until tender. The liquid can be used when calling for stock or broth.

METHOD 2: Boil meat until tender in chicken broth seasoned with hot sauce. The broth can be canned, cubed or powdered. To be used when your recipe uses chicken soup or seems to be mild in flavor.

METHOD 3: Boil in beef broth until tender using canned, cubed or powdered and a little hot sauce. Use in dishes calling for a heartier flavor. The meat can then be ground, chopped or flaked according to the cuts used. The stocks can be used in other recipes as called for.

ALLIGATOR BOULETTES

3½ lbs alligator meat, boiled
5 c onions, minced
1 c scallions, minced
⅓ c black pepper
¼ c salt
2 12 oz cans evaporated milk
2 lbs bread crumbs
corn flour
3 c olive oil
2 c bell pepper, minced
¼ c garlic powder
⅓ c cayenne pepper
6 eggs
1 c parsley, minced
peanut oil
plain flour

In a deep covered pot boil the meat until tender. Remove from heat, drain and chop in a food processor. Set aside. In a saucepan saute the onions, and bell peppers in olive oil until tender. Remove from the heat and put in a separate pan. Mix well all other ingredients except peanut oil, corn flour, flour, eggs and milk. Using a ½ ounce scoop roll boulettes into little meat balls. Mix the eggs and milk and blend well. Dredge in flour and then the egg batter and then the corn flour. Fry until golden in 350°F peanut oil. Makes 2½ gallons.

GATOR BOULETTES

2 lbs alligator
2 egg yolks
½ bell pepper
oil
2 stalks celery
salt and pepper to taste
milk
3 slices bread
2 onions
1 c water

Grind together the alligator, onions, celery and bell pepper. Soak bread in milk and press out. Add to ground mixture. Add egg yolks and season to taste. Spoon into hot oil and brown. Remove. In another pot add the water and bring to a boil. Put the boulettes in the water and steam on low heat for 35 minutes with a cover on.

BOYD'S SEAFOOD BOULETTES

4 c cooked seafood
1 c celery, minced
4 drops tabasco sauce
2 tbsp mustard
¼-½ tsp cayenne pepper
4 eggs, beaten
oil
1 c onion, minced
1 c bell pepper, minced
2 tbsp lemon juice
2 tbsp Season-All
2-2½ c italian bread crumbs
bread crumbs

The cooked seafood can be either ground or coarsely chopped consisting of crawfish, shrimp, fish, crab or alligator or a combination. Grind the shrimp, crawfish or alligator; flake and chop fish. Place the seafood in a large mixing bowl and add the vegetables and toss. Stir in the seasonings and mix well. Mix in the italian bread crumbs and then the eggs and blend thoroughly. Roll into balls about the size of golf balls and coat with the bread crumbs. Drop into hot oil and fry until golden brown, turning to brown evenly. Serve hot.

ALLIGATOR VERACRUZ

1 lb alligator fillets
¼ tsp salt
1 sm tomato, peeled, chopped
¼ c bell pepper, chopped
1 tbsp lemon juice
1 tsp parsley, dried
1 4½ oz can tiny shrimp, rinsed, drained
¼ c dry white wine
⅓ c water
dash black pepper
1 sm onion, chopped
2 tbsp chili sauce
1 tbsp butter
½ tsp thyme, dried, crushed
1 2½ oz jar mushrooms, sliced, drained

Slice the fillets on the bias into ¾ inch thick slices. Grease a 10 inch skillet and add the alligator, water, salt and pepper. Bring to a boil and then reduce the heat. Cover and simmer about 10 minutes or until the meat flakes easily. Drain well. In a medium saucepan combine the tomato, onion, bell pepper, chili sauce, lemon juice, butter, parsley and thyme. Bring to a boil and then reduce the heat to medium low. Cover and cook about 5 minutes or until the vegetables are tender. Stir in the shrimp, mushrooms and wine and return to a boil. Boil gently uncovered about 3 minutes or until of the desired consistency. Arrange alligator on a serving platter and spoon the tomato mixture over it. Serves 4-6.

ALLIGATOR A LA FISSI

2 lbs alligator meat, sliced thin
¼ c + 2 tbsp white vinegar
2 c oil
2 tbsp cayenne pepper
1 eggplant, diced
salt and pepper to taste
2 c + 1 c red wine
1 tbsp juniper berries
1 tbsp shallots, minced
3 c brown veal demi-glaze sauce*
butter

Combine 2 cups red wine, ¼ cup vinegar, garlic, berries and oil and mix well. Add the meat and marinate for 2 days in the refrigerator, turning occasionally. Combine the shallots, 1 cup red wine, 2 tbsp vinegar and cayenne pepper and boil together until reduced by half. Then add the brown veal demi-glaze sauce and salt and pepper to taste. Get a skillet very hot and saute the marinated alligator on both sides. Saute the eggplant in butter and when sauteed put in a serving dish and put the alligator meat on top of the eggplant and then pour the sauce over the top. Serves 6.

ALLIGATOR DE CYPREMORT

6-8 alligator tail fillets
milk to cover
½-¾ c + ½ tsp plain flour
1 stick butter
½ tsp + tabasco sauce
½ c dry white wine
½ c scallion tops, chopped
¼ tsp salt
1 sm white onion, chopped
1 sm bell pepper, minced
½ c scallion bottoms, chopped

Soak meat in enough milk to cover. Season the milk with ½ tsp salt and a dash of tabasco sauce per cup of milk. Soak for at least 1 hour. Drain and dredge in the ½-¾ c flour and shake to remove the excess. Melt half the butter and ½ tsp tabasco sauce in a glass baking dish under the broiler. Remove and add the scallions and pepper and top with the onion. Lay the fillets on top and dot with the remaining butter. Place the dish about 8-10 inches from the broiler and cook for 15-20 minutes or until the meat is brown on top. Baste with the butter sauce several times during cooking. Remove meat to a warm platter. Mix the ½ tsp flour and wine and add to the pan drippings and stir well. Transfer to a small saucepan and heat for 5 minutes. Pour sauce over fillets and serve. Serves 6-8.

ALLIGATOR BOUILLABAISE MARSEILLAISE

2-3 lbs alligator fillets
1 c shrimp, raw, peeled, whole
2 doz oysters, raw
½ c lobster meat, cooked, chopped
½ c crawfish tails
1 c whole tomatoes
½ tsp cayenne pepper
1 lemon, sliced, seeded
½ c butter
1 c white onions, minced
½ c scallions, minced
1 tsp garlic, minced
1 tbsp plain flour
1 tsp salt
2 c fish stock
pinch saffron

In a large saucepan melt the butter and saute the onions, scallions and garlic until tender. Add the shrimp, oysters, lobster and crawfish (crabs may be substituted) and continue cooking a few minutes more. Stir in the flour and cook 3-5 minutes longer. Add the tomatoes, salt, pepper and stock and cook slowly for 20 minutes. Add saffron and simmer 5 minutes more. Remove from the heat. While the sauce is cooking cut the fillets into 4 pieces and place in a baking pan. Bake in a preheated oven at 350°F for 15 minutes or until the alligator is done. Put the alligator in a soup bowl and pour sauce over it and garnish with sliced lemon. Serves 4.

GATOR FLORENTINE FILLETS

4 alligator fillets, 1 in thick
1 tsp salt
¼ stick + 2 tbsp butter
1 tsp white pepper
1 sm onion
¼ c evaporated milk
½ c parmesan cheese, grated
2 pkgs frozen chopped spinach
½ c water
½ c wine
2 tbsp plain flour
¼ tsp cloves, ground
½ tsp lemon juice

Cook the spinach according to package directions and drain. Mix the spinach with the salt and ½ stick of butter and place in the bottom of a buttered shallow baking dish and set in a warm oven. Simmer the fillets in the water, wine, pepper, onion and cloves until tender but do not let them fall apart. Remove fillets with a slotted spoon and place over the spinach and return to the oven to keep warm. Simmer the stock from the meat for 8-10 minutes and set aside. In a small saucepan melt the remaining butter and add the flour stirring constantly and then gradually add the stock (strained) and cook on a low fire until thickened. Add the milk and lemon juice and pour over the fillets and sprinkle with the cheese. Place in a 400°F oven for 5-7 minutes or until the cheese is lightly browned. Serves 4.

ALLIGATOR COURT BOUILLON

- 5 lbs alligator meat
- 2 cloves garlic, minced
- ¼ c scallion tops, chopped
- 1½ c onions, minced
- 2½ tsp salt
- 1 c plain flour
- 4 qts water
- ¼ c parsley, chopped
- cooked rice
- ½ bell pepper, chopped
- 2 tsp seasoning salt
- 3 stems celery, chopped
- 2½ tsp cayenne pepper
- 1 c oil
- 1 8 oz can tomato paste

Brown oil and flour until golden. Add onions, celery, bell pepper and garlic and cook until wilted. Add tomato paste and cook 25 minutes or until oil appears on the top of the mixture. Skim it off. Add water, salt, cayenne pepper and seasoning salt and simmer for 15 minutes. Add scallions, parsley and alligator which has been cut into pieces. Cook 25 minutes or until meat is tender. Serve over rice. Serves 10-12.

ALLIGATOR AMERICAN

- 4 lbs alligator tail, cut up
- 2 tbsp olive oil
- 2 tbsp celery, minced
- 2 carrots, minced
- 1 lge onion, minced
- salt and pepper to taste
- 1 lb can tomatoes, chopped
- 1½ c white wine
- 1 tbsp parsley, chopped
- 1 tsp thyme

In a large skillet heat the oil and add celery, carrots, onion and thyme. Saute this mixture, stirring occasionally, until it is lightly browned. Add the tomatoes and wine and simmer for 10 minutes. Add alligator, salt and pepper and combine well with the sauce. Cover and simmer 30-40 minutes, or until tender. Serve with chopped parsley.

LOUISIANA COURT BOUILLON

2-3 lbs alligator, cubed
1 1.2 oz pkg meat marinade mix
6 tbsp olive oil
2 lge bell peppers, chopped
1 clove garlic, minced
4 15 oz cans tomato sauce
6 c liquid (water, or marinade)
1 8 oz can mushrooms, sliced
1 med bay leaf
salt and pepper to taste
steaming rice
2 c italian salad dressing
water (if needed)
2 lge onions, chopped
6 ribs celery, chopped
1 10 oz cans Ro-Tel tomatoes
1 6 oz can tomato paste
1 lemon slice, quartered
1 tsp sugar
3-4 drops liquid crab boil*
Season-All to taste

Place alligator in a container with a cover and add the italian dressing and the marinade mix to completely cover adding water if necessary and refrigerate overnight. Drain and reserve the marinade. Heat oil in a heavy pot and saute the onion, bell pepper and celery until wilted. Stir in tomatoes, tomato sauce, tomato paste and liquid and simmer for 5 minutes. Add lemon, mushrooms, sugar, bay leaf, crab boil and seasonings blending well. Cover and simmer for 45 minutes. Add the alligator and simmer 2-2½ hours stirring as little as possible to retain chunks of meat. To serve mound steaming rice on a plate and spoon court-bouillon over it. (Note: You can add more alligator meat if desired without increasing the sauce.) Serves 6.

LOUISIANA COURT BOUILLON 2

3 lbs alligator tail, in pieces
2 6 oz cans tomato paste
3 c onions, chopped
1 c bell pepper, chopped
1 c scallions, chopped
2 4 oz cans mushroom steak sauce
salt and cayenne pepper to taste
1½ c oil
1 gal hot water
1 c celery, chopped
1 c garlic, chopped
1 c parsley, chopped
3 bay leaves

In a large pot heat the oil over a medium heat. Salt and pepper the alligator. Cook in the oil approximately 20 minutes. Add all seasonings and stir well and cook for approximately 10 minutes. Add the mushroom sauce and tomato paste and simmer over low heat approximately 45 minutes. Add 2 quarts of water and blend well. Cook over medium heat for 1 hour adding water to retain volume. Season again to taste.

FRANKLIN COURT BOUILLON

- 2 lbs alligator meat, cubed
- water to cover
- dash tabasco sauce
- 3 ribs celery, chopped
- 1 8 oz can tomato sauce
- 1 tbsp soy sauce
- ½ c white wine
- ¼ c parsley, chopped
- cayenne pepper to taste
- ½ c oil
- 1 med onion, chopped
- 1 sm bell pepper, chopped
- 1 6 oz can tomato paste
- 2 bay leaves
- pinch of oregano
- ½ c scallion tops, chopped
- salt to taste

Boil the meat in water to cover with a dash of tabasco sauce for 30 minutes. Drain and reserve 3 cups of the stock. Saute the onion, bell pepper and celery in hot oil until soft and then add the paste, sauce, bay leaves, soy sauce, oregano and 2 cups of the reserved stock. Bring to a boil and then simmer for 2-4 hours over a medium to low heat. Add more of the stock if it gets too thick. Add the meat and cook for 45 minutes and then add the wine, scallions, parsley and seasoning 10 minutes before serving. Serve over rice. Serves 4-8.

TECHE COUNTRY COURT BOUILLON

- 3 lbs gator fillets, white and pink meat mixed
- ½ c oil
- 1 onion, chopped
- 2 ribs celery, chopped
- 1 14½ oz can stewed tomatoes
- 3 bay leaves
- ¼ c worcestershire sauce
- salt and pepper to taste
- ½ c red wine
- 2 qts water
- ⅓ c plain flour
- 1 bell pepper, chopped
- 2 qts stock from meat
- 1 6 oz can tomato paste
- 4 cloves garlic, minced
- 1 lemon, wedged
- parsley, chopped

Boil the meat in the water for 30 minutes and then drain and reserve the stock. Heat the oil and add the flour and stir over a low heat until browned and then add the onion, bell pepper and celery and cook until soft. Add 1 quart of the stock and stir and then add the tomatoes, paste, bay leaves, garlic and worcestershire sauce. Bring to a boil and then simmer for 1-2 hours or until thick. Add stock now and then until all used and then add the meat after the sauce has cooked for 1 hour. Just before serving add the wine, parsley and season to taste. Serve over rice with the lemon wedges. Serves 10-12.

BELL ISLE COURT BOUILLON

3 lbs alligator meat, cubed
2 tbsp plain flour
1 28 oz can tomatoes
3 bay leaves
¼ tsp thyme
½ c red wine
salt and pepper to taste
2 tbsp oil
2 lge onions, chopped
1 qt hot water
¼ tsp marjoram
¼ tsp dry mustard
1 lemon, sliced, seeded
½ c hot cooked rice/person

Brown the flour in the oil until the color of caramels stirring constantly and then add the onions and cook until tender. Add the tomatoes and the meat and cover and cook for 10-15 minutes. Add the hot water slowly keeping the mixture at a boil and stirring all the time. Reduce the heat and add the bay leaves, half the wine and the lemon slices. Cover and cook over a low heat for several hours, stirring often. During the last hour of cooking add the seasonings. Remove from heat and add the remaining wine. Serves 8-10.

GRANNY'S GATOR HASH

3-4 c alligator meat, cooked, diced or cubed
¼ c plain flour
1 med onion, chopped
salt and pepper to taste
hot baked biscuits
¼ c oil
2 c hot water
1 4 oz can mushrooms, sliced
hot cooked grits

In a heavy saucepan heat the oil and make a roux* with the flour until very brown. Then add the onions and cook until tender. Add the meat and half the water and bring to a boil and then add mushrooms. Lower the heat and simmer for about 1 hour adding more water as necessary. Season to taste. (Bell peppers, scallions or fresh tomatoes may be added if desired.) Serve over the grits with biscuits. Serves 6-8.

GATOR HASH

3 c gator leg meat, cubed
2 tbsp plain flour
3 med potatoes, cubed, peeled
1 onion, diced
creole seasonings* to taste
2 tbsp butter
2 c chicken broth or stock
4 ribs celery, chopped
1 bay leaf

Melt the butter and then add the flour and brown stirring constantly and when browned add the broth or stock slowly while stirring. Add all the other ingredients and a little water if necessary to cover the meat. Bring to a boil and then lower the heat and simmer for 1-2 hours. Add more broth or water if needed and stir often. Add the seasonings during the last half hour. Serves 8-10.

ALLIGATOR SAUCE PIQUANT

5 lbs alligator meat
½ c celery, chopped
cayenne pepper
½ c bell pepper, chopped
salt and pepper to taste
4 tbsp butter
1 c water
1 8 oz can mushrooms, stems and pieces
2 med onions, chopped
8 cloves garlic, chopped
1 8 oz jar salad olives, rinsed
½ tsp sugar
¼ c parsley, chopped
1 6 oz can tomato paste
¼ c scallions, chopped
cooked rice

Saute onions in butter until brown. Add tomato paste and sugar and cook about 5 minutes. Add bell pepper, celery, garlic and mushrooms and stir well. Add water and cook 1 hour over low heat. Add scallions, parsley, alligator (cut in small pieces and preferably meat other than from the tail), salt, pepper and cayenne to taste. Cover pot and cook slowly for 30 minutes or until meat is tender. Add olives and cook a few minutes longer. Serve with rice.

WILD GAME* SAUCE PIQUANT

20 lbs alligator or turtle meat
1 qt oil
2 lbs plain flour
9 lbs onions, chopped
½ stalk celery, chopped
1 10 oz can Ro-Tel Tomatoes
2 8 oz cans tomato sauce
4 bell peppers, chopped
worcestershire sauce
tabasco sauce
garlic salt or powder
1 32 oz bottle ketchup
1 qt scallions, chopped
salt and pepper to taste
2 10¾ oz cans cream of mushroom soup

Cover meat with water and parboil for 30 minutes. Season with salt, pepper and garlic powder. Remove meat and save stock. Make a roux of the oil and flour until brown. Allow to cool 10-15 minutes. Add onions, bell peppers, and celery to roux and cook for 30 minutes. Add diced Ro-Tel tomatoes and cook for another 20 minutes, stirring often. Add tomato sauce, soup, ketchup and scallions. Cook for another 20 minutes stirring often. Add meat to gravy and add approximately 2 quarts stock and stir gently until meat is tender. Season to taste with tabasco sauce, worcestershire sauce and garlic powder. Serves 60. *Rabbit, deer, squirrel or other wild game can be used.

ALLIGATOR AND ANDOUILLE* SAUCE PIQUANT

5 lbs alligator meat, 1 in pieces
cajun seasoning*
10 ozs tomato sauce
⅓ c dark roux*
4 c spanish onion, chopped
1 c celery, diced
2 tbsp jalapeno pepper, diced
2 tbsp garlic, chopped
2 qts water
mixture of cornstarch and water for thickening
½ c + 1 tbsp olive oil
1¼ lbs andouille, diced
⅓ c butter
¼ c Wylers Chicken Granules
1 c bell pepper, chopped
1 tsp cayenne pepper
1 tsp sugar
3 c fresh mushrooms, sliced
½ c parsley, chopped
½ c scallions, chopped

Rub well the alligator meat with cajun seasoning and if possible allow to marinate overnight in a refrigerator. Using a high heat brown the alligator in ¼ cup olive oil and remove from the heat. Saute the andouille in the same oil for 5 minutes and remove from the pot. Pour tomato sauce into the pot with the remaining oil. Still on high heat stir the sauce until it is very brown. Keep stirring until a thick ball of paste forms. To this add the butter roux, chicken granules, onions, bell pepper, celery, peppers and sugar. Saute until the onions are clear. Return the alligator and andouille to the pot. Add garlic, mushrooms and 3 cups water and bring to a boil and then reduce heat to medium. Cook for 1 hour adding water as needed. Once the alligator is tender add the scallions and parsley. If desired add a mixture of cornstarch and water to thicken. Serve over rice. Serves 12-14.

ACADIAN ALLIGATOR SAUCE PIQUANT

2½ lbs alligator meat, mixed
1½ c roux*
4 cloves garlic, minced
2 lge stalks celery, chopped
2 tbsp tomato paste
1 tsp finely ground black pepper
1 tsp basil, dried
1 tsp thyme, dried
½ c burgundy
6 scallions, chopped
salt and pepper to taste
2 med onions, chopped
1 lge bell pepper, chopped
3 med tomatoes, peeled, chopped
2 tsp worcestershire sauce
1½ tsp chili powder
1 tsp oregano, dried
½ tsp cayenne pepper
2 c beef broth
¼ c parsley, minced

Cut alligator in strips and season with salt and pepper. Brown meat and mix well with hot roux. Stir in onions, garlic, bell pepper, celery and tomatoes. Stir until the vegetables are transparent. Add tomato paste, worcestershire sauce, peppers, oregano, basil, thyme, scallions, parsley and salt. Cook for 5 minutes. Stir in the wine and broth and simmer until the meat is tender and add water as needed. Serve over rice. Serves 4-6.

ALLIGATOR SAUCE PIQUANT A LA CHACHERE

3 lbs alligator meat, cubed
1 10 oz can Ro-Tel Tomatoes
1 8 oz cans tomato sauce
1 c burgundy
water
2 tbsp worcestershire sauce
2 sticks, butter, melted
2 8 oz cans mushrooms, sliced
3 onions, chopped
1½ bell peppers, chopped
6 ribs celery, chopped
3 cloves garlic, chopped
¼ c brown roux*
creole seasoning* to taste
½ tsp sugar
bay leaves

Season the meat generously with creole seasoning and then fry in the butter. Add the vegetables, except mushrooms, and saute for about 5 minutes until wilted. Add mushrooms, tomato sauce, Ro-Tel tomatoes and sugar and cook another 5 minutes while stirring. Mix roux in enough hot water to cover meat and add to mixture. Add all other ingredients and let come to a boil and then reduce heat to a simmer and cook for 3-4 hours until meat is tender and desired thickness is reached. Season to taste with creole seasoning and serve over rice.

V. J.'S ALLIGATOR SAUCE PIQUANT

4 lbs alligator meat, bite size
1 lb pork, cut up
2 lbs onions, diced
2 10¾ oz cans french onion soup
2 10¾ oz cans cream of celery soup
2 ribs celery, chopped
1 10 oz can Ro-Tel Tomatoes
½ c parsley flakes
1 c roux*
2 10¾ oz cans cream of mushroom soup
2 bu shallots, diced
2 bell peppers, chopped
3 tbsp garlic, minced
1 lge can V-8 Juice
1 c oil
1 c plain flour

Boil alligator in water, skimming off the foam from the top. Make roux* with the flour and oil stirring until dark brown. Add the vegetables and cook about 30 minutes until wilted. Add all remaining ingredients, except meats and parsley, and simmer for about 1 hour,. Add the meats and simmer about 2 hours. Add the parsley about 30 minutes before done. Serves 15.

BAYOU ALLIGATOR SAUCE PIQUANT

50 lbs alligator meat, cubed
1 qt oil
10 lbs celery, chopped
20 lge bell peppers, chopped
6 8 oz cans mushrooms, sliced
10 10½ oz cans cream of onion soup
6 10 oz cans Ro-Tel Tomatoes
3 bottles Louisiana Hot sauce
½ lb sugar
water (optional)
5 lbs plain flour
15 lbs onions, chopped
10 bu scallions, chopped
1 qt olives, sliced
10 10½ oz cans cream of mushroom soup
2 gals tomato sauce
1 can creole seasoning*
6 lemons, sliced, seeded
6 4 oz cans mushroom steak sauce

Get a big, really big pot. Prepare a roux by heating the oil and then adding the flour and cook over a medium heat until mixture is a dark brown stirring constantly. Mix and brown all the chopped vegetables and cook until the onions are clear. Add the alligator meat and cook about 1 hour. Add the remaining ingredients and cook for 2 more hours. Add water to obtain consistency desired and serve over hot rice. Serves 100+.

FESTIVAL GATOR SAUCE PIQUANT

2½-3 lbs alligator meat
2 c onions, chopped
2 c celery, chopped
4 cloves garlic, minced
3 8 oz cans mushrooms, sliced
1 qt Ragu Extra Thick with Mushrooms Sauce
cayenne pepper to taste
½ qt water
2 c scallions, chopped
1 c bell pepper, chopped
2-3 tsp basil
1 c oil
salt to taste

Heat ¾ cup of oil and add onions, scallions, celery and bell pepper and saute on a medium heat until wilted. Add the Ragu, basil and mushrooms with liquid. Rinse Ragu jar with ½ qt water and add to the sauce. Bring to a boil and then reduce heat and simmer for about ½ hour. While above is simmering season the alligator meat cut into chunks with salt and cayenne pepper. Brown meat with the remaining oil. Add to sauce. Cook down until meat is tender (adding more water if necessary). Add garlic and lemon. Cook until oil forms on top. Skim it off. Season to taste with salt and cayenne.

BIG BILL'S ALLIGATOR SAUCE PIQUANT

3 lbs alligator meat, boneless
1 lge Bermuda onion, chopped
1 clove garlic, minced
2½ tbsp Louisiana Hot Sauce
1 2-lb jar spaghetti sauce, regular
1 envelope french salad dressing
1 med bell pepper, chopped
½ tsp Season-All
2 8 oz cans tomato sauce
oil

Cut meat into cubes and mix with the salad dressing and marinate in the refrigerator overnight. In a 4 quart saucepan lightly brown the alligator in 3 tbsp oil over a medium heat. Pour off any excess water. While the meat is browning begin to saute the onions in oil in another 4 quart pot. After a couple of minutes add the bell pepper and after a few more minutes add the garlic. Saute the mixture until seasonings are soft. Add the tomato sauce and simmer for 30 minutes. Add the spaghetti sauce, Season All and hot sauce and simmer 15 more minutes over a medium heat. Add the alligator meat and cook over a medium heat for 30 minutes. Serve over rice. Serves 10.

JUSTIN'S ALLIGATOR SAUCE PIQUANT

5 lbs alligator meat, cubed
1 c scallions, chopped
1 c parsley, chopped
2 tbsp garlic, minced
6 drops Peychaud's Bitters
4 8 oz cans tomato sauce
4 c dry white wine
1 tbsp tabasco sauce
1 c green olives
1 c oil or bacon drippings
4 c onions, chopped
1 c bell pepper, chopped
½ c celery, chopped
1 sm lemon, chopped fine
2 c plain flour
1 tsp mint, dried, crushed
3 tbsp worcestershire sauce
cold water, enough to cover
1 lb fresh mushrooms

Make a roux of the oil or drippings and flour until dark brown. Then add the onions, bell pepper and celery and continue cooking. When the onions become clear add the scallions and parsley and continue cooking for a few minutes. Add a bit of cold water to the roux so that it becomes a thick liquid. Add garlic and the rest of the seasonings including the tomato sauce and wine. Add the alligator meat and cover with water about 2 inches over the rest of the ingredients. Bring to a boil stirring frequently being sure you have enough water that the sauce doesn't stick. Cook for several hours or until the meat is tender. Serve over spaghetti or rice.

SAUCE PIQUANT A LA LOUISIANA

20 lbs alligator meat, cubed
1 16 oz beer
8 ozs Coca Cola
1 tbsp cayenne pepper
4 ozs parsley, chopped
2 bell peppers, chopped
1 stick butter
6 10 oz cans brown mushroom gravy
1 29 oz can tomato sauce
1 7 oz jar stuffed olives
1 lb plain flour
2 ozs liquid crab boil*
2 lemon slices, seeded
6 oz jalapeno peppers, chopped
10 ozs white wine
1 tbsp salt
1 tbsp garlic salt
1 lb celery, chopped
6 ozs scallions, chopped
1 c oil
6 8 oz cans mushrooms, stems and pieces
½ gal tomatoes, whole
1 3 oz bottle Kitchen Bouquet
10 ozs creole seafood sauce
1 gal water
2 tbsp worcestershire sauce

Marinate the meat with the beer, wine, Coca Cola, salt and jalapeno peppers for 24 hours in the refrigerator. Drain the liquid from the meat and reserve. Add oil to a large pot and heat on a low flame. Add the onions and saute for about 15 minutes. Add the alligator to the oil and onions and raise the heat a little higher. Stir constantly to keep from sticking. When the meat starts to brown add cayenne, garlic salt, parsley, celery, bell peppers and butter. Cook for about 30 minutes. Dissolve the flour in the reserved marinade, water, crab boil, seafood sauce, worcestershire sauce and Kitchen Bouquet. Then add to the alligator meat and lower the heat and then add the mushrooms, mushroom gravy, tomato sauce, tomatoes, olives (drained and rinsed) and lemon slices. Continue to cook on a low fire stirring frequently. Keep a cover on the pot and cook for about 3 hours. Serve over rice. Serves 30-35.

ALLIGATOR SAUCE PIQUANT WITH PEAS

1 lge or 2 sm alligator leg(s)
½ c oil
1 14½ oz can stewed tomatoes
1 4 oz can mushrooms, sliced
1 15 oz can petit pois peas
salt to taste
water to cover
½ c plain flour
1 clove garlic, sliced
4 lge onions, chopped
cayenne pepper to taste

Separate the legs at the joints and boil in the water for ½ hour and then drain and rinse. Brown the flour in the oil in a heavy pot over a medium heat stirring constantly. Add the tomatoes and cook for 3-4 minutes. Add the meat and all remaining ingredients except the seasonings and cook for 2-4 hours stirring often. Remove the bones and discard. Season to taste and serve over rice. Serves 8-12 people.

PIQUANT BAKED GATOR

6-8 alligator tail fillets
½ c soy sauce
¼ c water
1 tsp oregano, ground
4 tbsp brown sugar
3 c dry red wine
½ c salad oil
6 cloves garlic, sliced
4 tsp ginger, ground

Arrange fillets in a shallow baking dish. Combine all remaining ingredients mixing well and pour over the meat. Cover and refrigerate for 24 hours. Remove from the refrigerator about 1 hour before baking and do not uncover but let come to room temperature. Preheat the oven to 375°F and then bake uncovered for 30 minutes and then lower the heat to 350°F and cover and bake for 30 minutes more. Serve with hot rice available for the pan drippings. Serves 6-8.

ALLIGATOR SAUSAGE JAMBALAYA

3 lbs alligator meat, cut up
½ bell pepper, chopped
¾ c oil
5 c water
1 lb smoked pork sausage
cayenne pepper
2 lbs onions, chopped
salt and pepper to taste
3 cloves garlic, chopped
2½ c rice
½ bu scallions, chopped

Brown alligator meat and sausage in hot oil (if not using smoked alligator). Remove meat and most of the oil and add onions, garlic, bell peppers and scallions and brown well. Add a little water, if necessary, to keep from sticking. Put meat back into the pot along with the water. Add salt and pepper to taste. Wash rice and add when mixture comes to a boil. When this comes to a boil again, lower heat and let all the water boil out. Stir well, lower the heat and cover for 15 minutes. Stir again, cover, and leave on low, low fire for 45 minutes. Watch that the rice does not burn and stick to the pot. (If you have smoked turtle or alligator sausage you can use this in place of the smoked pork sausage.)

BROWNED RICE CASSEROLE A LA ALLIGATOR

2 lbs gator meat, cubed
1 tsp tabasco sauce
2 cloves garlic, minced
1 8 oz can mushrooms, sliced, drained
½ bell pepper, chopped
1 14½ oz can stewed tomatoes
oregano to taste
1 14½ oz can chicken broth
1 14½ oz can water
1 tbsp olive oil
1 lge onion, chopped
1 c raw rice, long grain
pinch of rosemary
salt and pepper to taste

Boil the meat in the broth, water, tabasco sauce and garlic for 1-1½ hours until tender. Drain reserving the stock. Heat the oil in a heavy skillet and add onion and bell pepper and saute until soft. Stir in the rice and stirring constantly cook until rice is golden brown. Add the tomatoes, mushrooms and seasonings and simmer for 5 minutes. Place the meat in a deep greased casserole and top with the tomato mixture. Pour in enough reserved stock to cover about 2 inches. Bake at 350°F for 45-60 minutes or until the rice is done. Serves 6.

EASY GATOR RICE CASSEROLE

1 lb alligator meat, cooked*, diced
1 11 oz can cheddar cheese soup
1 lge onion, chopped
1 tsp salt
1 c raw rice
1 soup can water
1 lge bell pepper, chopped
½ tsp black or cayenne pepper

In a large bowl combine all the ingredients and mix well. Pour into a large well-greased casserole and cover tightly and bake at 375°F for 1 hour. Check to see if the rice is done, if not, return for 10-15 minutes longer. Serves 6-8.

GREEN ALLIGATOR CASSEROLE

2 lbs gator meat, cooked, diced
2 ribs celery, chopped
¼ c parsley flakes
1 10½ oz can cream of mushroom soup
½ lb american cheese, grated
½ c salad olives with juice
½ stick butter
1 lge onion, chopped
1 sm bell pepper, chopped
1 10¾ oz can cream of asparagus soup
1 4 oz can mushrooms, sliced
1 5 oz pkg green egg noodles, cooked

Melt the butter and saute until soft the celery, onion, bell pepper and parsley and then add the soups. Heat to near-boiling and add the cheese. Reduce the heat and cook slowly until the cheese is melted and then add the mushrooms and olives and cook for 2-3 minutes. Stir in the noodles and pour into a lightly greased casserole and bake for 1 hour at 350°F. Sprinkle with a mixture of parsley flakes and bread crumbs before baking, if desired. Serves 8-10.

ALLIGATOR STROGANOFF

1 lb alligator meat, cubed
water to cover
¼ c plain flour
¼ c corn oil
1 sm onion, minced
1 4 oz can mushrooms, stems and pieces
1 clove garlic, minced
1 beef bouillon cube
1 c sour cream
salt and pepper to taste
1 lge pkg noodles cooked, buttered

Put the meat in the water and boil for 30-60 minutes or until tender. Drain and reserve 1 cup of the stock. Heat the oil and the flour in a saucepan and stir until a dark brown and then add the onion and cook on low until soft. Add the mushrooms, garlic, reserved stock and the bouillon cube. Bring to a boil and then simmer for 15 minutes. If too thick add a little more stock and add the meat and continue cooking until very hot. Season to taste and just before serving add the sour cream and stir in gently but well. Serve over the noodles. Serves 4-6.

ALLIGATOR PILAF

1½-2 c alligator, cooked*, diced
1 c rice, cooked
1 14½ oz can stewed tomatoes
1 lge onion, chopped
½ bell pepper, chopped
2 ribs celery, chopped
1 tsp salt
¼ tsp cayenne pepper
bread crumbs
butter

Mix meat, rice and tomatoes in a heavy pot and heat well and then simmer for 15 minutes. Add onion, celery, bell pepper and season to taste. Place in a buttered casserole and sprinkle with bread crumbs and dot with butter. Bake at 350°F for 1 hour. Serves 4-6.

GATOR AND WILD RICE

1 c raw wild rice, cooked
3 c alligator meat, cooked*, diced
2 tbsp parsley, chopped
1 tsp salt
½ tsp cayenne pepper
1 8 oz can mushrooms, sliced
1 sm jar pimento, chopped
1½ c milk
½ c butter
1 sm onion, chopped
¼ c plain flour

Saute onion in the butter until soft and then add the flour and cook until a light brown. Add the liquids from the mushrooms, pimentos and milk and heat over a low fire until almost boiling and remove from the heat. Don't boil. Add the meat, mushrooms, pimento and parsley and season adding rice. In a 2 quart greased casserole place the rice mixture and bake for 30-45 minutes at 350°F. Serves 6.

GATOR JAMBALAYA

2 lbs gator meat, cooked*, diced
1 med onion, chopped
1 c celery, chopped
1 stick butter
4 c chicken or alligator broth
1 clove garlic, minced
1 c raw rice, long grain
2 tsp salt
1 tsp tabasco sauce
1 c parsley, chopped
1 scallion, chopped
1 4 oz can mushrooms, sliced

Saute the onion and celery in butter until soft and then add the meat, broth, rice mushrooms, garlic and seasonings and bring to a boil. Simmer 15 minutes and then add the scallions and parsley. Cover and simmer 15-30 minutes stirring occasionally. Serves 6-8.

ALLIGATOR SPAGHETTI

3 lbs alligator meat, cut up
3 6 oz cans tomato paste
4 tsp salt
1 c bell pepper, minced
¼ c parsley, chopped
2 c onions, minced
½ c worcestershire sauce
1 lb mushrooms, stems and pieces
3 35 oz cans tomatoes
4 tsp oregano
3 cloves garlic, minced
2 tbsp sweet basil, dried
½ lb sliced bacon, diced
1½ c water
¼ tsp tabasco sauce
2 tbsp sugar

In a 4 to 6-quart pot fry the bacon until crisp. Remove the bacon and all but 3 tbsp bacon grease. Add alligator meat and brown. (If using smoked alligator omit this stage.) Remove meat when browned and set aside. Saute onions, drained mushrooms, bell pepper and garlic for 10 minutes. Stir in the tomatoes, tomato paste, water, worcestershire sauce, parsley, basil, oregano, sugar, salt and tabasco sauce. Bring to boiling point, reduce heat and simmer, uncovered, stirring occasionally, for 3 hours or longer. Add reserved alligator meat and simmer until meat is tender. Serve over spaghetti with parmesan cheese. sauce yield is approximately 5 quarts.

ALLIGATOR SCALLOPINI

6 alligator fillets, sliced thin or pre-marinated
3 c scallopini or spaghetti
salt and pepper to taste
1 c shallots, minced
1 tsp dry mustard
4 cloves garlic, mashed to paste
1 c plain flour
meat sauce (your favorite)
¼ c butter
1½ c white vermouth
2 tsp paprika
1 lemon, sliced thin, seeded

Thoroughly blend the vermouth, mustard, paprika, garlic and lemon. Cover the fillets in a shallow bowl with the marinade and marinate for 6 hours turning hourly in a refrigerator. Drain and flour the fillets and salt and pepper to taste and saute about 1 minute on each side in the butter. Place the cooked slices on a dish and pour scallopini sauce over them. Top with shallots.

ALLIGATOR PRIMAVERA

1½ lbs fettuccine, pre-cooked (al dente)
4-5 oz roux*, white*
1 onion, julienne-cut
1 qt chicken broth
6 mushrooms, sliced
2 carrots, julienne-cut
1 rib celery, julienne-cut
1 lb alligator meat, diced
2 ozs butter
3 tbsp tomato paste
2 bell peppers, julienne cut
4 cloves garlic, chopped
salt and pepper to taste

Poach the alligator meat by slowly boiling in chicken broth for 20 minutes. Remove the meat and add the tomato paste, garlic and roux to the chicken broth until a medium-thick consistency is obtained. Add the vegetables and meat and bring to a boil slowly for 3-4 minutes. Reheat the pasta in hot water and top it with alligator primavera. Sprinkle with the parsley and place a small ball of butter on top. Serves 6.

GATOR NOODLE BAKE

1 lb gator meat, sliced thin
1 c milk
1 c chicken broth
1 lge bag egg noodles, cooked
½ c cracker crumbs
1 4 oz can mushrooms, sliced
dash tabasco sauce
salt to taste
¼ stick + butter

Marinate the meat in the milk for an hour or so and then drain and reserve the milk. Bring the meat to a boil in the broth and boil until tender. Add more broth (or water) to keep the liquid level about the same. When the meat is tender add the noodles and stir gently. Then add the mushrooms and season to taste. Butter a deep casserole and pour the noodles in. Heat the milk and add the ½ stick butter to it and heat until the butter melts and then pour over the noodles. Top with cracker crumbs and dot with butter. Bake at 325°F for 30-45 minutes. Let stand for 5-10 minutes before serving. Serves 6-8.

GATOR AND SHELL BAKE

1 c gator meat, cooked*, ground
6 ozs shell macaroni, boiled
1 10¾ oz can cream of celery soup
1 c potato chips, crushed or bread crumbs
butter
1 c milk
1 15 oz can petit pois peas
2 eggs, slightly beaten
salt and pepper to taste

Mix the soup, peas, milk and eggs in a heavy saucepan and heat on medium until just about to boil and then remove from heat. Mix in the meat and macaroni and stir in gently. Season to taste. Pour into a buttered casserole and top with the chips and bake at 350°F for 30-35 minutes. Let stand for 5 minutes before serving. Serves 4-6.

ALLIGATOR BUSY DAY CASSEROLE

1 c gator meat, ground, cooked*
1 6 oz pkg elbow macaroni, cooked
3 tbsp butter
3 tbsp plain flour
¼ c scallion, chopped
1½ c milk
1 c mild cheddar cheese, grated

Melt the butter and add the flour and cook stirring for 2-5 minutes on a low heat. Add the scallions and saute until wilted. Slowly add the milk and cook on medium to low heat until slightly thickened. Add the meat and heat and then remove from the heat and add ½ of the cheese and stir until the cheese begins to melt. Add the macaroni and mix well. Place this mixture in a buttered casserole and sprinkle the remaining cheese on top. Bake at 350°F for 30-40 minutes. Serves 6.

ALLIGATOR TALLERINI

1 lb gator meat, cooked*, ground
2 tbsp butter or corn oil
1 lge onion, chopped
3 cloves garlic, minced
1 can enchilada sauce
1 6 oz can tomato paste
12 ozs egg noodles, uncooked
1 15 oz can cream style corn
1 4¼ oz can chopped ripe olives
1 lb cheddar cheese, grated
½ tsp chili powder
salt and pepper to taste

Saute the onion in the butter until very soft and add the garlic, sauce, paste and noodles for 10 minutes. Add a little water if too dry or sticks. Remove from the heat and add the remaining ingredients and stir well until the cheese has started to melt. Place in a deep greased casserole and if desired sprinkle with bread crumbs or paprika. Heat at 300°F for 30-45 minutes. Serves 6-8.

FAVORITE ALLIGATOR CASSEROLE

2 lbs gator leg meat, boneless
1 4 oz can mushrooms, buttons
1 tbsp onion, minced
½ stick butter
¼ c plain flour
toasted almonds, sliced
1 pt heavy cream
1 pt alligator broth
1 tsp salt
¼ tsp black pepper
4 ozs egg noodles, cooked
water to cover

Place the meat in enough water to cover well and boil for 20 minutes. Drain and change the water and boil until the meat is tender. Drain reserving the liquid. Saute drained mushrooms and onion in butter and stir in the flour and add the cream and broth. Season and stir in the meat which has been cut into bite size pieces. Stir in the noodles. Place all in a buttered casserole and sprinkle with the almonds and bake at 300°F for 45 minutes. Serves 6-8.

LASAGNA A LA ALLIGATOR

2 lbs pink gator meat, sliced thin
½ c water
¼ tsp soy sauce
½ c parsley, chopped
broad lasagna noodles, cooked
¾ lb mozzarella cheese, grated
parmesan cheese, grated
½ tsp oregano powder
salt and pepper to taste
¼ c plain flour
2 c chicken broth
1 14½ oz can stewed tomatoes
2 cloves garlic, minced
1 bay leaf
pinch rosemary
pinch thyme
½ tsp sugar
water to cover

Boil the meat in water to cover for about 20 minutes and then drain and rinse. Now put the meat into a large pot with the broth, tomatoes, garlic, bay leaf and remaining seasonings and cook for 1 hour or until the meat is tender and the sauce cooked down. Mix the flour in the ½ cup water, soy sauce and parsley and stir well and add to the hot mixture and mix well. Simmer 10 minutes longer. In a well oiled lasagna pan layer the bottom with sauce, then a layer of noodles with sauce on top and a layer of cheese and repeat. Top with more cheese and bake at 350°F for 30 minutes. Serves 8-10.

GATOR A LA NONC

1 c gator meat, cooked*, ground
1 4 oz pkg egg noodles, cooked
1 10¾ oz can cream of mushroom soup
tabasco sauce to taste
¾ c cashew nuts, chopped
½ c milk
½ c mayonnaise
1 sm onion, chopped
1 tsp salt
1 10¾ oz can cream of chicken soup
¼ c bread crumbs

Mix the soups, onion, milk, mayonnaise and seasonings and heat slowly until very hot but not boiling and then add the meat and noodles. Place into a buttered 2-quart casserole and sprinkle with the cashews and bread crumbs. Bake at 325°F for 1 hour. (Any cream style soups may be used.) Serves 8.

CONFETTIED GATOR

1 lb gator meat, cooked*, ground
1 sm jar pimento, chopped
1 sm onion, chopped
½ bell pepper, chopped
2 lge tomatoes, chopped
1 8-12 oz pkg egg noodles, cooked
1 15 oz can cream style corn
¼ c ripe olives, chopped
salt and pepper to taste
½ stick butter
bread crumbs

Combine all the ingredients except the noodles and crumbs mixing well in a heavy saucepan. Heat on medium until the butter is melted and season to taste. Stir in the noodles and adjust the seasoning if needed and pour into a buttered casserole and top with the bread crumbs. Bake covered at 325°F for 1 hour and then remove the cover and continue baking for 15-20 minutes. Serve hot or warm. Serves 6-8.

ALLIGATOR EMERGENCY SUPPER

1-2 c gator meat, cooked, diced
½ stick butter
1 sm onion, chopped
½ tsp garlic powder
1 sm bell pepper, chopped
½ tsp cayenne pepper
¾ c milk
½ lb monterey jack cheese, grated
1 6 oz can tomato paste
1 paste can water
1-2 tsp salt
1 4 oz pkg egg noodles, cooked

Saute the onions in melted butter until soft and add the tomato paste, water, salt, pepper and meat and heat to boiling. Reduce the heat and simmer for 5-10 minutes. In a separate pot heat the milk but don't boil. Stir in the cheese, garlic powder, bell pepper and noodles. In a 2-quart casserole put a layer of the meat mixture and then a layer of the noodle mixture and repeat until all is used with a meat mixture on top. Sprinkle the top with more cheese and bake at 325°F for 15-20 minutes. Serves 6-8.

GOLDEN GATOR

1-2 c gator meat, cooked*, diced
1 6 oz pkg egg noodles, cooked
1 c milk
2 eggs, slightly beaten
salt and pepper to taste
1 c cheddar cheese, grated
1 tbsp vinegar or lemon juice
bread crumbs
butter, melted

Stir the vinegar into the milk and let stand for 5 minutes or more. Mix the meat, milk, eggs, cheese and seasonings to taste. Stir in the noodles and add the milk mixture and stir well. Place the mixture into a greased casserole and top with bread crumbs and drizzle the butter over the top. Bake at 350°F for 35-45 minutes until the top is lightly browned. Serves 4-6.

ALLIGATOR SKILLET DINNER

1 lb gator meat, cooked*, ground
1 lge onion, chopped
1 med bell pepper, chopped
1 14½ oz can stewed tomatoes
½ c tomato juice
2 c uncooked elbow macaroni
1 tsp salt
2 tbsp sugar
2 tsp chili powder
1 pt sour cream

Mix well all ingredients except sour cream in a large, heavy saucepan. Bring almost to a boil and then simmer for 25 minutes stirring often. Stir in the sour cream and adjust seasonings. Serve hot. Serves 4-6.

HURRY-UP ALLIGATOR SPAGHETTI

1 lb gator meat, boneless, cubed
2 16 oz jars mushroom spaghetti sauce
1 bay leaf
1 8 to 10 oz pkg spaghetti, cooked
cooked
½ tsp oregano flakes
½ tabasco sauce
½ tsp salt
parmesan or romano cheese
water

Boil the meat in water for 20 minutes and then drain and rinse. Return the meat to the kettle and add all the remaining ingredients except the spaghetti and cheese. Bring to a boil and cook over a medium heat covered lightly for 30 minutes or longer. Stir frequently. Remove the bay leaf. Serve over hot, buttered (if desired) spaghetti and sprinkle with the cheese. Serves 4-6.

SWAMP ALLIGATOR SPAGHETTI

3 lbs gator meat, boneless
water to cover
oregano, flakes or ground to taste
½ stick butter
2 14½ oz cans stewed tomatoes
2 tsp plain flour
¼ c red wine
1 8 oz can tomato sauce
dash tabasco sauce
2¼ c water
2 lge onions, chopped
½ tsp brown sugar
salt and pepper to taste

Boil the meat in water to cover with the bay leaf and a dash of tabasco sauce for 30 minutes. Remove meat and discard water. Saute the onions until tender in the butter and then add the tomatoes, sauce, 2 cups water and boiled meat. Heat to boiling and reduce heat to very low. Add the seasonings and cook until the meat is tender and liquid reduced by half. It should cook about 1-2 hours. Add more water if needed. Then add the sugar and mix the flour with the ½ cup of water and add to the gravy mixing well. Season to taste and simmer an additional 15 minutes. Add the wine and serve over buttered spaghetti. Serves 8-10.

SLOW POKE GATOR SPAGHETTI

1 lb gator meat, ground
1 14½ oz can beef broth
dash tabasco sauce
water as needed
½ c butter
2 med onions, chopped
1 rib celery and leaves, chopped
1 lge carrot, diced
1 lb spaghetti, cooked
¼ c parsley flakes
1 8 oz can mushrooms, sliced
1 8 oz can tomato sauce
1 14½ oz can stewed tomatoes
1 c dry red wine
1 lemon, seeded, sliced
salt and pepper to taste
oregano to taste

Boil the meat with the broth and tabasco sauce adding enough water to cover if needed and boil for 15-20 minutes. In another kettle heat the butter and saute the onions, celery, carrot and parsley and cook for 5 minutes stirring constantly. Add the mushrooms, sauce, tomatoes and wine and bring to a boil. Add the meat and broth and mix well. Add the lemon and reduce the heat to simmering. Cover lightly and simmer for 4 hours stirring often. In the last half hour add the seasonings to taste. Serve over the spaghetti which has been cooked at the last minute. Serves 8-10.

ALLIGATOR WHITE SAUCE SPAGHETTI

1 lb alligator meat, cubed
water
1 10 oz can chicken broth
2 10¾ oz cans cream of celery or cream of chicken soup
½ c milk
½ c water
¼ c white wine
1 lge jar pimento, chopped
2 hard boiled eggs, chopped
salt to taste
white pepper to taste
1 lb pkg thin spaghetti, cooked

Boil the meat in water for 30 minutes and then drain and discard the water. Rinse the meat. Put the meat in the broth and ½ cup of water and cook until tender. Now add the soup and milk and stir until smooth and cook for 15 minutes on low heat. Remove from the heat and add the wine, pimento and eggs and mix well. Season to taste. Serve over hot buttered spaghetti. Serves 8-10.

GATOR MEATBALL SPAGHETTI SAUCE

1 lb gator meat, ground
½ c parmesan cheese, grated
1 tsp salt
½ c italian bread crumbs
2 eggs, beaten
½ tsp tabasco sauce

Mix all above ingredients well and shape with wet hands into small balls about the size of ping-pong balls and place on an oiled baking sheet and bake at 325°F about 30 minutes or until set.

3 scallions, chopped
1 c parsley, chopped
4 c hot water
oregano to taste
3-4 cloves garlic, minced
2 8 oz cans tomato sauce
salt and pepper to taste

Combine all above ingredients and cook on a low heat for 1-2 hours. Add the meatballs during the last hour of the cooking time and after the sauce has been reduced by half. Check and adjust the seasonings 15 minutes before serving. Serve with pasta of your choice. Serves 6-8.

TEXAS-STYLE ALLIGATOR

2 lbs alligator meat, cubed
2 tbsp butter
1 lge red onion, chopped
1 17 oz can whole kernel corn
1 4 oz can mushrooms, pieces
1 lb pkg spaghetti, cooked
bread crumbs
1 10½ oz can tomato soup
½ tsp cumin powder
½ tsp celery seed
¼ tsp chili powder
1 jalapeno pepper, chopped
salt and pepper to taste
parmesan cheese, grated

Fry the onion in butter until soft and add the meat and liquids from the corn and mushrooms and cook covered until the meat is tender. Add the corn, mushrooms, soup and seasonings and simmer for 20 minutes. Stir in the spaghetti and adjust the seasonings if necessary and pour into a casserole or baking pan and top with the bread crumbs and parmesan cheese. Bake at 350°F until the top is browned. Serves 6-8.

SAUSAGE STUFFING

- 10 lbs alligator meat
- 1½ lbs ham fat
- 1½ lbs pork fat
- 3 lge onions, minced
- 1 tbsp garlic powder
- ½ c pepper sauce
- ½ c lard
- 6 c water
- salt and pepper to taste
- sausage casings

Finely grind meats, combining ham with alligator. Combine remaining ingredients well and stuff into sausage casings. Grill 15-20 minutes or bake at 375°F for 30 minutes.

ALLIGATOR SAUSAGE

- 10 lbs alligator meat, ground
- 1½ lbs lean ham, ground
- 1½ lbs pork fat, ground
- 3 lge onions, minced
- 1 tbsp garlic powder
- 1 c pepper sauce
- ½ c lard
- 6 c water
- seasonings to taste
- sausage casings

Mix all ingredients well and stuff into casings and cook as any other sausage.

ALLIGATOR BURGER

- 4 lbs alligator, ground
- 1½ c onions
- 1 bu scallions
- 1 c bell pepper
- 1 stalk celery
- 3 tbsp salt
- 1 tbsp black pepper
- 3 tbsp garlic powder
- 1 egg
- ½ c plain flour
- oil
- plain flour for coating

Grind together onions, scallions, celery and bell pepper and mix with flour, egg and all other seasonings along with the meat and mix thoroughly. If needed, add a little flour for texture. Make patties thin and coat with flour. Get cooking oil hot in a deep-fat frier and drop patties in. Patties will remain on bottom until almost done and then they will rise to the top. Allow to cook until they turn slightly brown and then remove from fryer. Drain well on paper towels.

GATOR BURGERS

- 5 lbs alligator meat, ground
- 2 onions, minced
- 3 potatoes, diced small
- 3 slices bread, diced small
- salt and pepper to taste
- garlic powder to taste

Mix all ingredients well and form into patties. Pan fry until golden brown.

BURGERS A LA ALLIGATOR

2 lbs alligator meat, ground
1 tsp tabasco sauce
2 tbsp bell pepper, diced
1 clove garlic, ground
salt and pepper to taste
1 boiled potato, mashed

Combine all ingredients in above order. Form into patties and brown in a greased skillet over medium heat for 4-6 minutes on each side.

TECHE ALLIGATOR BURGERS

2 lbs alligator meat, ground
½ c scallions, chopped
¼ c parsley, chopped
2 eggs
2 tbsp lemon juice
¼ c plain bread crumbs
salt and pepper to taste

Mix the ingredients together well and shape into patties and roll in the bread crumbs. Fry or grill.

ALLIGATOR MEAT PATTIES

1 lb lean beef, ground
½ lb alligator meat, ground
1 egg
½ c cracker crumbs
½ tsp worcestershire sauce
½ tsp garlic powder
½ tsp soy sauce
oil

Combine all ingredients (except oil) in above order. Make sure each patty is firm. Pan-fry over medium heat until golden brown using a small amount of oil.

PATTY-GATOR

1 lb gator meat, boiled*, ground
1 stack saltine crackers, crushed
1 egg, slightly beaten
1 sm onion, minced
cornmeal
oil
½ sm bell pepper, minced
dash tabasco sauce
salt to taste
little evaporated milk
plain flour
seasonings to taste

Add the meat to the crushed crackers mixing well and add the remaining ingredients except the salt and milk. Add enough milk to make the mixture hold together in small patties. Add salt to taste. Combine equal amounts of cornmeal and flour and season to taste. Using wet hands shape the mixture into patties and dip each into the seasoned cornmeal-flour* mixture and fry in hot oil until brown. Serve on hamburger rolls with the usual hamburger condiments. Makes about 6 patties.

ALLIGATOR MEATLOAF

5 lbs alligator meat, ground
2 lge irish potatoes, ground
2 slices bread
2 eggs, lightly beaten
½ c ketchup
seasonings to taste

Preheat oven to 350°F. Remove crusts from bread and soak in water. Squeeze out excess water. In a large bowl combine the meat, potatoes, bread, eggs, ketchup and seasonings. Place into 2 loaf pans and bake for 45 minutes.

FESTIVAL CUTLETS

1 lb alligator meat, cooked*, ground
½ stick butter, melted
2 eggs, 1 beaten
salt and pepper to taste
bread crumbs
3 slices white bread, crust removed
1 c milk
¼ c parsley flakes
1 spoon water
butter

Soak bread in the milk and then squeeze dry reserving the milk. Combine the meat, bread, butter, unbeaten egg, reserved milk, parsley and seasonings. With wet hands shape the mixture into cutlets and dip into the beaten egg which has the water added. Then dredge in bread crumbs and place on waxed paper and chill for several hours or overnight. Melt enough butter to cover the bottom of a skillet and fry each cutlet until brown on each side turning carefully so as not to break them. Serve hot with any sauce you desire. Makes 4-6 cutlets.

GATOR CAKES

3 c gator meat, cooked*, ground
3 c mashed potatoes, cooled
1 sm onion, minced
plain flour
seasonings to taste
1 scallion, chopped
¼ stick butter
2 eggs, slightly beaten
oil

Saute the onion and scallion in butter until softened. Add to the meat along with the eggs and potatoes and mix well. Season to taste. Add dry potato flakes if too soft to shape or a little milk if too stiff. Shape into hamburger size patties and dip in flour and fry on both sides until golden. Serve as hamburgers or main dish. Makes 6-8 cakes.

LEMON GATOR PATTIES

1 lb alligator meat, cooked*, ground
salt and pepper to taste
2 tbsp water
½ c bread crumbs
2 tbsp lemon juice

Mix well all the above ingredients and shape into 6 patties about 1 inch thick. Place in a shallow baking dish and bake at 400°F for 6-8 minutes. Set aside and lower the oven to 350°F and then make the sauce with the ingredients below.

¼ c ketchup
¼ tsp prepared mustard
pinch ground cloves
1 tbsp lemon juice
¼ tsp salt
2 lemons, sliced thin, seeded

Combine all ingredients except lemons and spoon over the patties. Return to the oven and bake for 15 minutes. Serve with sliced lemons on top of each. Serves 6.

ALLIGATOR CAKES

1 c alligator meat, cooked*
1 lge onion, chopped
2 tbsp plain flour
¾ c evaporated milk
½-1 c potato flakes
oil
1 egg
½ tsp salt
¼ tsp cayenne pepper
½ tsp garlic salt
plain flour for dredging

Combine the meat, onion and seasonings in a large bowl and set aside. In a smaller bowl stir the flour into the milk and then add the egg and beat to blend. Add to the meat mixture and then add the potato flakes a little at a time until the mixture holds its shape. Make into balls or patties and coat with additional flour and cook in hot oil until browned well.

GATOR IN A BUN

½ c alligator meat, cooked*, ground
worcestershire sauce
½ c chili sauce
onion, minced (optional)
½ c american cheese, grated (optional)
4-6 hot dog buns

Cut the buns in half and remove some of the center leaving enough of the bun so that it doesn't fall apart when filled. Combine the meat, cheese, chili sauce, onion and worcestershire sauce mixing well. Fill the buns with the mixture and put the halves together. Wrap each in waxed paper and then in foil. Bake at 300°F for 20-25 minutes or until hot. Serves 4-6.

GATOR MEATLOAF

2 lbs alligator meat, ground
2 lge eggs
½ c celery, minced
2 c milk soaked bread, drained
½ pkg soup mix (your choice)
butter
4 slices bacon, diced
½ c onion, minced
3 cloves garlic, minced
½ c boiling water
bacon slices
salt and pepper to taste

Combine the alligator, diced bacon, eggs, onions, celery, garlic, and bread and blend thoroughly. Form into a loaf and place in a buttered pan and place in a preheated 350°F oven and bake for approximately 1½ hours. Remove the meatloaf and place on a cookie sheet. Raise the temperature to 450°F. Wrap the meatloaf in bacon and add a little butter to the pan and return to the oven. Allow the bacon to get crisp (about 10 minutes). Baste at intervals with the water and soup mix.

GATOR LOAF SUPREME

3 c gator meat, cooked*, diced
1½ c soft bread crumbs
2 eggs, slightly beaten
⅛ tsp rosemary
pinch of marjoram
salt and pepper to taste
few grains nutmeg
2 ribs celery, minced
1 c sour cream
½ c mayonnaise
¼ c parsley, minced
paprika
bell pepper rings, optional
pimento strips, optional

Combine all ingredients except pepper rings, pimento and paprika in a large bowl and toss lightly until well mixed. Put into a greased loaf pan and place in a preheated oven with the pan set in another pan of hot water. Bake at 350°F for 45 minutes and then reduce the heat to 325°F and bake 30 minutes longer. Cool for 5-10 minutes and then slide a knife around the edges to loosen and then invert onto a platter. Garnish with the pepper slices, paprika and pimento slices. Serves 6.

GATOR LOAF

1 lb alligator leg meat, ground
1 lb bulk sausage
1 can bean with bacon soup
2 eggs, slightly beaten

Combine the alligator and sausage and cook for 20-30 minutes stirring often. Drain in a colander and cool slightly. Combine the soup, eggs, cooled meat and mix well and put into a loaf pan and bake at 350°F for 75 minutes. Remove from the oven and cool and then invert to a cutting board. Slice and place on a serving platter and pour more hot soup over the top if desired. Serves 8-10.

ALLIGATOR STUFFED BELL PEPPERS

4-6 sm to med bell peppers
2 c gator meat, cooked*, ground
1 egg
1 c bread crumbs
1 c cooked rice
½ tsp poultry seasoning
1 sm onion, chopped
1 tsp salt
dash tabasco sauce
¼-½ c milk
1 12 oz can tomato sauce
½ sauce can water

Remove the tops from the peppers and remove the seeds and pulp from centers. Place in a pot of salted water and boil for 5-10 minutes. Remove and drain. Mix the remaining ingredients using half the crumbs except sauce and water and mix well adding enough milk until no longer dry but not mushy. Fill the peppers with the mixture and cover the tops with the remaining crumbs. Place in a deep dutch oven or casserole depending if you plan to bake or cook on the stovetop. Combine the tomato sauce and water and pour over the peppers. On the stove cook at a low heat for 30-45 minutes basting often. If baking, bake at 375 F for 30-45 minutes also basting often. Serve with extra sauce on rice.

STUFFED BELL PEPPERS WITH GATOR

2 c gator meat, cooked*, ground
1 c cooked rice
1 10½ oz can cream of mushroom soup
1 tbsp lemon juice
salt and pepper to taste
2 tbsp + butter
1 c monterey jack cheese, grated
6 bell peppers
bread crumbs
water

Cut off the tops of the peppers and remove the seeds and pulp. Boil for 6-8 minutes and then drain. Mix the meat, rice, soup, lemon juice, 2 tbsp butter and seasoning and heat over a medium fire until hot. Stir in the cheese and lower the heat stirring until the cheese melts. Then pack the peppers with the filling. Top with bread crumbs and dot with butter. Place into a shallow casserole and add about 1 inch of water to the bottom of the casserole. Bake uncovered at 225°F for 45-60 minutes. Serves 6.

ROLLED ALLIGATOR ROAST

1 lge alligator tail fillet
bell peppers, minced
4 oz butter
1 c white wine
onions, minced
garlic, minced
bacon slices
creole seasoning*

Try to get a fillet about 3 feet long, 1 inch thick and 6 inches wide. Season generously and sprinkle onions, bell pepper and garlic on top. Roll the fillet and tie with string to hold in place. Melt the butter in a dutch oven and place roast in oven and cover with slices of bacon. Heat oven to 500°F and cook until the bacon is crisp. Then pour the wine over the roast and lower oven to 300°F. Cover and bake for 2 hours or until tender. Serves 8.

ALLIGATOR COURTABLEU

2 6 oz alligator fillets
3 oz turkey tasso*
1 oz mild cheddar cheese, shredded
⅛ tsp black pepper
1 egg, beaten
1 tbsp plain flour
vegetable cooking spray
garlic red pepper sauce
1 oz monterey jack cheese, shredded
3 ozs sm shrimp, poached
¼ tsp sugar
⅛ tsp salt
⅛ cayenne pepper
⅛ garlic powder
1 tbsp milk

Preheat oven to 400°F. Pound fillets to ½ inch thick. Combine salt, sugar, peppers and garlic powder and set aside. Prepare egg wash with milk and set aside. Lay out the fillets and spread the remaining ingredients and seasoning mix over the center of fillet leaving a ¾ inch border around the edges. Put egg wash onto each fillet around border and sprinkle with the flour. Roll tightly and secure the ends with toothpicks. Place rolls in a casserole sprayed with a vegetable cooking spray. Spray rolls lightly. Bake for 20 minutes in preheated oven. Remove the alligator from the oven and cool for 20 minutes before slicing. Remove toothpicks and slice into ¾ inch thick medallions. Pour the sauce into the bottom of a serving platter and arrange the medallions on it. Serves 4.

GARLIC RED PEPPER SAUCE

8 ozs clam juice, bottled
¼ c onion, minced
1 tbsp cornstarch
¼ c evaporated skim milk
½ tsp cayenne pepper
1 tsp garlic, minced
1 tbsp butter
⅛ c cold water

Combine the clam juice, cayenne pepper and onion in a small saucepan and bring to a boil. Mix the cornstarch with the water and add to the mixture. Stirring constantly cook for 1 minute. Remove from heat and fold in the butter and milk. Cool.

ROASTED ALLIGATOR

2 3 in slices alligator tail
¼ c butter
¼ c garlic, diced
½ stick butter
½ c worcestershire sauce
few sprigs parsley

Stuff alligator slices with garlic and onions in small pockets. Pour melted butter and worcestershire sauce over top of roast and let stand a few minutes. Place in a castiron dutch oven with a small amount of water. Cook in a slow oven (225°F) for 2 hours. Add a few potatoes and carrot strips for extra zest if desired. Garnish with parsley before serving.

STUFFED ROASTED ALLIGATOR

1 3-4 lb alligator tail
½ c bell pepper, minced
3 tsp salt
oil
1 c onions, minced
4 cloves garlic, minced
2 tsp cayenne pepper
water

In a small bowl mix together the onions, bell pepper and garlic with the 2 tsp salt and 1 tsp pepper. With a sharp pointed knife make slits in the roast and with your finger stuff the mixture into the slits. Rub the roast well with remaining salt and pepper. Place meat in a deep roasting pan with a little oil and cook at 375°F. Add a little water to make a gravy and baste with the pan juices. Cook until the juices run clear.

GARLIC STUFFED ROASTED ALLIGATOR

1 5 lb alligator roast
salt and cayenne pepper
½ c oil
1 med bell pepper, coarsely chopped
¼ c parsley, minced
6 cloves garlic, thinly sliced
½ c plain flour
2 med onions, thinly sliced
2 c chicken broth

Make several slits in the roast and stuff with garlic. Then season well with salt and pepper to taste. Sprinkle all sides with flour. In a heavy pot heat the oil and brown the roast on all sides and reduce the heat. Add the onions and bell pepper and chicken broth. Cover and place in a preheated 350°F oven. Roast for 2-2½ hours basting occasionally with broth and pot drippings. Add water if it becomes too dry. Slice and garnish with minced parsley. Serves 8-10.

ALLIGATOR SHELLS

1 lb alligator meat, chopped
½ lb Velveeta Cheese
1 10½ oz can cream of mushroom soup
½ c cream
½ stick butter
garlic powder
salt and pepper to taste
paprika
patty shells

Combine the first 5 ingredients in a saucepan and simmer for 15 minutes. Sprinkle with garlic powder, salt and pepper. Pour in scooped patty shells and sprinkle top with paprika.

GATOR SHISH-KA BOBS

1 lb alligator meat
1 c italian salad dressing
bell pepper
med onions, quartered

Cut meat into 1 inch cubes. Marinate meat in salad dressing overnight. Cut bell peppers into 1 inch pieces. String meat and vegetables alternately on skewers. Cook over hot coals turning often until brown.

SEAFOOD SHISH-KA BOB

- 1½ c milk
- 1 lb shrimp wrapped in bacon
- cherry tomatoes, cut in half
- bell pepper, cut in 1 in squares
- pear onions, peeled
- 1 bottle italian dressing
- 5 lbs alligator, 1 in squares
- fresh mushrooms, whole
- fresh pineapple, cut 1 in squares
- 1 sm bottle sweet and sour sauce

Marinate the alligator meat in milk and italian dressing overnight in a refrigerator. Light your coals. Coals should be low. Put all ingredients on skewers alternating them. When the coals are ready place skewers on a rack and cook 25 minutes and baste with the sweet and sour sauce. Serves 8.

GINGER PLUM ALLIGATOR

- 4 lbs alligator fillets, sliced thin
- 1 c Karo Dark Corn Syrup
- ⅔ c scallions, minced
- 2 tsp ginger, ground
- 1 12 oz jar Damson Plum Preserves
- ½ c soy sauce
- 2 cloves garlic, minced

In a saucepan stir in all the ingredients except alligator. Stirring constantly bring to a boil over medium heat and boil for 5 minutes. Remove 1 cup of mixture and set aside. Place alligator meat in a shallow baking dish and pour the remaining mixture over the meat and cover and marinate in the refrigerator for at least 4 hours. Place the alligator on rack in a broiler pan and broil 6 inches from heat, turning and basting every 5 minutes until done. Heat reserved mixture and serve with the alligator.

ALLIGATOR ROLLATUMS

- alligator fillets
- bacon bits
- 2 c water
- onions, chopped
- dill pickle, sliced
- oil

Cut fillets into 2 inch wide by 4 inch long strips. Place onion, bacon bits and pickle inside strips. Roll the strips, secure with a toothpick, and brown in a skillet in a little oil. Add water. (If desired, use ¼–½ cup red wine for a richer flavor.) Cover and simmer until tender, about 40 minutes.

GATOR BAKE DE JONGHE

- 1 lb alligator meat, pounded
- oil
- cracker meal, seasoned*
- 4 ozs de jonghe butter

Cut well pounded meat into julienne strips and coat well with the cracker meal which has been well-seasoned. Deep-fry until golden brown. Place browned alligator into 4 small ramekins and top each with 1 oz De Jonghe butter. Finish under the broiler until golden brown. Use as an appetizer or a main course.

DE JONGHE BUTTER

2 lbs butter, softened
¾ c parmesan cheese, grated
¾ c bread crumbs, coarse ground
½ c sherry
½ c parsley, chopped
1½ tsp salt
½ tsp white pepper
4 cloves garlic, minced

Blend all ingredients well and be sure the mixture is evenly distributed. Spread the butter in a loaf pan and refrigerate.

WINE BAKED ALLIGATOR FILLETS

1 lb alligator fillets
½ tsp oregano, dried, crushed
½ c swiss, mozzarella or fontina cheese, shredded
¼ c dry white wine
salt and pepper to taste
2 med carrots, julienne-cut
2 ribs celery, julienne-cut

Place the fillets in a greased 10 x 6 inch baking dish and season with salt and pepper. Pour the wine over the fillets and sprinkle with oregano. Bake covered at 350°F until the alligator flakes when tested. Sprinkle fillets with the cheese and bake uncovered 5-10 minutes more. In a medium saucepan cook the carrot and celery in a small amount of boiling salted water about 10 minutes or just until tender. Drain. Cut the fillets into 4 equal portions and transfer to a serving platter and spoon some of the alligator juices from the pan over them. Arrange the carrot and celery strips around the fillets on the platter. Serves 4.

BROILED ALLIGATOR AMADINE

1 lb alligator fillets
½ tsp lemon peel, grated fine
few dashes tabasco sauce
½ c almonds, sliced
½ c butter
½ c lemon juice
salt and pepper to taste
1 tbsp parsley, minced

Melt the butter in a small saucepan and stir in the lemon juice, peel and tabasco sauce. Cut the alligator crosswise into 4 portions and place them on an unheated rack in a broiler pan. Brush with some of the lemon butter mixture and sprinkle with a little salt and pepper. Broil 4 inches from the heat for 12 minutes. Brush again with lemon butter mixture and turn and brush again. Broil 10-12 minutes more or until the fillets flake easily. Top with sliced almonds the last 2 minutes of broiling. Transfer fillet portions to a serving platter. Combine the remaining lemon butter mixture with the parsley and drizzle over the portions. Serves 4.

ALLIGATOR CABBAGE BAKE

1 lb alligator meat, cooked*
1 sm head cabbage
1 c cooked rice
½ stick butter
bread crumbs
1 10¾ oz can cream soup, any kind
1 med onion, chopped
salt and pepper to taste
2 cloves garlic, minced
parmesan cheese

Coarsely chop cabbage and boil in salted water until tender. Drain and reserve the liquid in case it is needed. Wilt the onions in melted butter and then add the soup garlic and meat and simmer for a few minutes. Add the rice and cabbage and mix thoroughly. If too dry add a little of the reserved liquid. Season to taste. In a greased 3 quart casserole top with bread crumbs which have equal amounts of parmesan cheese and dot with butter if desired. Bake at 300°F for 30-45 minutes or until hot and the top is browned. Serves 8.

ALLIGATOR HOLLANDAISE

4-6 alligator steaks, 1 inch thick
hollandaise sauce
salt to taste
butter
1 lemon, sliced, seeded
1 pkg crab boil*, in bag
water to cover meat

Combine all ingredients except butter in a heavy pot and bring to a boil and cook until the meat is tender. Remove the steaks with a spatula and place in a buttered casserole dish. Strain liquid and reserve 1 cup. When the hollandaise sauce is made, pour over the steaks and sprinkle with more paprika and parsley. Serves 4-6.

HOLLANDAISE SAUCE

1 tbsp butter, softened
3 egg yolks
1 tsp cornstarch
1 tbsp lemon juice
1 cup reserved stock
paprika parsley flakes

Use a double boiler and add water to the bottom section and beat the egg yolks in the top section. Add the butter, cornstarch and lemon juice and place over the boiling water but do not let the water touch the bottom of the top section. Slowly add the boiling stock and stir constantly until thickened. Add a little paprika and parsley.

CHEESY BAKED GATOR

- 2 lbs alligator meat, cubed
- ½ c french salad dressing
- 2 tbsp lemon juice
- oil
- ¼ tsp salt
- 1 3 oz can fried onions
- ¼ c parmesan or romano cheese, grated

Combine the meat, dressing, lemon juice and salt in an air tight container mixing well and refrigerate for 6-24 hours. Shake several times while marinating. Drain the meat from the marinade and reserve the marinade sauce and place the meat in a shallow well oiled baking pan and cover. Bake at 325°F for 30-45 minutes or until meat is tender. Add some of the reserved marinade if the meat gets too dry or the bottom of the pan gets too dry. Remove the cover and sprinkle the onions, crushed, and cheese over the top. Bake again at 350°F for 5-7 minutes or until bubbly and hot on top. Serves 6-8.

ALLIGATOR TETRAZZINI

- 2 lbs alligator meat, cubed
- 1 c milk
- 1 14½ oz can chicken broth
- 1 4 oz can mushrooms, drained, reserve liquid
- ¼ c bread or cracker crumbs
- ¼ tsp poultry seasoning
- ½ tsp paprika
- ¼ lb cheddar cheese, grated
- 4 c cooked rice
- 1 tbsp parsley flakes
- 4-6 slices bacon, cooked, crumbled
- 3 tbsp plain flour
- ½ tsp garlic salt
- ¼ tsp white pepper
- butter

Put the meat in the milk and refrigerate for a few hours or overnight. Drain reserving the milk. Mix the mushroom liquid and broth and heat and then add the meat and cook on low until the meat is tender. Drain the meat and put the broth back in the saucepan. Set meat aside. Mix the flour with the reserved milk and stir into the hot broth and season with salt and pepper and cook until slightly thickened. Remove from the heat and add the cheese. Mix the meat with the mushrooms and rice. In a buttered casserole alternate layers of the meat mixture with the sauce with sauce on top. Sprinkle the parsley, bacon, crumbs and remaining seasonings over the top. Bake at 400°F for 30 minutes. Serves 8-10.

FANCY GATOR

- 4 alligator tail fillets, 1 inch thick
- 1 envelope dry onion soup mix
- ½ tsp salt
- dash tabasco sauce
- ½ c dry sherry
- ½ c raw rice
- 2 10½ oz cans cream of celery soup

Marinate the fillets in the sherry for a few hours and then drain and reserve the sherry. Combine the sherry, rice, soup, salt and tabasco sauce mixing well and put into a greased 2-quart shallow baking dish. Put the fillets on top and press down into the rice mixture slightly. Sprinkle the soup mix on top. Cover tightly and bake at 325°F for 2½-3 hours or until the rice is cooked and the meat is tender and all the liquid is absorbed. Serves 4.

GATOR QUICHE

½ lb alligator meat
2 tbsp plain flour
½ c milk
water
1 sm-med onion, chopped
1½ lge bell pepper, chopped
creole seasoning*
½ c mayonnaise
2 eggs
8 ozs mozzarella cheese, shredded
¼-½ bu parsley, chopped
½-¾ bu scallions, chopped
salt to taste
1 9 in frozen pie shell

Marinade

½-¾ bottle worcestershire sauce
½ c wine vinegar
¼ c wine or Amaretto
2 cloves garlic, minced
4-5 tbsp mustard (regular or dijon)
½ c onions, chopped
¼ c bell pepper, chopped
¼ c celery, chopped

Mix marinade ingredients well and place alligator to marinate overnight in the refrigerator. Boil alligator in the marinade with enough water to cover the meat for 20 minutes. Allow to sit for 30 minutes and then drain and allow to cool. Grind the meat and the boiled vegetables together. Blend the mayonnaise, flour, eggs and milk and mix well. Stir in the alligator mixture and the cheese. Add the freshly chopped vegetables and mix well. Place the mixture in the defrosted pie shell and bake at 350°F for 40 minutes until golden brown. Allow to cool before cutting and serve while still warm.

CRUNCHY GATOR CASSEROLE

2-3 c gator meat, cooked*, diced
1 10¾ oz can cream of celery soup
½ c sharp cheddar cheese, grated
paprika
½ tsp salt
3 c potato chips, crushed
1 soup can cream
butter

Combine meat, soup, cream and salt and heat to boiling and then remove from heat. In a 2 quart buttered casserole dish line the bottom with a little more than half the potato chips and pour the meat mixture over them. Cover with the remaining chips and sprinkle with the cheese and dust with paprika. Bake at 350°F for 30 minutes. Serve hot. Serve 4-6.

ALLIGATOR MOUSSE WITH FENNEL BUTTER SAUCE

½ lb alligator tail
2 c alligator or fish stock
2 c whipping cream
lettuce leaves
butter
1 tsp fennel seed
2 tsp unflavored gelatin
1 tsp salt
beurre blanc fennel sauce
ice

Line the bottoms of 8½-cup ramekins or soufflé dishes with parchment paper and butter and set aside. Bring the stock and fennel seed to a boil in a medium saucepan over a medium-high heat. Boil for 3 minutes. Let cool to room temperature and then strain into a medium saucepan. Stir in alligator and gelatin and cook over a medium-high heat until the mixture is reduced by ⅕ and the alligator flakes easily. Transfer mixture to a processor or blender and puree about 3 minutes. Transfer to a large bowl. Set this bowl into a larger bowl filled with ice cubes, stirring puree constantly until room temperature. Remove from the ice and set aside. Combine the cream and salt in a medium bowl and beat until soft peaks form. Fold the whipped cream into the puree. Spoon mixture evenly into the prepared ramekins. Refrigerate covered overnight. To serve arrange lettuce leaves on 8 inch salad plates. Unmold each serving into the center of each plate. Gently remove the parchment and spoon beurre blanc fennel sauce over each mousse and serve immediately. Serves 8.

BEURRE BLANC FENNEL SAUCE

6 black peppercorns
2 tbsp dry vermouth
½ c well-chilled butter, cut into 16 pieces
2 tbsp fennel vinegar
1 lge scallion, chopped

Combine peppercorns, vinegar, vermouth and scallion in a small heavy saucepan and cook over a medium high heat until reduced to a glaze, about 3-5 minutes. Remove from heat. Whisk in 2 pieces of butter 1 piece at a time and stir until mixture is smooth. Place pan over a low heat and whisk in remaining butter in the same manner. Serve at once. Makes ½ cup.

FENNEL VINEGAR

¼ c white wine vinegar
1 tsp fennel seed

Combine the vinegar and fennel in a small saucepan and bring to a simmer over medium heat. Let simmer for 5 minutes and strain.

ALLIGATOR MOUSSE

1 lb alligator, finely chopped
salt and pepper to taste
1 tsp lemon juice
1 egg + 2 yolks
3 tbsp + butter
1 c half and half cream
1 c bechamel sauce*, thick
1 tsp nutmeg
1 tbsp orange zest, grated, no white
½ c whipping cream
3 tbsp plain flour

Add to bechamel sauce 3 tbsp butter, flour and half and half and mix thoroughly and rub through a sieve. Add alligator, salt, pepper, nutmeg, lemon juice, zest, egg, yolks and cream and mix well. Place into a well buttered 1 quart mold, or individual ½ c molds, set in a pan of hot water and bake in a preheated 325°F oven until knife inserted comes out clean in about 30 minutes. Remove from oven and water bath. Unmold on serving plate(s). Serves 8.

ALLIGATOR SOUFFLE

1 c alligator, cooked, flaked
1 c dry vermouth
1 c alligator stock
3 tbsp plain flour
¼ c carrots, minced
1 tsp lemon juice
pinch allspice
2 c water
3 eggs, separated
3 tbsp butter
¼ c celery, minced
¼ c scallions, minced
salt and pepper to taste

Poach enough boneless alligator tail in the vermouth and water to get 1 cup of cooked, flaked meat. Save the stock. Melt the butter over a low heat and stir in the flour until it is well- blended. Slowly add the alligator stock while stirring with a wire whip. When this reaches the boiling point stir in the alligator and vegetables and lower the heat and simmer for a few minutes. Add beaten egg yolks while stirring and allow to thicken. Do NOT boil. Season with the salt, pepper, lemon juice and allspice. Cool this mixture in the refrigerator for an hour or so. Whip the egg whites until stiff and fold them gently into the cooled mixture. Bake the souffle in an ungreased high-sided baking dish at 325°F until it is firm.

ALLIGATOR STUFFING

1 lb alligator, body meat
1 bu scallions, minced
2 tbsp brandy
salt and pepper to taste
1 c bread crumbs
2 eggs, beaten
1 bu parsley, chopped

Marinade

1 c red wine
3 cloves garlic, mashed
¼ c extra virgin olive oil
1 lemon, sliced thin

Mix all marinade ingredients thoroughly and cover the alligator meat with it for 6 hours, turning every 30 minutes. Remove the meat and grind it. Then mix in the eggs, bread crumbs, brandy, salt and pepper. This can then be used to stuff mushroom caps, cabbage rolls, egg rolls, tortellini shells or any other item that you may wish to stuff and cook accordingly.

ALLIGATOR DRESSING

5 lbs alligator, ground
1 bu scallions, chopped
3 tbsp parsley, chopped
chicken bouillon
1 loaf stale bread
3 onions, chopped
butter
salt and pepper to taste

Saute scallions, onions and parsley in butter until onions are clear. Brown alligator in the vegetable mixture. Mix well. Break the bread in enough bouillon to moisten. Combine well the bread and meat mixture, season with salt and pepper, and bake at 350°F for 30 minutes.

SIMPLE SIMON ALLIGATOR DRESSING

1-2 c alligator meat, cooked*, minced
½ tsp salt
½ tsp tabasco sauce
1 tbsp soy sauce
bread crumbs, optional
1 4 oz can mushrooms, sliced
2 tbsp butter
2 c cooked rice
1 10¾ oz can cream of mushroom soup
parmesan cheese, optional

In a large oven-proof skillet melt the butter and add all the remaining ingredients mixing well and cook on a medium heat until hot. Cover tightly and place in a 225°F oven for 45 minutes. A little more or less time may be needed to cook. Remove cover the last 15 minutes. Sprinkle with bread crumbs and/or parmesan cheese if desired. Serves 6.

ALLIGATOR EGGPLANT DRESSING

1 lb alligator meat, ground
½ stick butter
1 med eggplant, peeled, chopped
1 med onion, chopped
1 rib celery, chopped
½ med bell pepper, chopped
1 c raw rice, cooked
2 cloves garlic, chopped
⅛ tsp cayenne pepper
2 tsp Season-All
10 scallion tops, chopped
1 4 oz can mushrooms, stems and pieces, drained, chopped

Saute meat in butter and then add the onion, celery, bell pepper and garlic and cook until done. Add the eggplant and cook until the egg plant is tender. Add seasonings, mushrooms and scallions. Add the rice to the mixture and mix well. Serve as a dressing or use as a stuffing.

DIRTY RICE A LA ALLIGATOR

2 lbs alligator meat, diced
1 lge onion, chopped
2 cloves garlic, minced
3 tbsp oil
1 tbsp plain flour
3 c + water
1 c raw rice, long grain
salt and pepper to taste

Parboil the meat in enough water to cover for 30 minutes and then drain and rinse. Set meat aside. Heat the oil in a heavy pot with a lid and stir in the flour and brown while stirring constantly. Add the onion, garlic and meat and cook for 15 minutes. Add the 3 cups of water and bring to a boil. Season lightly and add the rice mixing well and cook covered on the lowest heat for 20-30 minutes. Serve hot. Serves 6.

FISHERMAN'S ALLIGATOR STOVETOP SUPPER

2 c gator meat, cooked*, diced
2 tbsp butter
1 box Rice-A-Roni, chicken flavor
1⅔ c boiling water
½ tsp onion salt
¼ c celery and leaves, chopped
½ c bell pepper, diced
¼-½ tsp tabasco sauce

In a covered skillet melt the butter and add the Rice-A-Roni and saute until a light brown. Stir in the hot water and add the flavor pack from the Rice-A-Roni and the remaining ingredients mixing well. Cover and simmer for 15-18 minutes or until the rice is tender and liquid is absorbed. Serves 4-6.

STUFF AND PUFFED GATOR

2 c gator meat, cooked*, cubed
1 8 oz pkg cornbread stuffing mix, any flavor
¼ stick butter, melted
paprika
6 eggs, beaten
1 lb monterey jack cheese, grated
3 c milk

Place the butter in a 2-quart casserole and grease sides with it and then sprinkle the stuffing mix in the bottom. Combine the meat, eggs, cheese, milk and seasoning pack mixing well and pour over the stuffing mix and sprinkle with paprika. Bake for 1 hour at 325°F or until puffed and golden brown. Serves 6-8.

GATOR BEAST FEAST

1 lb alligator meat, cubed, boiled
1 tbsp lemon juice
1 tbsp cornstarch
½ tsp sugar
½ c almonds, slivered or bread
¼ tsp curry powder
¾ c orange juice
2 c bread cubes or torn sliced bread crumbs
1 orange, peeled, chopped
½ stick butter, melted
oil

Oil a shallow baking dish and place the boiled meat in the bottom. Make the next layer by combining the chopped orange with juice, bread cubes, butter and curry powder and when well mixed, spread over the meat and place in a 325°F oven for 25-35 minutes or until top begins to brown. Make a sauce by combining the orange juice, cornstarch, lemon juice and sugar in a small saucepan and cook on medium heat until smooth and thickened slightly. Pour over the casserole and return to the oven and bake 8-10 minutes longer. Top with the almonds before serving. Serves 4-6.

ALLIGATOR STIR FRY

14 ozs alligator meat
salt
pinch of white pepper
1 tbsp cornstarch
½ egg white
½ tbsp oyster sauce*
1½ tbsp sherry
1 c cooked rice
1 pepper, crushed
1 lb mushrooms, sliced
½ tsp soy sauce
¼ tsp sesame oil
½ bu white scallions, cut in 1 inch pieces
1 c white wine
1 c peanut oil
1 clove garlic, minced
toasted almonds, sliced
1 c chicken stock

Slice alligator meat into 1½ inch long and ⅛ inch thick pieces. Marinate meat in mixture of salt, white pepper, egg white, cornstarch and white wine from 2 hours to overnight. In a heavy skillet or wok heat the peanut oil until very hot. Drain and add alligator and brown for about 20 seconds, separating and stirring lightly until meat is sealed. Remove from skillet. Pour off all but ¾ tbsp of the peanut oil. Reheat oil to hot. Add garlic, crushed pepper, mushrooms and alligator meat. Sprinkle on soy sauce, sesame oil, scallions, oyster sauce and sherry and stir fry quickly for about 20 seconds. Add chicken stock and continue stir frying until stock comes to a boil and thickens. Stir frying should not take longer than 2-3 minutes. Make sure vegetables are still crisp. Serve with rice and garnish with almonds. Serves 2.

STIR FRIED GATOR

- 1 lb alligator, ½ inch strips
- ½ tsp white pepper
- ½ tsp chili powder
- 1 tbsp worcestershire sauce
- 2 c beef stock or water
- 1 sm yellow squash, in julienne strips
- 1 c broccoli florets, fresh
- 4 red bell pepper rings, thinly sliced
- ¼ tsp salt
- ½ tsp cayenne pepper
- 3 tsp soy sauce
- 3 tsp cornstarch
- 1 sm zucchini, in julienne strips
- ½ med onion, thinly sliced
- 1 c cauliflower florets, fresh
- vegetable cooking spray

Spray a large skillet with vegetable spray and place over high heat. Add the meat and saute, stirring for 5 minutes. Add the salt, white pepper, cayenne pepper, chili powder, soy sauce and worcestershire sauce and cook for 10 minutes stirring often. Dissolve the cornstarch in stock and add along with all remaining ingredients. Cook 10 minutes or until sauce thickens stirring occasionally. Serves 4.

SZECHUAN SPICY ALLIGATOR

- ¾ lb alligator meat
- ⅓ c carrots, julienne strips
- 2 shallots, julienne strips
- ½ tsp garlic, minced
- 1 tsp sesame oil
- 2 tbsp sugar
- 1 tbsp vinegar
- ¼ tsp white pepper
- 1 tbsp + ½ c oil
- 2 tbsp soy sauce
- ⅓ c celery, julienne strips
- ⅓ c onions, julienne strips
- 3 hot peppers, dried
- ½ + ¼ tsp salt
- ½ tsp szechuan peppercorns, crushed (optional)
- 1 tbsp sherry
- 1 tsp cornstarch
- ½ egg white

Slice the alligator into strips ⅛ inch thick by 2 inches long. Sprinkle with cornstarch, 1 tbsp soy sauce, 1 tbsp oil, ½ tsp salt and white pepper. Mix thoroughly. Coat with the egg white and let these marinate for 20 minutes. Heat a wok or skillet until very hot and add the ½ cup of oil for 45 seconds over a high heat. Drop alligator in the hot oil and stir gently to separate and cook until about 70% done. Remove the meat and drain. To the remaining oil in the pan break the dried peppers in half and cook until they are brown. Add the garlic and then the vegetables and stir fry for 1-2 minutes. Return the alligator back to the pan and add the sherry, vinegar, 1 tbsp soy sauce, ½ tsp salt, pepper corns, sugar and sesame oil. Stir for 30 seconds. Remove from pan to a serving platter.

ALLIGATOR STRIPS—LUAN STYLE

2 lbs alligator tail
2 onions, sliced thin
1 13½ oz can pineapple chunks, in syrup
1 4 oz can mushrooms
sweet and sour sauce
1 tbsp oil or bacon drippings
½ c water
1 bell pepper, julienne strips
2 tomatoes, peeled, quartered

Cut the alligator into finger-size strips and brown in the oil or drippings. Add the onion and water. Seal the pot and cook over a low heat until the meat is tender in about 20 minutes. Drain the pineapple chunks and reserve the syrup. Make the sweet and sour sauce. Combine the sauce, pineapple, bell pepper and mushrooms to the meat. Cover and cook for 5 minutes. Fold in the tomatoes and heat through. Serve over hot rice. Sliced celery, water chestnuts, chinese pea pods and broccoli make delicious additions. Serves 6-8.

SWEET AND SOUR SAUCE

½ c brown sugar, firmly packed
2 tbsp cornstarch
2 tbsp vinegar
1 tsp ginger, ground
⅓ c soy sauce
reserved pineapple syrup

Mix well and cook until thick over a low heat.

ALLIGATOR STIR FRY WITH HOISIN SAUCE

1 lb alligator meat
2 tbsp hoisin sauce*
1 tsp cornstarch
1 lge onion, sliced and in rings
1 8 oz can bamboo shoots, drained
⅓ c water
2 tbsp soy sauce
2 tbsp oil
1 lge bell pepper, 1 in pieces
hot cooked rice

Thinly slice alligator into bite size strips and set aside. In a small bowl combine the water, Hoisin sauce, soy sauce and cornstarch and set aside. Preheat a large skillet or wok over high heat and add the oil. Stir fry the onion rings and bell pepper in hot oil for 2 minutes. Remove from the skillet or wok. Add more oil if necessary and add ½ of the alligator strips and stir fry for 2-3 minutes. Remove and stir fry the remaining alligator. Return all meat to the skillet or wok. Stir in the Hoisin sauce mixture and cook and stir until thickened and bubbly and cook 2 minutes more. Stir in the drained bamboo shoots and the onion-pepper mixture. Cover and cook 1 minute. Serve over hot rice.

ALLIGATOR AND PEPPERS

1 lb alligator fillets
3 tbsp soy sauce
1½ tsp cornstarch
2 tbsp oil
1 lge carrot, shredded
⅓ c peanuts
3 tbsp lime juice
1 tbsp sugar
¼ tsp ginger, ground
2 lge onions, sliced and in rings
1 4 oz can green chili peppers, drained, chopped

Thinly slice the alligator into bite size pieces and set aside. In a small mixing bowl combine the lime juice and soy sauce and stir in the sugar, cornstarch and ginger and set aside. Preheat a large skillet or wok over high heat and add the oil. Stir fry the onion and carrot for 2-3 minutes and then remove. Add more oil if necessary and add ½ of the alligator and stir fry for 2-3 minutes and then remove. Repeat with the remaining alligator. Return all alligator to skillet or wok and the soy mixture and cook until thickened and bubbly and continue cooking for 2 minutes more. Stir in the onion, carrot, chili peppers and peanuts and cover and cook for 1-2 minutes or until heated through. Serves 4.

ALLIGATOR SUKIYAKI

1-2 lbs alligator meat, sliced very thin across the grain
2 tbsp butter
1 med onion, sliced thin
½ c soy sauce
3-4 bell peppers, cut in strips
1 8 oz can mushrooms, sliced
pinch nutmeg
½-1 tsp white pepper
2 8 oz cans water chestnuts, sliced, drained
4 sm tomatoes, cut in wedges
1 can bean sprouts, drained

Melt butter and saute onion until softened and then stir in the meat slices and cook 15-20 minutes until almost done. Add the soy sauce, mushrooms, pepper, nutmeg and water chestnuts. Cover and simmer for 30-60 minutes. If it becomes too dry add liquids from chestnuts and bean sprouts. Add the tomatoes, peppers and bean sprouts and cover and simmer for 8-10 minutes. Vegetables should be tender and crisp. Serve with white or wild rice. Serves 4-6.

ORIENTAL GATOR

2 c gator meat, cooked*, diced
2 10¾ oz cans golden mushroom soup
1 c cashew nuts, chopped
bell pepper rings, optional
pimento strips, optional
3 ribs celery, diced
4 scallions, chopped
½ c milk
2 lge cans chow mein noodles

Combine all the ingredients except the optional ones. Lightly oil a shallow casserole and fill with the mixture and bake at 350°F until very hot and bubbly, about 25-30 minutes. If desired before serving garnish with pepper rings and pimento strips. Serves 6-8.

BEAN SPROUT AND GATOR BAKE

1 lb gator meat, cooked*, ground
2 tbsp butter, melted
1 sm onion, chopped
dash soy sauce, optional
6 eggs, beaten with 1 tsp salt
dash tabasco sauce
1 can bean sprouts, drained
paprika

Combine onion and butter and saute until onion is soft. Add the meat, eggs, tabasco sauce and bean sprouts and mix well. Soy sauce can be added if desired. Place in a 1½ quart oiled casserole and sprinkle with paprika and bake at 350°F for 20-25 minutes. Serves 6-8.

FIVE COUNT ALLIGATOR CASSEROLE

1 c gator meat, cooked*, minced
1 10¾ oz can cream of mushroom soup
1 can chow mein noodles
1 12 oz can evaporated milk
1 10 oz can chicken noodle soup

Combine and heat the soups in a heavy saucepan. Add the meat and the noodles and then place in a greased casserole and bake uncovered for 1 hour at 350°F. Lower the heat if it begins to brown before the time is up. Serves 4-6.

ALLIGATOR CHOP SUEY

1 lb gator meat, cooked*, ground
¼ stick butter
2 lge onions, chopped
2 ribs celery, chopped
½ c celery tops, chopped
salt and pepper to taste
1 8 oz can mushrooms, buttons
2 bell peppers, cut in strips
3 c beef broth, canned or cubes
½ c plain flour
soy sauce, optional

Saute the meat in butter and then add the onions, celery, mushrooms, bell peppers and broth and cook over a low heat until very hot and the vegetables are tender but not soft, about 15 minutes. Take a small amount of the liquid and mix in the flour mixing well. Add back to the pot and stir well and cook until thickened. Add soy sauce if desired and season to taste. Serve over rice or chinese noodles. Serves 6.

STIR FRY ALLIGATOR A LA TECHE

2 c alligator meat, sm cubes
1 c italian salad dressing
1 15 oz can pineapple chunks, drained, reserve liquid
1 tbsp oil
3-4 c cooked rice
1 4 oz can mushrooms, sliced, drained
¼ c soy sauce
1 lge onion, sliced
1 lge bell pepper, cut in 1 inch pieces
1 tbsp cornstarch

Make a marinade of the salad dressing and soy sauce and place the meat in it mixing well and let stand in the refrigerator for several hours or overnight. Drain and reserve the marinade. Saute the onion, pepper and mushrooms in oil until tender-crisp. Remove with a slotted spoon and set aside. Brown the meat slightly in the drippings and then cover and cook on low until meat is tender adding reserved marinade if needed. Remove the meat from the skillet and add the pineapple juice, cornstarch and any remaining marinade and heat to boiling and then reduce heat and cook until thickened. Remove from the heat and add the meat, vegetables and pineapple chunks and stir gently. Cook again on medium to heat throughout. Serve at once over hot rice. Serves 4-6.

TROPICAL GATOR STEAKS

6-8 alligator tail steaks, cut 1 inch thick
1 lemon, sliced, seeded
dash tabasco sauce
1 c dry white wine
½ c almonds, slivered
1 20 oz can pineapple, sliced
1 bell pepper, in rings, chopped
water to half-cover meat
¼ c brown sugar
1 stick butter, melted

Steam the steaks in the water with the lemon slices and tabasco sauce added. Steam covered for 15-20 minutes. Remove and discard the water. Preheat the oven to 350°F. Place the butter in the bottom of a casserole and carefully place the steaks in it. Reserve 1 pineapple slice for each steak and chop the remainder and mix with the juice. Combine the chopped pineapple, pepper, wine and sugar and pour over and around the steaks. Cover and bake for 15 minutes. Remove the cover and continue baking for 15-30 minutes longer. Place a pineapple slice on each steak and sprinkle with almonds. Spoon a little pan sauce over each steak and broil for a few minutes until the almonds are browned. Serves 6-8.

ONION BUTTERED GATOR

3-4 lbs gator fillets, cut 1 in thick
2 tsp worcestershire sauce
½ tsp dry mustard
½ tsp tabasco sauce
½ sm onion, chopped
½ c butter, melted
¼ c parsley, chopped

Combine the parsley and onion in a small saucepan with the butter and heat. Pour over the meat and place in a shallow baking dish. Mix the worcestershire sauce, mustard and tabasco sauce and pour over the top of the meat. Bake covered at 350°F for 30 minutes. Uncover and bake 30 minutes longer basting often with the pan drippings. Serve with pan drippings in a sauce boat to be used with other vegetables. Serves 6-8.

SWEET AND SOUR ALLIGATOR

1½ lbs alligator tail meat, ¾ in cubes
¼ c + 2 tbsp plain flour
1 tsp salt
1 c pineapple juice
3 tbsp cornstarch
⅓ c cider vinegar
½ tsp garlic salt
1 8 oz can unsweetened pineapple chunks
1 med onion, thinly sliced
hot cooked rice
2 eggs, beaten
1 tbsp milk
4 c oil
⅓ c brown sugar, firmly packed
1 8 oz can tomato sauce
⅓ c light corn syrup
¼ tsp black pepper
1 med bell pepper, 1 in squares
2 stems celery, sliced diagonally

Combine eggs, flour, milk and salt and mix well. Add the alligator cubes and coat well. In a 2-quart deep fryer heat the oil to 350°F and deep-fry alligator a few pieces at a time until golden brown. Drain well. In a 4 quart saucepan combine pineapple juice, brown sugar, cornstarch, tomato sauce, vinegar, corn syrup, garlic salt and black pepper and stir well. Cook over medium heat stirring constantly until thickened. Stir in the fried alligator chunks, pineapple chunks, bell pepper, celery and onion. Cover and simmer 10 minutes. Serve over hot rice. Serves 6.

GATOR SWEET AND SOUR

2 lbs alligator meat, cubed
1 c plain flour
4 carrots, sliced in thin strips
2 10 oz jars sweet and sour sauce
salt and cayenne pepper
2 c oil
1 no. 2 can pineapple chunks
1 med bell pepper, sliced thin
3 c water

Lightly coat alligator cubes with flour and brown in oil and drain well. Combine carrots, onion, bell pepper, pineapple chunks, water and sweet and sour sauce and simmer for 30 minutes. Add the alligator and cover and simmer for 30 minutes. Salt and pepper to taste. Serve over white rice. Serves 6.

SWEET SOUR ALLIGATOR AND PEPPERS

1 lb alligator fillets, cubed
1 med red or green bell pepper
1 tbsp cornstarch
¼ c soy sauce
¼ c catsup
3 tbsp dry sherry
hot cooked rice
3 tbsp cornstarch
1 14¼ oz can pineapple chunks in own juice
dash pepper
¼ c honey
3 tbsp vinegar
3 tbsp peanut oil

Coat the alligator cubes with 2 tbsp cornstarch and set aside. Cut the bell pepper into ½ inch squares and set aside. Drain the pineapple and reserve the juice and set aside. In a small mixing bowl combine the pineapple juice, 1 tbsp cornstarch, soy sauce, honey, catsup, vinegar and sherry. Stir well and set aside. In an electric wok or skillet cook alligator in hot oil for 2-4 minutes or until the alligator flakes easily and remove. Add bell pepper and stir fry for 2 minutes. Remove peppers. Stir the pineapple juice mixture and add to the skillet or wok and cook and stir until the mixture is thickened and bubbly. Cook and stir 2 minutes more. Add pineapple chunks, alligator and pepper and heat about 1 minute more or until heated through. Serve over hot rice. Serves 4.

SWEET AND SOUR GATOR

2 lbs alligator fillets
1 8 oz bottle russian style dressing
1 envelope dry onion soup
1 8-10 oz jar pineapple or apricot preserves

Mix the preserves, soup mix and dressing and pour over the fillets in a shallow buttered baking dish and let stand in the refrigerator for 8-12 hours covered. Let come to room temperature and then place in a 350°F oven and bake for 1 hour. Uncover and continue baking until tender basting often. Serves 4-6.

HONG KONG ALLIGATOR

1 lb alligator meat, cubed
½ c cider vinegar
2 lge onions, sliced
1 8 oz can water chestnuts, sliced, drained
salt and white pepper
hot cooked rice
2 lge bell peppers, cut in 1 in squares
1 lge can pineapple chunks, drained
¼ c brown sugar
water as needed
2 16 oz cans mandarin slices,

Mix all the juices from the drained ingredients with the vinegar and marinate the meat in it for 24-36 hours refrigerated. Place the meat and marinade in a heavy saucepan and bring to a boil. Lower the heat and add the water chestnuts and onions. Cook for 20-45 minutes or until the meat starts to get tender. Add the sugar and remaining ingredients and stir carefully with a wooden spoon and cook for 8-10 minutes. Adjust seasoning if needed. Serve over rice.

PINEAPPLE GATOR

2 lbs alligator tail meat, cubed
1 lge can pineapple chunks, drained, reserve juice
¼ c water
3 tbsp brown sugar
1 med onion, sliced
cooked rice
¼ tsp ground ginger
¼ c cider vinegar
3 tbsp cornstarch
2 tbsp soy sauce
1 bell pepper, cut in 1 inch pieces
salt and white pepper

Marinate the meat in the pineapple juice for several hours and then place in a heavy pot and bring to a boil. Simmer covered until the meat is beginning to flake. Mix the water, sugar, ginger, vinegar, cornstarch and soy sauce and stir well. Add to the meat and bring back to a boil stirring gently until thickened. Add the bell pepper and onion and cook until tender crisp. Season to taste and serve over rice. Serves 4-6.

ALLIGATOR HAWAIIAN

2 c gator meat, cooked*, cubed
1 lge can pineapple chunks, drained, reserve liquid
2 tbsp cornstarch
1 tbsp butter
1 8 oz can water chestnuts, sliced drained, reserve liquid
1 bell pepper, cut in 1 in squares
1 tbsp soy sauce
1 jar cocktail onions, drained, reserve liquid
1 cube chicken bouillon
salt and pepper to taste
hot cooked rice

Combine the saved liquids and place in a saucepan with the cornstarch and heat until slightly thickened. Add the butter and stir until melted. Add the water chestnuts and bell pepper and cook for 10 minutes. Add the meat and cook for 10 more minutes. Add the soy sauce and onions and cook for 5 minutes stirring carefully. Carefully stir in the pineapple and simmer until hot again. Season to taste. Serve over rice. Serves 4-6.

CHINESE LETTUCE STEW A LA ALLIGATOR

2 c alligator meat, cooked*, diced
½ med head lettuce
1 tsp salt
½ tsp cayenne or black pepper
1 c chicken broth
1 clove garlic, crushed
paprika, optional
1 sm onion, minced
½ stick butter
1 15 oz can petit pois peas, drained
¼ tsp ground ginger
½ tsp sugar
hot cooked rice

Combine all the ingredients except the rice in a large, heavy saucepan. Bring to a boil and reduce the heat and simmer gently, stirring carefully for 1-2 hours. Let stand 15 minutes before serving. Serve in rice nests and sprinkle with paprika if desired. Makes about 4 cups.

PEPPERY FRIED RICE WITH ALLIGATOR

1 lb alligator, boneless
2 eggs, beaten
1 6 oz pkg frozen pea pods
1 clove garlic, minced
3 tbsp soy sauce
¼ tsp cayenne pepper
1½ c quick-cooking rice
2 tbsp oil
1 sm red or green bell pepper
⅓ c water chestnuts, sliced
¼ tsp ginger, ground

Thinly slice the alligator and bell pepper into bite size strips. Prepare rice according to package directions omitting the salt. In a 10 inch skillet cook the eggs in 1 tbsp oil without stirring until set. Invert the skillet over a baking sheet to remove the eggs and then cut the eggs into short, narrow strips. In the same skillet cook the frozen pea pods, bell pepper and garlic in remaining oil about 1 minute or until the pea pods are thawed. Remove the vegetables from the skillet. Add more oil if needed and add ½ the alligator to the skillet and stir fry 2-3 minutes or until done. Remove from the skillet and repeat with the remaining alligator. Return all the alligator and stir in the rice, ginger, eggs, vegetables, water chestnuts, soy sauce and cayenne pepper. Heat through. Pass additional soy sauce if desired. Serves 6.

ALLIGATOR A L'INDIENNE

2 lbs alligator meat, cubed
2 lemons
1½ tbsp salt
3 tbsp butter
½ c celery, chopped
1 pinch thyme
1 pinch mace
½ tsp curry powder
¼ c coconut milk
1 c cream
1 c long grain rice
1 jar chutney, top quality
4 c water
1 sm bag crab boil*
1 tsp white pepper
½ c onion, minced
½ c parsley, minced
½ bay leaf
2 tbsp plain flour
2 c chicken stock, consomme or bouillon
1 c white wine
2 tbsp lemon juice
1 ripe mango (optional)

Braise alligator for 1 hour in a court-bouillon made with the water, 2 lemons cut in half, crab boil, a tsp each of salt and white pepper. Keep the pot cover cracked so that water will reduce to nothing by the end of the cooking time. Discard the lemon and crab boil and remove the meat to a deep platter and set aside. Prepare cream curry sauce by cooking onions in butter until transparent and then add celery, parsley, thyme, bay leaf and mace. When onions become pale golden add flour and curry powder. Stir over high heat 2-3 minutes until flour begins to color. Pour in the stock and coconut milk. (Coconut milk is made by blanching ½ cup shredded coconut in milk and straining.) Blend with a wire whisk as you bring to a boil. Quickly reduce heat to a low simmer and cook for 45 minutes. Add wine and blend. Strain the sauce through a medium sieve using the back of a wooden spoon to press pulp against sides and bottom of the sieve. Discard the pulp. Before serving reheat the sauce. When warm mix in cream and lemon juice and then pour over the meat. Serve immediately. Accompany with ample servings of chutney. A tropical adaptation is to add the chopped meat of a ripe mango to the chutney.

INDIAN RICE

In India, rice is prepared by bringing 2 quarts of water and 1 tbsp salt to a boil in a 3-quart pot. Pour in the rice, stirring immediately. Bring back to a boil. Reduce heat to maintain a gentle boil. Cook uncovered for 15 minutes. Drain rice into a colander and rinse with cold running water. Remove rice to the center of a napkin or large kitchen cloth. Fold corners over to cover rice and place in a warm place to dry out for 15-30 minutes.

CURRIED GATOR

6-8 c gator meat, diced
2 c half and half cream
1 14½ oz can chicken broth
1 broth can water
2 tbsp curry powder
¾ c plain flour
1 c butter
dash tabasco sauce
1 med onion, chopped
1 c celery, chopped
2 med apples, chopped, unpeeled
1 sm bell pepper, chopped
2 cloves garlic, minced
½ tsp ginger
1½ tsp salt

Marinate the meat in the cream for 6-12 hours refrigerated. Heat the broth and water to boiling in a large saucepan and reduce the heat to a simmer. In a separate skillet melt the butter and add the flour and curry powder and cook on low until well blended and then add the onion, celery, apples and bell pepper and saute until all are tender but not browned. Add the garlic and ginger and cook uncovered on low for 15 minutes. Now add the meat and cream and increase the heat stirring constantly until boiling is reached and then reduce heat and cook for 1 hour, stirring often. Season with the salt and tabasco sauce. Serve over rice. Serves 10-14.

QUICK GATOR 'N GRAVY

1 lb gator leg meat, boneless
½ tsp salt
1 bay leaf
1 12 oz jar or can brown gravy
water to cover
½ tsp tabasco sauce
¼ c vinegar
1 16 oz can Veg-All, drained
seasonings to taste

Boil the meat in the water with the salt, tabasco sauce, bay leaf and vinegar until it begins to fall apart. Drain and rinse. Cut into bite size pieces. Return the meat to a pot and add the vegetables and gravy mix and mix and heat completely. Season to taste. Serve as a soup with bread or crackers or over rice or hot buttered noodles. Serves 4-6.

SWISS STYLE ALLIGATOR

2 lbs alligator pink meat
water to cover
1 bay leaf
1 carrot, scraped, halved
1 tsp tabasco sauce
¼ stick butter
seasonings to taste
1 med onion, chopped
½ bell pepper, chopped
2 ribs celery, chopped
½ tsp lemon juice
1 8 oz can tomato sauce
2 tomato sauce cans water

Boil meat in water to cover along with the carrot, bay leaf and tabasco sauce until tender. Drain. In another saucepan heat butter and add the onion and celery and saute for 5 minutes. Add the bell pepper, lemon juice, sauce and water and simmer gently. Add the meat and cover the pot and let simmer slowly for 1 hour. Season in the last 20 minutes. Serve over rice. Serves 4-6.

BLACKENED ALLIGATOR

alligator tail steaks, pounded
drawn butter*
blackened seasoning

Dip steaks in butter and then coat with seasoning on both sides. Cook in white hot black iron skillet until done, turning once. Serve with lemon and butter. (Have fan running over stove if inside or it will set off your smoke alarm.)

DILLY CORN CASSEROLE A LA GATOR

3 c alligator meat, cubed
2 14½ oz cans chicken broth
2 broth cans water
1 tsp tabasco sauce
4 tbsp butter
4 tbsp plain flour
1½ c bread crumbs
2-3 c reserved stock
½ tsp celery salt
½ tsp dry mustard
½ tsp dill seed
2 17 oz can whole kernel corn, drained, reserve liquid
½ c butter, melted

Boil the meat in the broth, water and tabasco sauce until tender. Drain and reserve the boiling liquid. Melt the 4 tbsp butter in a dutch oven and add the flour and cook slowly stirring constantly until lightly browned. Slowly add the corn liquid and adding the reserved stock using about 1½ cups to start with. Add the celery salt, mustard and dill seed stirring all the while. If too thick add more stock. The sauce should be as thick as a heavy pancake batter. Remove from the heat. Pour 1 can of the corn into a buttered casserole and cover with half the meat sauce and repeat with the remaining corn and sauce. Top with the bread crumbs and melted butter and bake at 350°F for 30 minutes. Serves 6-8.

GATOR ETOUFFÉ

2 lbs alligator meat, bite size
2 lge onions, chopped
½ c butter
salt and pepper to taste
1 clove garlic, chopped
¼ c bell pepper, chopped
2 tbsp tomato paste

Saute the onion, bell pepper and garlic in butter. Add tomato paste and saute until onions are clear. Add seasoned meat. Cover and steam until meat is tender. Serve meat and sauce over rice.

ALLIGATOR ETOUFFÉ

1 lb alligator meat, cut in pieces
cayenne pepper to taste
4 stems celery, chopped
2 sticks butter
¼ c parsley, minced
2 onions, chopped
2 cloves garlic, minced
1 14½ oz can stewed tomatoes
½ c scallions, chopped
salt and black pepper

Saute onions, garlic, scallions, parsley and celery in butter until soft. Add tomatoes and simmer for 20 minutes in a covered pot. Add alligator meat and let cook over a low fire until tender, approximately 1 hour. If gravy is too thick add a little hot water. Serve over rice.

CROCODILE ETOUFFÉ

2 lbs crocodile** tail
½ c + oil
½ c onions, diced
½ c bell pepper, diced
1 jalapeno pepper, chopped
½ tsp cumin
1 tsp salt
1½ c fish or shrimp stock
1 c plain flour
2 tbsp tomato paste
½ c celery, diced
1 tbsp garlic, chopped
¼ tsp thyme
½ tsp cayenne pepper
½ tsp black pepper

Make a dark roux* of ½ cup each oil and flour. When a dark brown add the tomato paste and stir well until mixture is smooth and browned. Then add the vegetables and seasonings and saute until the vegetables are soft. Dredge the crocodiles strips in remaining flour and saute in a little oil until golden brown. Add to the vegetable mixture and also add the stock. Simmer for 30-40 minutes. Serve over rice.

**Alligator can be used.

ALLIGATOR ETOUFFÉ A LA TECHE

2 lbs gator meat, cooked*, diced
1 stick butter
1 bay leaf
½ tsp basil
1 8 oz can tomato sauce
1 c white wine
1 tbsp lemon juice
french bread
1 rib celery with leaves, chopped
¼ c plain flour
1 c scallions, chopped
1 lge onion, chopped
1 sm bell pepper, chopped
1 c water or alligator stock
hot cooked rice
garlic butter

Make a roux* with the butter and flour until the color of caramel. Add the scallions, onion, bell pepper and celery and simmer until tender. Add the bay leaf, basil and seasonings and simmer for 20 minutes. Add tomato sauce, wine, water or stock and cook for 30 minutes. Add the meat and cook 30 minutes more. Stir in the lemon juice and adjust seasonings if needed. Serve over rice with french bread with garlic butter on the side.

GARLIC BUTTER

½ c butter
2 garlic cloves, minced
1-2 drops tabasco sauce
½ tsp salt
dash of black pepper

Bring butter to room temperature and add other ingredients and cream thoroughly. Refrigerate.

ALLIGATOR STUFFED EGGPLANT

3 eggplants, cut in half
2 tbsp butter
3 slices bread, cubed
3 tbsp bacon fat or butter
½ c onion, minced
3 cloves garlic, minced
¼ c scallion tops, chopped
¼ c celery, chopped
1 egg, slightly beaten
2 tsp salt
½ tsp black pepper
½ tsp cayenne pepper
1 c alligator meat, ground
1 tsp fresh parsley, minced
plain bread crumbs
parmesan cheese, grated

Scoop out pulp of eggplant which has been cut lengthwise. Cover dish with plastic wrap and cook pulp in butter on high in microwave 15 minutes. Cook shells on high 5 minutes. Moisten bread with water, squeeze and set aside. Melt fat or butter in a 2-quart dish. Saute onion, garlic, scallions, celery and alligator 6 minutes, stirring often. Put in food processor a few seconds to make mixture fine. Add egg, salt and pepper, cooked eggplant pulp and moistened bread. Cook on high 5 minutes. Stir in parsley. Stuff mixture in eggplant shells, sprinkle with bread crumbs and parmesan cheese and place on a flat plate. Microwave 2-3 minutes until heated through.

STUFFED EGGPLANT WITH ALLIGATOR

1 lb alligator meat, ground
1 med onion, chopped
1 rib celery, chopped
½ med bell pepper, chopped
2 cloves garlic, chopped
10 scallion tops, chopped
½ c + italian bread crumbs
water
1 med eggplant
2 tsp Season-All
1 4 oz can mushrooms, stems and pieces, drained, chopped
⅛ tsp cayenne pepper
1 egg, well beaten
½ stick + butter
lemon juice

Saute the meat in the ½ stick butter until browned and then add the vegetables and garlic and cook until done. Cut eggplant lengthwise and remove the pulp. Place the eggplant shells in enough water to cover and add the lemon juice to keep them from discoloring. To the meat mixture add the Season-All, cayenne pepper, egg, mushrooms and scallions and stir in the ½ cup bread crumbs mixing well. Spoon into the shells and sprinkle more bread crumbs over the tops and dot with butter. Bake at 375°F for 20 minutes. Serves 2.

ALLIGATOR STUFFED SQUASH

1 c gator meat, boiled, ground
3-4 sm white scalloped squash
½ c onion, chopped
½ c scallions, chopped
1 sm bell pepper, chopped
3 tbsp parsley flakes
3 cloves garlic, minced
butter
1 14½ oz can stewed tomatoes
½ stick butter
1 tsp salt
½ tsp black pepper
½ tsp thyme
½ c water
bread crumbs

Cut the tops off the squash and place in a shallow pot with a cover and a little water to steam with. Bring water to a boil and simmer until the squash is tender. Scoop out the pulp of the squash leaving enough shell so they don't fall apart. Melt the ½ stick butter and add the vegetables and cook until they are tender. Then add the meat and squash pulp and mix well and add the remaining ingredients except the bread crumbs and remaining butter. Simmer until the liquid is absorbed and is thick. Fill the shells with the mixture and top with the bread crumbs and dot with butter. Place in a shallow casserole and bake 30-45 minutes at 375°F. Small eggplants, mirletons or yellow squash may be used in place of the white squash. Serves 4-6.

ALLIGATOR SQUASH CASSEROLE

4-6 med yellow squash, sliced
1 lge onion, chopped
3 ribs celery, chopped
1 sm bell pepper, chopped
½ stick butter
bread crumbs
2 c gator meat, cooked*, diced
4 eggs, beaten
½ c milk
1 sm box cornbread mix
1 c parmesan cheese, grated

Boil the squash until tender and drain. Melt the butter and saute the onion, celery and bell peppers until soft. Add the squash and continue cooking until the squash is heated through again and then remove from the heat. Mix the cornbread mix with the eggs and milk and add to the vegetable mixture. Stir well and add the meat and mix thoroughly. Season with salt and pepper and add the grated cheese reserving a little for the top. Pour into a buttered casserole dish and top with the bread crumbs and the reserved parmesan cheese. Bake at 325°F for 45-60 minutes until lightly browned. Serves 8-10.

ALLIGATOR "AL E. GATOR'S"

4 5 oz alligator filets
½ c pecans, minced
1 c plain bread crumbs
oil
1 c plain flour
1 c milk
1 egg
butter

Pound alligator with a meat mallet until alligator is ½ inch thick. Dust with flour. Dip in beaten egg and milk. Blend pecans and bread crumbs and bread alligator with this mixture. Fry in a mixture of oil and butter. Serve with plantation sauce.

PLANTATION SAUCE

1 cleaned fresh mango, diced
1 cleaned fresh papaya, diced
3 tbsp brown sugar
3 tbsp butter
3 tbsp pecans, chopped
½ oz rum
1 c water

Melt butter in pan and saute pecans. Add sugar and stir for 2 minutes. Add mango and papaya and mash. Add water and bring to a boil. Add rum after the water boils. Cook 15 minutes on simmer.

ALLIGATOR AND SCALLOPS

4 alligator steaks
1 egg, beaten
1 c milk
½ c butter

Cut and pound steaks until 1 inch thick. Combine egg and milk. Dip each piece into the egg mixture. In a skillet heat butter and saute steaks on each side 2-4 minutes. Serve scallop sauce over alligator steaks.

SCALLOP SAUCE

scallops
white wine
cornstarch
water

Poach scallops in water and reserve the broth. Add equal amounts of white wine and the reserved broth in a saucepan. Slowly boil down this liquid until it is half of the original. Add the poached scallops. You may want to add cornstarch to thicken and make an attractive garnish.

ALLIGATOR SCAMPI

alligator (white meat)
fresh garlic, chopped
butter
mushrooms, chopped
tomatoes, chopped
shallots, chopped
white wine
salt and pepper to taste

Cut alligator into fillets and tenderize until double in size. Season with salt and pepper. In sauce pan heat the butter and garlic; saute meat and then add mushrooms, tomatoes and shallots. Cook over low heat until meat is tender. Just before serving add a small amount of white wine.

SCAMPI A LA GATOR

18-24 alligator meat strips,
  1 inch diameter, 2-4 inch long
cayenne pepper
1 stick butter mixed with
  garlic puree to your taste
parsley or chives, chopped
2 tbsp lemon juice
salt
1 tbsp paprika
5 cloves garlic, minced
1 12 oz pkg vermicelli, cooked
2 c corn oil

Mix well the meat, oil, paprika, garlic, lemon juice salt, cayenne pepper in a bowl and refrigerate for 12 hours or more. Drain the meat discarding the marinating liquid. Heat the garlic butter in a skillet and add the meat strips and stir fry and then cover and let the meat steam. Remove the cover and let the remaining liquid cook down. Remove the meat carefully and pour the remaining liquid in the skillet over the vermicelli and toss lightly. Add the meat to the vermicelli and mix. Heap on a platter and sprinkle with the parsley or chives. Serves 6.

ALLIGATOR AU GRATIN

2 lbs alligator
3 ribs celery, chopped
2 scallions, chopped
½ lb butter
4 tbsp plain flour
1 12 oz can evaporated milk
2 egg yolks
1 lge onion, chopped
10 ozs mild cheddar cheese, grated
salt and pepper to taste

Saute the vegetables in butter. Add the flour and blend. Season alligator meat cut into small pieces with salt and pepper. Saute meat with the vegetables. Blend the milk and eggs and then add to the vegetables and meat and heat through. Remove from heat and place in serving dishes and sprinkle with cheddar cheese.

ALLIGATOR AU GRATIN A LA TECHE

2 c gator meat, cooked*, chopped
1 c mayonnaise
1 c cheddar cheese, grated
2 tsp prepared horseradish
½ tsp tabasco sauce

Combine all ingredients except cheese mixing well. Put into a shallow buttered baking dish and top with the cheese. Bake about 10 minutes at 300°F or until the cheese is melted. Serve at once. Serves 6.

ALLIGATOR THERMIDOR

1 lb alligator meat, cubed,
  boiled in seasoned water
½ tsp salt
¼ tsp cayenne pepper
½ c dry golden sherry
2 egg yolks, beaten
paprika
1 4 oz can mushrooms, sliced, drained
½ stick + butter
1 c scallions, minced
2 tbsp plain flour
1½ c half and half cream
buttered bread crumbs

Melt the ½ stick of butter in a heavy saucepan and saute the scallions and then stir in the flour and cook for 5 minutes on low heat until the flour is lightly browned. Add the cream stirring until smooth. Add the salt, cayenne pepper and mushrooms and heat through. Beat the egg yolks with the sherry and add to the mixture stirring quickly to keep smooth. Add the meat and heat through. Fill a casserole or individual serving shells with the mixture and cover with buttered bread crumbs and dot with butter. Sprinkle with paprika. Place in broiler until browned, about 5 minutes. Serves 2-4.

ALLIGATOR NEWBERG

1 lb alligator meat, cubed
½ stick butter
3 tbsp plain flour
½ tsp salt
cayenne pepper to taste
1 pt heavy cream
2 egg yolks, slightly beaten
2 tbsp white wine

Put the meat and butter in a heavy pot and saute until the meat flakes and is tender but not falling apart. If it begins to dry add a little more butter. Stir the flour into the cream and add to the meat slowly stirring constantly. Stir a small amount of the hot liquid into the egg yolks and then add to the mixture stirring to prevent lumping. Season to taste and add the wine. Remove from the heat and let stand for a few minutes. Serve over toast points, english muffin halves, hot biscuits, or slices of fresh french bread. Serves 4-6.

ALLIGATOR RAREBIT

1 c gator meat, boiled, minced
2 tbsp butter
2 tbsp plain flour
3 c mild cheddar cheese, grated
english muffin halves or toast
paprika or parsley flakes
½ tsp salt
1 tsp dry mustard
2 tsp worcestershire sauce
1¼ c half and half cream
dash white pepper

Over a very low heat melt the butter and add the flour mixing well. Add the cheese and stir until it is melted. Add the salt, mustard, worcestershire sauce and a dash of white pepper mixing well. Add the cream slowly stirring all the time on a very low heat. Stir in the meat and cook until heated through. Serve over the english muffins or toast. Sprinkle with paprika or parsley. Can also be used as a dip. Serves 6.

SCALLOPED ALLIGATOR

2 c gator meat, cooked*, diced
2 c chicken broth
3 ribs celery, diced
1 c Velveeta Cheese, cubed sm
salt and pepper to taste
1 med onion, chopped
1 egg, beaten
1 10¾ oz can cream of chicken soup
½-1 c + bread crumbs

Mix well all the ingredients except the bread crumbs. Add bread crumbs until the mixture is mushy but not runny. Place in a casserole and top with bread crumbs and bake uncovered for 45-60 minutes at 325°F. Serves 6-8.

ALLIGATOR POTATO SCALLOP

1 c gator meat, cooked*, diced
butter, melted
5 tbsp mayonnaise
bread crumbs
1 5 to 6 oz box scalloped potatoes
1 c celery, onion, bell pepper, total
pinch poultry seasoning

Prepare potatoes as directed on box in a buttered 2-quart casserole. Then add all ingredients except bread crumbs and butter and mix well. Bake at 375°F for 45 minutes and then top with the bread crumbs and drizzle the melted butter on top and bake 10 minutes longer. Serve hot. Serves 4-6.

ALLIGATOR COQUILLE DE ST. MARIE

3 c alligator tail meat, cubed
1 stick + butter
1 scallion, chopped
4 tbsp plain flour
2 c water
½ c white wine
1 4 oz can mushrooms, sliced, drained, reserve liquid
2 egg yolks
2 tbsp parmesan cheese, grated
1 tbsp paprika

Boil the meat in the water, wine and mushroom liquid until tender. Drain and reserve the liquid. Saute the onions in the stick of butter until soft and add the flour and mix well. Add the reserved boiling stock to the onion mixture and heat on a low fire until almost boiling. Beat egg yolks in a measuring cup and slowly add a little of the hot broth to the eggs mixing well and then pour this into the pot stirring to prevent lumping. Cook on low for 5 minutes and then add the meat and mushrooms and cook for 10 minutes more stirring often. Put into baking shells and sprinkle with a mixture of cheese and paprika and dot with butter. Bake at 350°F until brown. Serves 4-6.

ALLIGATOR CASSEROLE

2 lbs alligator, ground
4 cloves garlic, chopped
1 stick butter, softened
8 slices toast
salt and pepper to taste
bread crumbs
6 scallions, chopped
4 tbsp parsley, chopped
2 eggs, beaten
half and half cream
cayenne pepper

Saute alligator, scallions, garlic and parsley in some butter. Butter bread crumbs. Pull apart toast and mix with the seasonings. Add eggs and half and half to soften and season to taste. Cover with enough buttered crumbs and bake in a casserole for 30 minutes at 375°F.

GATOR AND ARTICHOKE CASSEROLE

1 lb gator meat, diced, boiled in seasoned water, diced
1 4 oz can mushrooms, sliced, drained, reserve liquid
1 lge can artichoke hearts, drained
½ stick butter
½ c milk
½ c heavy cream
½ c sherry
salt and pepper to taste
parmesan cheese, grated
paprika
¼ c plain flour

Oil a 2-quart casserole and layer the meat, mushrooms and artichoke hearts. Make a sauce with butter and flour and brown lightly over a low heat and then add the milk, reserved mushroom liquid and cream. Cook, stirring constantly, until thickened and then add the sherry. Remove from heat and season to taste. Pour over the layered items in the casserole and sprinkle with the parmesan cheese and dust lightly with paprika. Bake for 25-30 minutes at 375°F. Serve over hot cooked rice or egg noodles.

ELEGANT GATOR CASSEROLE

2 lbs alligator meat, boneless
water to cover
¼ c vinegar
1 clove garlic
¼-½ c parmesan cheese, grated
1 14½ oz can chicken broth
1 tsp + dash tabasco sauce
3 pkgs frozen broccoli, cooked as pkg directions, drained
mock Hollandaise sauce

Boil meat in the water with the vinegar, garlic and dash tabasco sauce added until tender. Drain and rinse and dice the meat. Then boil the meat for 15 minutes in the chicken broth and remaining tabasco sauce. Cool and drain reserving the stock for other recipes. In a shallow buttered casserole layer the broccoli in the bottom. Canned drained asparagus can be substituted, if desired. Top with the meat and sprinkle with the parmesan cheese. Pour mock hollandaise sauce over the top and bake at 325°F until hot. Then place under the broiler to brown slightly. Serves 6.

MOCK HOLLANDAISE SAUCE

½ c mayonnaise
2 tbsp lemon juice
½ c heavy cream
dash paprika

Mix well all ingredients.

RITZY GATOR

2 c gator meat, cooked*, diced
2 tbsp butter
1 tbsp bell pepper, minced
2 tbsp onion, minced
½ c dry sherry
Ritz Cracker crumbs
½ tsp brown mustard
2 c milk
1 c monterey jack cheese, grated
4 tbsp plain flour
1 sm jar pimento, chopped
paprika

Saute bell pepper and onion in butter until soft and then add the flour a little at a time stirring constantly. Add milk slowly stirring to insure a smooth sauce. Add the mustard, cheese, sherry and pimento blending well and when heated through add the meat and cook slowly for 5 minutes. Pour into a greased casserole and top with the crumbs and paprika and bake at 350°F for 20-25 minutes. Serves 4-6.

ALLIGATOR EGGPLANT CASSEROLE

1-2 c gator meat, cooked* ground
1 lge eggplant, peeled, diced
½ c water
1 stick butter
1 lge onion, chopped
1 egg, beaten
½ lb cheddar cheese, grated
2 c italian bread crumbs
salt and pepper to taste

In melted butter saute the eggplant, onion and water and bring to a boil. Simmer, covered lightly, until the eggplant is tender. Remove from the heat and add all remaining ingredients and mix well. In a 1-quart greased casserole add the eggplant mixture. Top with additional bread crumbs and a little parmesan cheese, if desired. Bake at 325°F for 30-45 minutes. Serve hot. Serves 4-6.

LAZY WAY GATOR CASSEROLE

2 c gator meat, white, cooked*, diced
2 c milk
½ c lemon juice
2 sticks butter, melted
1 tbsp worcestershire sauce
1 tsp salt
4 hard boiled eggs, mashed
2 eggs, slightly beaten
4 c white bread, torn into sm pieces
2 scallions, chopped
4 tbsp dijon mustard
½ tsp cayenne pepper
½ tsp black pepper

Mix well all the ingredients in a large bowl. Grease a 2-quart shallow casserole or individual baking shells. Spoon the mixture into the casserole or baking shells and bake at 400°F for 30 minutes. Dust with paprika and parsley flakes if desired. Serves 8.

SWIFT ALLIGATOR SUPPER

2 c alligator meat, cooked*, ground
1 10¾ oz can cream of mushroom soup
tartar sauce
5 6 oz bags potato chips
lemon wedges
paprika

Crush the potato chips and combine with all the other ingredients saving a few crumbs for the top. Place in a buttered casserole dish and top with the reserved crumbs and sprinkle with paprika. Bake 30 minutes at 300°F. Serve with lemon wedges and tartar sauce. Serves 6-8.

DAVID'S GATOR SUPPER

1 lb alligator meat, cooked*, ground
½ stick butter
1 sm onion, chopped
1 6 oz can tomato paste
1 paste can water
dash tabasco sauce
8 ozs egg noodles, cooked
1 8 oz pkg cream cheese, softened
¾ c romano or parmesan cheese, grated, or combination
1½ tsp salt
½ bell pepper, chopped
1 c milk

Add meat and onion to butter in a heavy pot and cook for 5 minutes. Add the tomato paste, water, 1 tsp salt and tabasco sauce and cook on low heat for 5-10 minutes. In a separate pot heat the cream cheese, cheese(s), ½ tsp salt, bell pepper and milk over a low fire until well-blended and smooth stirring constantly. Add the noodles to this mixture. In a large buttered casserole alternate layers of the meat mixture and the noodle mixture beginning and ending with the meat mixture. Sprinkle with additional grated cheese and bake at 350°F for 20-30 minutes or until top is lightly browned. Serve with hot rolls or bread sticks. Serves 6-8.

GATOR MARSALA

1½ lbs alligator fillets
2 tbsp butter
1½ c dry marsala wine
salt and black pepper
½ c all-purpose flour
3 tbsp olive oil
½ c chicken broth

Season fillets with salt and pepper to taste and dust with the flour, shaking off the excess. In a large heavy skillet melt the butter with the olive oil over a moderate heat. When the foam subsides add the meat and brown for 2-3 minutes on each side. Transfer to a plate. Pour off and discard most of the fat from the skillet and add the marsala and the broth. Boil over high heat for 1-2 minutes, scraping the bottom to dissolve the brown bits. Return the meat to the skillet. Cover and reduce the heat to low and simmer 10-15 minutes. When the sauce has reduced to a thick syrupy glaze, remove from the heat. Serve next to garlic mashed potatoes.

CROCKED GATOR

10-14 alligator ribs
6 ozs barbecue sauce
1 tbsp garlic powder
1 tbsp salt
1 tbsp pepper

Season ribs. Fill crock pot with ribs. Pour barbecue sauce (your favorite) over the ribs. Cover and cook until tender.

CROCKED ALLIGATOR

2 lbs alligator meat, cubed
water to cover
1 10¾ oz can cream of mushroom soup
1 14½ oz can chicken broth
hot cooked noodles
1 4 oz can mushrooms, sliced
½ c red wine
1 clove garlic, speared on a toothpick
1 sm onion, chopped

Boil the meat in the water for 30 minutes and then drain and rinse. Discard the water. Put all remaining ingredients along with the meat except the noodles in a crock pot and cook on low heat for 8-12 hours or 5-6 hours at high heat. Serve over the noodles. Serves 6-8.

MICROWAVE ALLIGATOR

2 alligator tail chops, ½ inch thick
1 med. onion, sliced
1 tsp Season-All or
lemon pepper seasoning

Season chops with Season-All or lemon pepper seasoning. Put in a 1½ quart dish and microwave on high uncovered 5 minutes. Arrange onion over chops and cover with plastic wrap and microwave on simmer at 30% power for 20 minutes. Allow to sit for 5 minutes before serving. Serves 1-2.

MISTRAL'S WAY ALLIGATOR

½ lb alligator meat
tarragon
½ c white wine
½ c heavy cream
plain flour, seasoned*
butter

Tenderize and cut meat into bite size pieces. Coat meat with seasoned flour and tarragon. Saute in butter and brown on both sides. Add wine and cream to make a thick cream sauce. Serve.

ALLIGATOR CABBAGE MAINTENON A LA TECHE

2 lbs gator meat, cooked*, diced
1 sm head cabbage, about 4 lbs
½ stick butter
seasonings to taste
½ c plain flour
1 14½ oz can stewed tomatoes
boiling water

Cut up cabbage into bite size pieces and then melt the butter and stir in the tomatoes and add the cabbage and meat mixing well. Add enough water with the flour to cover the mixture. Bring to a boil and reduce to a simmer. Stir often and cook for 2-3 hours or until the liquid is almost gone. Season to taste. Serves 6-8.

SAUTEED ALLIGATOR PAWS

1 lb alligator tenderloins
1½ c plain flour
salt and pepper to taste
parsley
4 eggs, extra lge, beaten
½ c drawn butter*
lemon wedges
mustard sauce 2

Use either the tail or jowl tenderloin and cut into medallions and tenderize with a meat mallet until very thin. Roll in the flour which has been seasoned with salt and pepper. Dip each piece into the beaten eggs. Then quickly saute meat in hot saucepan with butter until golden brown on both sides. Drain and serve with mustard sauce. Garnish with lemon wedges and parsley.

MUSTARD SAUCE 2

1 c mayonnaise
1 tbsp prepared mustard
1 tsp soy sauce
1 tsp lemon juice

Blend all ingredients together well.

SMOTHERED ALLIGATOR A LA TERREBONNE

2 lbs alligator meat
¼ c oil
1 onion, minced
1 bell pepper, minced
¼ c parsley, minced
½ c celery, minced
¼ c scallions, minced
1-2 bay leaves
¼ tsp sweet basil
salt and pepper to taste

Saute onions in oil until golden brown and then add the bell pepper and celery. Saute until tender, then add the meat (cut into bite size pieces) and seasonings. Simmer for 40 minutes or until meat is tender. Add parsley and scallions about 5 minutes before serving.

CREOLE SMOTHERED ALLIGATOR

3 lbs alligator meat, boneless,
  pink meat from legs or torso
1 12 oz can beer + can water
¼ c oil
¼ c plain flour
1 8 oz can tomato sauce
hot rice cooked in salted water
1 10 oz can beef consomme
1 sm onion, chopped
½ bell pepper, chopped
1 rib celery, chopped
10 green stuffed olives, halved
salt and pepper to taste

Marinate the meat in the beer and water for 8-10 hours or overnight. Drain and reserve the liquid. Heat the oil and add the flour and cook stirring constantly over a medium heat until lightly browned. Add the meat, tomato sauce and consomme and bring to a boil. Add the onion, bell pepper, celery and some of the beer mixture to make a watery gravy. Cook for 20 minutes on medium-high and then reduce the heat and cook for 2-3 hours stirring often until it is thick. Add the olives and salt and pepper 10 minutes before serving. Serve over rice. Serves 6-8.

BAYOU BEND ALLIGATOR

3 lbs gator meat, cubed, boned, preferably pink meat
½ c + 1 tbsp plain flour
½ c oil
1 clove garlic, mashed
1 lge onion, sliced
cayenne pepper to taste
1 8 oz can mushrooms, sliced
1 c dry white wine
½ c brandy
¼ c ketchup
scallions, chopped
salt and pepper to taste

Roll meat in the ½ cup of flour and add to heated oil in a heavy pot and brown lightly. Add the garlic, onion, drained mushrooms and about half of the wine. Bring to a boil and cook until the meat is tender. Add the liquid from the mushrooms and remaining wine and cook for 10 minutes. Mix the tbsp of flour with the ketchup and add to the pot stirring to prevent lumping. Stir in the brandy and cook for 5 minutes or until thickened. Season to taste and add the scallions just before serving over hot cooked rice. Serves 6-8.

CREOLE GATOR EN BATEAUX

2 lbs gator meat, cubed, boned
1 tbsp tabasco sauce
juice of I lemon
1 tbsp + butter
1 tbsp plain flour
3-5 sm french breads
1 lge onion, chopped
1 rib celery, chopped
1 14½ oz can stewed tomatoes
1 tsp cornstarch
salt to taste

Sprinkle the tabasco sauce and lemon juice over the meat and let stand in the refrigerator for 4-6 hours stirring several times. In a heavy pot melt the tbsp butter and then add the flour, onion and celery and cook on low until soft and starting to brown. Add the tomatoes and cook for 5 minutes. Add the meat and enough water to cover. Cook, lightly covered over a moderate heat until the meat is tender and the liquid is reduced. Add cornstarch to thicken. Make the bateaus (boats) by slicing breads lengthwise and scooping out some of the centers. Butter well and place on cookie sheets and heat in a 250°F oven for 15 minutes. Fill with the mixture and serve very hot. Serves 6-10.

TEXICAN ALLIGATOR OR TURTLE CHILI

2 lbs alligator or turtle, coarse ground or diced
1 lge onion, chopped
2 jalapeno peppers, chopped
½ tsp oregano
½ c red wine
2 c red kidney beans, cooked, (optional)
4 slices salt pork, chopped
1 can Ro-Tel Tomatoes
1 clove garlic, minced
2 tbsp chili powder
½ tsp cumin
salt and pepper to taste

Brown the salt pork in a large, deep iron skillet, frying out all fat as liquid. Add alligator or turtle meat and onions and brown at a medium heat. Add tomatoes, seasonings, wine, pepper and a dash of salt. Mix well and reduce heat and simmer about 2 hours, adding small amounts of water if necessary. Refrigerate overnight and then reheat and add beans (if desired). Serves 5.

GATOR POT PIE

1-2 c gator meat, cooked*, flaked
paprika
parsley flakes
1 9 in pie shell or 4 individual patty shells
½ c Cheese Whiz
1 10 oz pkg frozen peas in cream sauce
cracker or bread crumbs

Cook the peas as directed on package. Add the gator meat and cheese stirring well and heat until cheese is completely melted. Pour into baked pie shell or patty shells and top with crumbs and sprinkle with paprika and parsley flakes. Place under a broiler for a few minutes to brown.

HUNTER ALLIGATOR PIE

1 lb gator meat, cooked*, ground
1 17 oz can green peas, drained
1 10¾ oz can cream of celery soup
butter
mashed potatoes

Mix the peas, meat and soup in a greased casserole and bake at 325°F until very hot stirring occasionally. Remove from the oven and spread the potatoes over the top and dot with butter. Return to the oven and bake for 8-10 more minutes or until the potatoes begin to brown on peaks. Serves 4-6.

ALLIGATOR STEW

1 lb alligator meat, cubed**
½ c oil
½ c onions, chopped
½ c bell or banana pepper, diced
½ c celery, chopped
2 tbsp parsley, minced
1 tsp garlic powder
1 12 oz can corn, undrained
1 10 oz can Ro-Tel Tomatoes
1 8 oz can tomato sauce
20 ozs water
salt and pepper to taste

In a skillet, brown the onions and drain. In a dutch oven combine the meat, chopped vegetables, tomatoes, sauce, seasonings and browned onions. Cover and cook over low heat for 1½ hours or until the meat is tender. **Use meat from legs or body preferably.

CREOLE STYLE ALLIGATOR

4 lbs alligator meat, cubed
1½ bell pepper, diced
1 tbsp lemon juice
2 tbsp brown sugar, packed
2½ c celery, chopped
½ c plain flour
½ c butter
2 tsp worcestershire sauce
1½ tsp tabasco sauce
½ tsp black pepper
1⅓ c onions, chopped
⅔ c white wine
3 bay leaves
8 cloves, whole
3 20 oz cans tomatoes

Melt the butter and add the bell peppers, onions and celery and saute about 10 minutes or until tender. Remove from the heat. Add flour and blend thoroughly and add the tomatoes gradually, stirring constantly. Add salt and pepper, sugar, bay leaves and cloves. Bring to a boil and then add the alligator meat to the mixture and bring to another boil. Reduce heat and simmer, uncovered, over low heat about 45 minutes, stirring occasionally. Remove from heat and stir in worcestershire sauce, tabasco sauce, lemon juice and white wine. Serve over hot rice. Serves 15.

ALLIGATOR NOIR

2 lbs alligator, tail or jowl
2 ribs celery, minced
3 cloves garlic, chopped
½ tsp marjoram
½ tsp cayenne pepper
½ c plain flour
1 qt veal stock (veal tail stock preferred)
lemon slices, seeded
sherry (optional)
2 med onions, minced
1 bell pepper, minced
1 tsp thyme
1 tsp basil
½ tsp black pepper
½ c oil + chicken broth
¼ c green salad olives, sliced
¼ c parsley, chopped

Cut meat into 1 inch cubes and fry in a little oil until browned. Remove from the oil and set aside. Discard the remaining oil. Make a roux* of the flour and ½ cup oil until medium browned. Add onions, bell pepper and garlic mixing well and simmer until the vegetables are well done. Add the seasonings mixing well and let simmer for 1 minute. Add the alligator and let simmer covered for 10 minutes. Gradually add the veal stock and dilute with chicken broth if stock flavor too strong to allow the alligator flavor to come through. Chicken broth can also be added to get the desired gravy consistency you want. Add the lemons and olives and let simmer on low heat until the meat is tender. Just before serving add the parsley and serve over rice with a little sherry if desired.

VEAL STOCK

- 2½ lbs veal knuckle, bones or tail
- water to cover
- 6 ozs onions
- 6 ozs carrots
- 1 stalk celery
- 1 bouquet garni*
- salt
- 4 white peppercorns

Cover veal with enough water to cover and bring to a boil. Remove any scum that rises. Add the vegetables and seasonings. Bring again to a boil and then lower the heat and simmer for 1 hour. Strain and then boil again until it is reduced to 1 quart. Strain again. For a concentrated form boil again and reduce to 2 ounces.

CREOLE ALLIGATOR

- 2 lbs alligator fillets, cubed
- 2 sm onions, minced
- 2 ribs celery, chopped
- 2 tbsp parsley, minced
- 2 cloves garlic, minced
- 2 sticks butter, melted
- 5 c tomatoes
- 1 c mushrooms, sliced
- 1 bay leaf
- 2 tbsp sugar
- 4 dashes tabasco sauce
- 1 c tomato paste
- salt and pepper to taste
- 2 tbsp lemon juice
- 2 tsp worcestershire sauce
- 2 tbsp parmesan cheese, grated

Saute onions, celery, parsley and garlic in half the butter until soft. Stir in the tomatoes, mushrooms, bay leaf, sugar, tabasco sauce, tomato paste and season to taste. Bring to a boil then reduce heat and simmer, stirring occasionally, for 15-20 minutes until slightly thickened. Combine lemon juice and remaining butter and pour over the alligator meat and sprinkle with salt and pepper. Add the alligator and worcestershire sauce to mixture. Cook over a low heat for 10 minutes or until the alligator is tender. Sprinkle with the cheese and serve over rice or noodles. Serves 10-12.

GATOR IN HOT SAUCE

- 3 lb alligator, cut in chunks
- ½ c oil
- 1 c onions, chopped
- ½ c scallions, chopped
- 2 8 oz cans tomato sauce
- 1 10 oz can Ro-Tel Tomatoes
- salt to taste
- 1 tsp sugar
- 1 c bell peppers, chopped
- ½ c plain flour
- 3 ribs celery, chopped
- ¼ c parsley, chopped
- 1 can whole tomatoes
- garlic powder to taste
- 1 8 oz can mushrooms, stems and pieces
- 1 qt water

In a heavy pot over medium heat make a roux* (brown) of the flour and oil. When browned add the onions, bell peppers and celery and cook until the onions are tender, stirring often. Add meat and brown about 15 minutes. Add remaining ingredients and cook until the meat is tender and sauce of desired thickness. If too thick add water as needed, Serve on hot rice. Serves 6-8.

GATOR STEW WITH TURNIPS

1-2 lbs alligator meat, diced
⅓ c oil
2 ribs celery, chopped
5 cloves garlic, chopped
4 c water
1 c flour
3 onions, chopped
1 med bell pepper, chopped
¼ c parsley, chopped
turnips, as desired, diced

Brown the flour and oil. Pour meat and vegetables into the finished roux and smother until the onions are clear. Add the water and cook about 2 hours or until the meat is tender. Serve over rice.

DEBBIE'S ALLIGATOR STEW

3-4 lbs alligator meat, cubed
2 onions, minced
8 tbsp butter
2 sprigs thyme (optional)
4 sprigs parsley (optional)
creole seasoning*
4 tbsp plain flour
2 14½ oz cans stewed tomatoes
4 cloves garlic, minced
1 bell pepper (optional)
salt and pepper to taste

Brown flour in butter and add onion and bell pepper (if using). Saute until the onion is transparent and tender. Add tomatoes, garlic and seasonings to taste. Simmer on a low heat for 30 minutes. Add the alligator and cook in the gravy until tender. If gravy becomes too thick add a little water to desired consistency. Serve over rice with garlic bread and a tossed green salad. Serves 6-8.

ALLIGATOR CREOLE

4 lbs alligator meat, cubed
¾ c plain flour
2 c yellow onions, chopped
1 c celery, chopped
¼ c scallions, chopped
4 med tomatoes, chopped
1 8 oz can tomato sauce
¼ tsp cayenne pepper
1 c olive oil
2 cloves garlic, chopped
½ c bell pepper, chopped
2 tsp salt
4 tbsp parsley, chopped
2 6 oz cans tomato paste
¼ tsp black pepper
3 c water

Boil the alligator meat as you would shrimp. In a large skillet heat olive oil and add the flour and make a roux* (brown). Add the onions, celery and bell peppers and cook until soft. Add the tomatoes, tomato paste and sauce and blend. Add water and mix and simmer for 30 minutes. Add the drained alligator meat and the remaining ingredients and mix well and cook slowly for 1 hour. Serve over hot rice. Serves 8-10.

ALLIGATOR KOTTWITZ

6 pieces alligator tail, 4-6 oz each, lightly pounded
19 oz artichoke bottoms, drained, sliced
2 tbsp dry white wine
salt and black pepper
8 tbsp butter + 1 stick
1 lb mushrooms, sliced
juice of 1 lemon

Season alligator with salt and pepper to taste. Melt the 8 tbsp butter in a large sauce pan and cook the meat over a moderate heat for almost 3 minutes on each side. Remove meat from pan and place in a warm oven until serving. Add the artichoke and mushrooms to the sauce pan. Add the wine and cook for 5 minutes over a medium heat. Blend in the stick of butter and add the lemon juice and warm gently. Serve the alligator topped with the artichoke-mushroom mixture.

CREAMED GATOR STEW

2 lbs alligator, cubed
1 c white wine
1 carrot, in chunks
¼ c tabasco sauce
½ c mushrooms, sliced
plain flour
1 14½ oz can chicken broth
1 white onion, minced
½ c celery, chopped (optional)
1 tsp salt
½ c heavy cream

Thoroughly coat alligator with flour and brown slightly. Place in a large casserole dish and cover with the remaining ingredients except mushrooms and cream and cover and cook 1-2 hours in a moderate oven until the meat is tender. Add the mushrooms and cream. Mix well and serve.

GATOR AND CABBAGE STEW

1 lb gator meat, cubed
1 slice salt meat, cubed
1 lge onion, chopped
2 cloves garlic, minced
1 10 oz can chicken broth
1 c water
1 chicken bouillon cube
1 tsp salt
¼ tsp soy sauce
1 sm head cabbage, shredded
3 tbsp plain flour
oil

In a large saucepan heat just enough oil to cover the bottom and add the gator and salt meat and cook on medium heat until lightly browned. Then add the onion, garlic, broth, bouillon cube and soy sauce and bring to a boil and cook for 1 hour on low heat stirring often. Add the cabbage and cook for an additional hour. Combine the flour and water stirring well and add to the stew and stir until thickened and then cook 20 minutes more. Add the salt cautiously as the salt meat may have been enough. Serves 3-5.

ALLIGATOR RED-EYE STEW

4 lbs alligator leg meat with bones
1 4¼ oz can black olives, chopped
1 lge carrot, scraped
1 lge potato, peeled
1 tbsp salt
1 c sherry or red wine
3 lemons, sliced, seeded
fresh or dried parsley
2 scallions, tops, chopped
3 ribs celery, chopped
cooked rice, ½ c/person
1 8 oz can mushrooms, buttons
water to cover
¾ c salad style olives with juice
1 bay leaf
2 tbsp tabasco sauce
¾ c oil
¾ c plain flour
1 14½ oz can stewed tomatoes
2 lge onions, chopped
1 bell pepper, chopped

Boil the alligator in the water with the carrot, potato, bay leaf and 1 tbsp tabasco sauce for 45 minutes. Remove meat and discard remainder. Brown the flour in the oil until darker than caramels and then add the tomatoes, onions, celery, bell pepper, olives mushrooms, salt, remaining tabasco sauce and half the wine and add enough water to make a thin gravy. Add the meat and bring to a boil and then reduce the heat and simmer for 2-4 hours stirring often. Add the lemons, parsley and scallions during the last 30 minutes and adjust seasoning if needed. It should be hot. Remove from heat and add remaining wine. Serve over rice.

SOUTH BEND ALLIGATOR STEW

legs from med alligator or 1 large leg
2 lge bell peppers, chopped
3 scallions, chopped
2 ribs celery with leaves, chopped
3 cloves garlic, halved
¾ c oil
¾ c plain flour
2 lge red onions, chopped
1 14½ oz can stewed tomatoes
water to cover meat
3 bay leaves
2 carrots, scraped
1 tbsp tabasco sauce
parsley
seasonings to taste
water as needed

Disjoint the legs removing any fat and membranes and boil in enough water to cover with 2 bay leaves, carrots and tabasco sauce for 30 minutes. Drain, rinse and discard the water. Mash the carrots. Bone and cube the meat. Brown the flour in the oil and then add the onions and tomatoes and cook until the onions are clear. Add the carrots and all remaining ingredients except parsley and meat using water as needed to keep desired thickness. Simmer for 2-5 hours adding some meat in half hour intervals. Add the last of the meat about 45 minutes before serving. Season to taste and add the parsley just before serving. Serve over rice. Serves about 10.

CIDER STEW A LA ALLIGATOR

1 lb alligator meat, cubed
1 c apple cider
2 lge onions. sliced
3 tbsp oil
1 tbsp cornstarch
¼ tsp thyme
3 lge potatoes, peeled, diced
4 carrots, scraped, cut in quarters
1 can lima beans, drained, save liquid
3 tbsp plain flour
1 tbsp chili sauce or ketchup

Soak the meat in the cider for several hours. Add the flour to the oil and stirring cook over medium heat until browned. Add the onions and saute until soft. Add the meat and cider and bring to a boil and then add the chili sauce or ketchup and cook slowly until the meat is tender. Add remaining vegetables during the last 30 minutes of cooking. Mix the cornstarch and lima bean liquid and add to the stew to thicken. Season to taste after adding the thyme. Can serve alone or with rice. Serves 6-8.

TECHE STYLE MULLIGAN STEW

2-3 lbs alligator meat, diced
water to cover
2 14½ oz cans stewed tomatoes
1 15 oz can creamed corn
1 tsp + dash tabasco sauce
2 sticks butter
salt to taste
3 med potatoes, chopped
additional pepper if needed
4 ribs celery with leaves, chopped
1 carrot, scraped
1 lge onion
1 bay leaf
1 tbsp sugar
1 c bread crumbs or instant potato flakes
3 med onions, chopped

Boil meat in water to cover with the large onion, carrot, bay leaf and dash tabasco sauce for 45 minutes. Drain retaining the meat and discard the liquid and vegetables. Melt the butter in a large kettle and add meat and enough fresh water to cover and then add the potatoes, onions and celery. Slowly cook until the vegetables are tender and then add the tomatoes, corn, seasonings and sugar. Simmer for 15-20 minutes and adjust seasonings if needed. Add the bread crumbs or potato flakes if gravy is too thin. Serves 8.

GATOR IRISH STEW A LA BAYOU

alligator leg meat, cubed
water to cover
¼ tsp dill seed
dash tabasco sauce
3 cubes beef bouillon
parsley flakes
3 c beef broth or water
8 sm onions, peeled, whole
3 carrots, scraped, cut in ½
3 med potatoes, peeled, cubed
½ c half and half cream
2 tbsp plain flour
salt and pepper to taste

Boil the meat in the water to cover for 20 minutes. Drain, rinse and discard the water. Place the meat in a large pot with the dill seed, tabasco sauce and beef cubes along with the water or broth. Bring to a boil and then reduce the heat to a simmer. Cook lightly covered about 30 minutes. Add onions, carrots and potatoes and increase the heat long enough to boil and then simmer until the vegetables are tender. Combine flour and cream well and add a little of the pot liquid to this and then add to the stew stirring constantly until slightly thickened. Season and add the parsley. Serve with hot biscuits. Serves 4-6.

FRANKLIN STYLE GATOR OVEN STEW

2 lbs gator meat, cubed, boiled
  45 minutes in seasoned water
1 10¾ oz can cream of mushroom soup
¼ c water
1 pkg onion soup mix
½-¾ c red wine
dash tabasco sauce
1 4 oz can mushrooms, sliced

Place the drained meat in a heavy greased roasting pan and mix the remaining ingredients and pour over the meat. Cover tightly and bake at 375°F for 1 hour. Reduce the heat and cook for an additional 2 hours. Do not remove the cover or stir while cooking. Remove from oven and let stand for 10 minutes before serving over rice, grits or noodles. Serves 4-6.

ALLIGATOR SOUP

6 slices alligator, diced
6 onions, chopped
6 tomatoes, chopped
1 tbsp parsley, chopped
1 glass wine
2 c oil
water
salt and pepper to taste
cayenne pepper to taste
2 c plain flour

Make a dark roux* of the oil and flour and then saute the onions in the roux. Add tomatoes and other seasonings along with the meat. Cook approximately 1 hour or until meat is tender.

SOUP DU CROCODILE

2 lbs alligator, cubed
2 tbsp oil
1 c brown roux*
2 c onion, chopped
1 c celery, chopped
⅔ c bell peppers, chopped
1 lb can whole tomatoes
4 tbsp parsley, chopped
1 lemon, sliced, seeded
1 tbsp salt
1 tsp garlic powder
1 tsp cayenne pepper
1 tsp black pepper
2 in fresh basil sprigs
2 qts water
dry sherry (optional)

Heat the oil and add the roux, tomatoes, onions, celery, bell peppers, lemon, peppers, salt, garlic and herbs. Stir well and then add the meat and water. Bring to a boil, cover, and simmer 2½ hours or until the meat is tender. Add parsley and simmer a few more minutes. Add sherry if desired when serving.

DOWN THE BAYOU 'GATOR SOUP

3 lbs alligator meat
1 c oil
2 8 oz cans tomato sauce
1 med onion, minced
6 ribs celery, minced
1 tbsp tabasco sauce
1 tbsp basil, dried
½ tsp thyme, powdered
2 tbsp + salt to taste
2 c plain flour
1 16 oz can tomatoes
½ lb fresh mushrooms, sliced
½ bell pepper, minced
4 tbsp butter
1 tbsp worcestershire sauce
2 tbsp parsley flakes
½ tsp white pepper
water

Boil the alligator in water with the 2 tbsp salt until tender. Reserve 1 quart of the broth. Bone and dice alligator. In a skillet saute the vegetables until caramelized. In a large pot make a roux* with the flour and oil until deeply browned. Add the reserved broth gradually and when smooth add the tomatoes, tomato sauce and alligator and stir well. Add all the seasonings and vegetables blending well. Cook approximately 1 hour until thickened and well blended, stirring occasionally. Makes approximately 3 quarts.

BAYOU GATOR SOUP

2 lbs gator meat, cubed
1 c celery, chopped
2 cloves garlic, chopped
1 14½ oz can tomatoes
3 tbsp flour
1 glass sherry
½ lemon, sliced, seeded
salt and pepper to taste
2 qts + 1 c water
2 lge onions, chopped
½ c scallion tops, chopped
4 med bay leaves
1 8 oz can tomato sauce
3 tbsp oil
1 tbsp worcestershire sauce
parsley
½ c bell pepper, chopped

Marinade

½ bottle worcestershire sauce
1 c wine vinegar
6 tbsp dijon mustard

Mix the marinade ingredients and add the cubed alligator meat and marinate overnight in a refrigerator. Drain the alligator and rinse in cool water. Season with salt and pepper and fry in oil until brown. Remove from the oil. Add onions, bell pepper, celery and garlic and cook until tender. Return the gator to the pot and add the tomato sauce, tomatoes and 1 cup water and cook about 30 minutes. Add the lemon, bay leaves and scallion tops and 2 quarts of water and simmer for 1 hour. Add parsley and worcestershire sauce to the soup and also wine to taste. Serves 8-10.

Prudhomme's Alligator Soup

Seasoning mix

- 1 tbsp + ½ tsp salt
- 1¾ tsp cayenne pepper
- 1 tsp onion powder
- 1 tsp sweet basil, dried
- ¾ tsp thyme leaves, dried
- ½ tsp oregano leaves, dried
- 2 tsp sweet paprika
- 1½ tsp garlic powder
- 1 tsp ground cumin
- ¾ tsp white pepper
- ½ tsp black pepper

Soup ingredients

- 1 lb alligator tail meat, cubed
- 1¾ c onions, minced
- ½ c plain flour
- 1 tsp garlic, minced
- 2 tbsp sherry
- ½ lb unsalted butter
- 1 c celery, minced
- 2 c tomatoes, peeled, chopped
- 6 c rich seafood stock
- 2 c cooked rice

Combine all the seasoning mix ingredients in a small bowl and blend thoroughly. Season the meat which has been pounded to ½ inch thick and then cubed with 1 tsp of the mix working it in with your hands. In a 5½-quart saucepan melt 1 stick of butter over high heat until about half-melted. Add the alligator meat and cook until the meat is browned and sediment on pan bottom is continuously building up, about 13 minutes, stirring occasionally and scraping pan bottom clean of browned sediments. Remove from heat and with a slotted spoon transfer the meat and any bits of brown sediment to a bowl and set aside. Return the pan to high heat and add the remaining butter and ½ cup of the onions and cook about 5 minutes scraping all brown sediment free from the pan bottom and then stirring occasionally. Stir in ½ cup of the celery, then gradually the flour, stirring until well blended. Cook about 2 minutes stirring and scraping frequently. Stir in the tomatoes, 2 tbsp of the seasoning mix and the garlic and cook about 4 minutes, stirring constantly, and scrape browned sediment from pan bottom as it develops. Add ½ cup of the stock and scrape until browned sediment on pan bottom is dissolved. Add ½ cup more stock and the remaining seasoning mix and cook about 2 minutes, stirring and scraping frequently. Now add the remaining stock and stir and scrape well. Add the meat and sherry and bring mixture to a boil, stirring and scraping frequently. Reduce heat and simmer about 15 minutes, stirring and scraping occasionally. Add the remaining onions and celery and continue simmering until meat is tender and flavors marry, about 30 minutes. Remove from heat and serve immediately in bowls. Serve 1½ cups soup over ½ cup rice for a main dish; for an appetizer serve 1 cup of the soup without rice.

RICH SEAFOOD STOCK

2 quarts cold water
1½-2 lbs rinsed shrimp heads and/ or shells, or crawfish heads and/or shells, or crab shells, or rinsed fish carcasses (heads and gills removed), or any combination of these. You can substitute oyster liquor for all or part of the seafood stock called for in a recipe.
vegetable trimmings or 1 med onion, 1 clove garlic, 1 rib celery

Combine all the ingredients being used with the water and bring to a boil and then reduce heat and simmer for 30 minutes. Strain and discard the solid material. Now continue simmering this strained stock until evaporation reduces the liquid by half or more.

ALLIGATOR SOUP A LA LOUISIANA

10 lbs alligator meat, cubed
2 sticks butter
8 c onions, minced
14 c tomato sauce
8 qts water
3 c parsley, lightly packed
24 hard boiled eggs
2 sticks unsalted butter
2 lbs spinach, finely chopped
4 c celery, minced
4 tsp garlic, minced
3 c plain flour
1 lemon, seeded, sliced
1½ c sherry

Seasoning mix

3 tsp garlic powder
1 tsp black pepper
1 tsp white pepper
10 bay leaves
1 tsp basil, dried
3 tsp salt or to taste
3 tsp cumin
1 tsp lemon pepper seasoning
3 tsp thyme
1 tsp cayenne pepper

Melt the unsalted butter in a pot and add the alligator meat and brown 6-8 minutes. Stir in the seasoning mix, spinach, onions and celery and cook for 15 minutes. Stir in the tomato sauce and cook for 10 minutes. Stir frequently. Add flour and garlic stirring well and cook for 5 minutes stirring constantly. Add 2 quarts water and stir and scrape bottom of pan. Stir in remaining water and bring to a boil. Boil 5 minutes and lower heat and bring to a simmer and cook 1 hour. Chop parsley and eggs and add to soup and cook 20 minutes more. Add sherry and remove from fire. Serve.

ALLIGATOR SOUP WITH CATFISH

- 1 lb alligator fillets, cubed
- 1 c onions, minced
- 4 c water
- ½ c potatoes, peeled, cubed
- ½ c Ro-Tel Tomatoes
- 1 tsp granulated garlic
- ½ tsp cayenne pepper
- 1 16 oz can cream-style corn
- 2 sticks butter
- 1 c bell pepper, minced
- ⅓ lb catfish fillets, cubed
- 2 c tomato sauce
- 1 tsp salt
- ½ tsp black pepper
- 1 c sliced carrots

In a large pot saute onions and bell peppers in melted butter. When vegetables are clear add all other ingredients except corn and let simmer for 2 hours or more. Remove from the heat and add the corn and stir. Serves 6-8.

ANTOINE'S ALLIGATOR SOUP

- 2¼ lbs alligator meat, diced
- 4 bu scallions, minced
- ½ oz parsley, minced
- 1 lemon, sliced, seeded
- thyme to taste
- 1 oz worcestershire sauce
- ½ gal beef stock (optional)
- caramel coloring
- 1 lge onion, minced
- 4-5 cloves garlic, minced
- 2½ ozs butter, melted
- 3 bay leaves
- 6 ozs sherry
- 16 ozs tomato juice
- salt and pepper to taste
- 3-4 ozs brown roux*

Fry the alligator in butter and set aside. Saute all the vegetables in the pan drippings. Add the roux, beef stock and caramel coloring and season to taste with salt and pepper and add worcestershire sauce, tomato juice, sherry, lemon and bay leaves. Simmer for 2 hours or until the meat is tender. Makes approximately 1 gallon.

SIMPLE ALLIGATOR SOUP

2½ lbs alligator meat, diced
1 carrot, diced
⅜ c oil
4 lge onions, chopped
½ bu celery, chopped
1½ 8 oz cans tomato sauce
1 bay leaf
1 sprig basil
1¼ lemons, sliced thin, seeded
salt and pepper to taste
¼ c parsley, chopped
5 qts water
½ stick butter
¾ c plain flour
¾ bell pepper, chopped
1 can whole tomatoes
1 6 oz can tomato paste
1 sprig thyme
1 sprig chervil
¼ c soy sauce
1 bu scallions, chopped

Boil together the alligator meat and carrot in the water for 1 hour. Set aside. Make a brown roux* with the butter, oil and flour. Add the onions and brown. Add all remaining ingredients except scallions and parsley to the roux mixture and then add to the alligator and cook for 2 hours. Add scallions and parsley and cook for 30-45 minutes.

CABBAGE AND GATOR SOUP

2 c alligator meat, cubed
2 14½ oz cans beef broth
4 broth cans water
½ tsp + dash tabasco sauce
2 lge potatoes, peeled, diced
cooked, crumbled bacon
3 med carrots, scraped, sliced
2 lge onions, chopped
3 cloves garlic, minced
1 sm head cabbage, shredded
salt to taste

Boil the meat in the water, broth and dash tabasco sauce until tender and then drain and reserve the stock. To the stock add the potatoes, carrots, onions and garlic and simmer for 45 minutes. Add the cabbage except 2 cups to be added later and cook for 30 minutes. Remove from the heat and allow the soup to cool. When cooled add a little at a time to a blender and blend until smooth. When blended return to the pot and bring to a boil and then add the meat and simmer stirring often for 45 minutes. Add the remaining cabbage and cook until tender. Add the salt and remaining tabasco sauce and serve with the bacon sprinkled on top. Serves 6.

GATOR AND SHELL SOUP

1 lb alligator meat, cubed
2 c water or to cover meat
2-3 c additional water
1 lge onion, chopped
1 c raw macaroni shells
1 soup can milk
2 carrots, sliced
1 tsp salt
½ tsp tabasco sauce
pinch of cloves
1 11 oz can bean and bacon soup

Boil the meat in the 2 cups of water or enough to completely cover for about 30 minutes. Drain and rinse discarding liquid. Return the meat to the kettle with the additional water, onion, carrots, salt, tabasco sauce and cloves and bring to a boil. Simmer over medium heat for 30-45 minutes and then add the macaroni and simmer another 20 minutes. Add the soup and milk mixing very well. Adjust seasonings if necessary. Serve with soda crackers or french bread. Serves 6-8.

BEER AND NOODLE ALLIGATOR SOUP

2 lbs alligator meat, cubed
1 c water
1 12 oz can beer
1 lge onion, sliced
1 12 oz pkg egg noodles
4 tsp cornstarch
1 lge apple, peeled, cubed
2 beef bouillon cubes
pinch of thyme
1 clove garlic, chopped
2 carrots, sliced
½ tsp poppy seeds
1 bay leaf, crushed
salt and pepper to taste

Combine the meat, water, beer and onion and let marinate in the refrigerator for several hours. Place meat mixture in a large dutch oven and bring to a boil and then simmer for 1 hour skimming any foam that forms. Add remaining ingredients except salt and pepper and simmer for 2-3 hours and then season. Serves 4-6.

TEXICAN ALLIGATOR SOUP

1 lb alligator meat, cubed
½ c butter
1 lge onion, chopped
1 sm bell pepper, chopped
2 cloves garlic, minced
2 c water
1 14½ oz can chicken broth
2 tbsp tomato paste
1 tsp seasoned salt
1 tsp chili powder
½ tsp cumin powder
2 bay leaves
dash tabasco sauce
parsley, chopped, optional

Saute the meat in the butter and then add the onion, bell pepper and garlic and saute until vegetables are tender. Do not brown. Add the water, broth, paste and all seasonings except parsley and bring to a boil and then simmer for 1 hour, uncovered, or until the meat is tender and soup slightly thickened. Serve over rice and sprinkle with parsley if using. Serves 6.

CREAMY ALLIGATOR SOUP

2 lbs gator meat, boiled, ground
1 med onion, chopped
½ stick butter
1 can cream of shrimp soup
½ c plain flour
3 hard boiled eggs, grated
1 tsp lemon pepper seasoning
1 tsp paprika
½ tsp salt
1 qt half and half cream
¼ c dry sherry
parsley or scallion tops, chopped

Saute the onion in butter in a large kettle until soft but not brown. Add soup, flour, lemon pepper, paprika and salt stirring constantly. When almost boiling add the cream very slowly stirring constantly. Reduce the heat to a simmer and cook for 5 minutes. Stir in the wine, meat and eggs and simmer until heated through. To serve sprinkle with parsley or scallions. Serves 6.

CORNY GATOR SOUP

1 lb gator meat, cubed
2 tsp butter
2 tsp plain flour
1 sm onion, minced
parsley or scallion tops, chopped, optional
½ tsp salt
½ tsp cayenne pepper
1 c milk
1 8 oz can whole kernel corn

In a dutch oven melt the butter and add the meat, cover, and cook for 15 minutes, stirring often. If liquid forms saute without cover until liquid evaporates. Mix the flour with the milk and pour over the meat. Add all remaining ingredients except parsley or scallions and stirring constantly bring to a boil. Reduce the heat and stir until just simmering. Cover and cook until the meat is flaking. Adjust seasonings if necessary. Serve with the parsley or scallions sprinkled on top. Serves 4.

CHEESY GATOR SOUP

1½ lbs alligator meat, diced
5 c water
½ c carrots, minced
½ c celery, minced
1 c scallions, minced
1 8 oz jar Cheese Whiz
1 med onion, chopped
1 tbsp prepared yellow mustard
½ c butter
1 c plain flour
4 c milk
¼ c sherry
½ tsp salt
1 tsp black pepper
¼ tsp cayenne pepper

For 45 minutes boil the meat, carrots, celery, scallions in the water and then set aside until the sauce is made. Saute the onion in the butter and then add the flour blending well. Boil the milk and then stir briskly into the onion mixture with a wire whisk. Add the cheese, salt and peppers stirring well. Add the mustard and then stir this mixture into the meat mixture blending well. Bring to a boil and add the wine and cook for 10 minutes.

GATOR SOUP AU SHERRY

- 3-4 lbs gator meat
- 2 bay leaves
- 2 tbsp salt
- 3 lge onions, chopped
- 1 6 oz can tomato paste
- 1 c sherry
- 3 hard boiled eggs, chopped
- 1 lemon, sliced thin, seeded
- 3 qts water
- 1 tsp cayenne pepper
- 2 sticks butter
- 1 c plain flour
- 3 qts stock from meat
- ½ c worcestershire sauce
- ¼ c parsley flakes

Boil the meat in the water with the bay leaves, cayenne pepper and salt until tender. Drain the meat and dice reserving the liquid and add enough water to the liquid to make 3 quarts of stock. In a large kettle melt the butter and saute the onions until soft. Stir in the flour and stir until browned. Add the paste and cook for 5 minutes. Add the stock and sherry to the mixture and cook slowly for 30 minutes. Add the meat and simmer 30 minutes longer. Add the egg and remove from the heat and stir in the parsley and lemon. Let stand on a very low heat until served. Makes about 3 quarts.

POTAGE DE CROCODILE A LA TECHE

- 3 lbs alligator meat, diced
- 4 ozs olive oil
- 3 ozs plain flour
- 2 bell peppers, minced
- 3 lge onions, minced
- 4 ribs celery, minced
- 4 cloves garlic, minced
- 1 10 oz can Ro-Tel Tomatoes
- 1 8 oz can tomato sauce
- 2 ozs rice vinegar
- 3 bay leaves
- 3 pts water
- 1 pt chablis
- 4 scallions, minced
- 1 tsp chervil
- 2 tbsp worcestershire sauce
- 2 tbsp cayenne pepper
- 4 tbsp salt
- 2 tbsp white pepper
- 1 lge lemon, seeded, cut in sm pieces

Make a dark roux* with the olive oil and flour and then add the bell peppers, onions, celery and garlic and cook until wilted. Add all other ingredients except the meat and slowly cook for 45 minutes. Saute the meat in a little oil for 15 minutes and then add to the soup and simmer 45 minutes longer.

ALLIGATOR GUMBO

2 lb alligator meat, diced
1 tsp garlic powder
1 tsp cayenne pepper
1 sm bell pepper, chopped
1 20 oz can whole tomatoes
½ lemon, sliced thin, seeded
4 tbsp parsley, chopped
½ c sherry
2 tbsp oil
2 tbsp plain flour
1 c onions, chopped
2 qts water
dash tabasco sauce
dash worcestershire sauce
salt to taste

Lightly brown the meat in the oil and then remove and add the flour and brown. Add the onions and saute until limp and then add the tomatoes and garlic powder while cooking. Return the meat to the pot and add the water, seasonings and lemon. Simmer and remove the lemon slices. Add the sherry when serving. Serves 6.

TURTLES

Preparation—In preparing a live turtle for the pot there are certain items you should have before you start: a good sharp knife (I use a skinning knife), preferably a short and long one, a stout hook of some sort, a hacksaw and/or hand axe, a large pot of boiling water, a container for the meat, an outside table, a hose with running water and towels. I use one of two methods of killing the turtle depending on size. With a large turtle such as the Alligator Snapping Turtle (*Macroclemys temminckii*) I use the hook to pull out the head and cut it off. Smaller turtles are dropped in the boiling water and as soon as dead removed. Then I cut off the head and, in either case, hang them up with the head end down to bleed. The boiling water does two things. It kills the turtle quickly and removes the old epidermis and shields* of the carapace* and plastron* easily. Also, the toe nails snap off. In the larger turtles that you may not have a pot big enough for, the water can be used for the same purpose when processing. After the turtle has drained, lay it on its back (carapace) and cut all around the bottom shell (plastron) evenly on the same plane. Turtles like the Alligator Snapping Turtle (*Macroclemys*), Sliders (*Trachemys*), River Cooters (*Pseudemys*), and Map Turtles (*Graptemys*) have the plastron connected to the carapace by a bony bridge. You need to saw through this bridge or use a hand axe. Turtles like the Snapping Turtle (*Chelydra*) have cartilage here and can be cut through with the knife at the proper seam. After severing the bridge, begin lifting up the plastron, cutting attached muscles as close as possible and remove. After the plastron has been removed start removing the carapace by cutting along the shell and the meat going completely around the turtle. The leg quarters can then be removed separately or left whole. Where the vertebral column attaches to the carapace at the head end and tail end may cause some problems in severing. The axe can be used here or with some effort of twisting and cutting these can be broken. Then the meat can be lifted out. Remove and retain any shelled eggs (if any), egg strings and liver, being careful not to break the gall bladder. Remove the gall bladder. There is a strip of meat along the vertebral column that can be either cut out with some difficulty or left in the shell to cook. The boiling water should have removed any of the old epidermis and left a nice clean skin. The carapace and plastron after having the shields removed with the boiling water can be chopped up and boiled to make a turtle stock. The meat along the back can then be removed and used. Cut off the toes and then cut and disjoint. Keep the skin and cook. In my estimation this is one of the best parts of the turtle. Commercial sources do not sell the skin. Some say do not use the fat but I use it. It is delicious in small quantities. If using soft shell turtles (*Apalone*), process the same way. The cartilagenous part of the carapace can be cut up and cooked along with the meat.

Elsewhere I have given recipes for prepared terrapin*. I prepare mine the same as given above. Commercially, in the United States, turtles of the genera *Macroclemys* (Alligator Snapping Turtle), *Chelydra* (Snapping Turtle), *Apalone* (Softshell Turtles) and *Malaclemys* (Diamondback Terrapin) are the most commonly used. *Trachemys* (Sliders), *Pseudemys* (River Cooters), and *Graptemys* (Map Turtles) are also eaten but to a lesser extent because the meat contents are low. Gopher Tortoises (*Gopherus polyphemus*) are also eaten in the United States, and in Mexico, Bolson Tortoises (*Gopherus flavomarginatus*). Sea turtles are commonly eaten everywhere except the Leatherback Seaturtle (*Dermochelys coriacea*) since it is not edible.

With the turtle heads, I make skulls. Sometimes the carapace and plastron are skeletonized for research specimens. I usually incubate the shelled eggs and release the young in the area the parent came from, when known. If the turtle was left in the boiling water too long, incubation then is not feasible. Then I eat the eggs.

TURTLE FINS WITH SAUCE MADERE

4 sea turtle fins
2 ozs butter

Brush the fins on all sides with butter, grill over moderate heat, turn after 10 minutes and continue cooking. Serve on a dish with sauce madere poured over them.

SAUCE MADERE

2 ozs butter
1 wine glass madeira
1 tsp chervil, chopped
salt and pepper to taste
1 wine glass white wine
1 tbsp tarragon, chopped
4 ozs onion, chopped
4 ozs mushrooms, chopped
1½ ozs plain flour
1 tbsp parsley, chopped

Melt the butter in a saucepan and add the onion and sauté until golden brown. Stir in the flour and continue cooking and stirring until the mixture is golden brown. Gently add the white wine a little at a time. Season with salt and pepper and continue to cook for 15 minutes. Add the chopped herbs and mushrooms. Five minutes before serving, add the madeira and allow the sauce to simmer and reduce a little. Pour the sauce over the turtle fins and serve.

TURTLE FINS

4 sea turtle fins
pinch of fennel
pinch of tarragon
salt and pepper to taste
plain flour
white wine
butter

Simmer turtle fins in boiling water until tender and then skinned. Dip the fins in seasoned flour and brown them in butter. When they are nicely browned add a little white wine, tarragon, and fennel. Simmer until tender.

TURTLE FINS A L'AMERICAINE

4 sea turtle fins
plain flour
salt and pepper to taste
¼ c white wine
butter

Simmer turtle fins in boiling water and wine until tender and then skinned. Dip in seasoned flour and brown in butter. When brown add sauce a l'americaine. Simmer until tender.

SAUCE A L'AMERICAINE

3 tbsp butter
1 clove garlic, chopped
1 tbsp tarragon, chopped
freshly ground black pepper
1½ c white wine
5 ripe tomatoes, peeled, seeded, chopped
3 tbsp parsley, chopped
1½ tsp thyme
3 tbsp tomato paste
6 scallions, chopped
salt
1 sm onion, chopped

Melt the butter and sauté the onions for a few minutes. Add the scallions, tomatoes, garlic and herbs and simmer for 1 hour. Season to taste and let cook down and blend thoroughly. Add the tomato paste. Blend with white wine.

TURTLE FINS MORNAY

4 sea turtle fins
plain flour
butter
¼ c white wine
parmesan cheese, grated
salt and pepper to taste

Simmer turtle fins in boiling water until tender and then skin. Dip the fins in seasoned flour and brown in butter. When they are browned, add white wine and simmer until nearly done. At the last, add a little sauce mornay*, sprinkle with grated cheese, and run under the broiler to brown.

FLIPPER TURTLE SOUP

2 lbs turtle flippers
1 14 oz can tomatoes, chopped
½ c barley
2 tbsp worcestershire sauce
2 qts water
4 lge potatoes, chopped
2 lge onions, chopped

Parboil the flippers and then remove the skin and dice into small pieces and then add to the water along with the vegetables and barley and cook slowly for 1½ hours. Season with the worcestershire sauce just before serving.

TURTLE EGGS IN A PUFF

12 turtle eggs
½ c oil
2 chicken egg whites, beaten stiff
plain flour

Mash the boiled yolks of the turtle eggs into a paste. Add egg whites. Roll into a ball and sprinkle with flour and fry in deep oil for 5 minutes. Serve hot.

TURTLE EGGS

shelled turtle eggs
butter
pepper
salt

Boil in salted water. The yolk will harden but the white will not, no matter how long boiled. Cut or tear off top of shell, holding egg in left hand while tearing or cutting off top with right hand. Add salt and pepper and a dab of butter to contents and pop yolk into mouth.

MOCK TURTLE EGGS

6 hard boiled egg yolks
1 tbsp parsley, minced
sherry
½ egg, slightly beaten
salt and pepper to taste
plain flour, seasoned*

Mash the yolks and combine with the ½ egg or less and add enough sherry to moisten into a paste. Add a little flour if needed for consistency. Form into small balls and dust lightly with the flour and bake at 275°F for 15 minutes. (Can be added carefully to a turtle sauce Piquant just before serving.)

MOCK TURTLE EGGS 2

hard boiled eggs
butter
boiling water
1 egg, beaten
cayenne pepper

Mash the yolks with a little butter and add a beaten egg with a little cayenne pepper and form into egg-size balls. Drop them briefly for 2 minutes into boiling water. (They also can be rolled in flour and sautéed in butter.)

TURTLE STEAK HOLSTEIN

1 lb turtle steak, ground
salt and pepper to taste
lettuce (for garnish)
6 eggs
butter
tomato (for garnish)

Combine the ground meat with 2 eggs and salt and pepper to taste. Form into 4 balls and flatten slightly. Fry in hot butter until lightly browned on the outside. Serve with a fried egg on the top and garnish with the lettuce and tomato. Serves 4.

TURTLE ROULADE

1 lb turtle steaks
1 onion
salt and pepper to taste
butter
2 c water
1 bay leaf
plain flour
4 slices bacon
1 spiced pickle
paprika
1 c red wine
2 cloves, whole
thyme

Cut the steaks into 8 equal slices across the grain and pound the meat slightly. Lay ½ slice bacon on each piece. Slice the pickle in 8 wedges and place a piece on the upper end of each steak. Slice the onion in half and lay a slice from 1 half on each piece of pickle. Then roll everything into tight rolls and fasten with a toothpick. Season with salt, pepper and paprika and then roll in flour and brown in hot butter. Chop the remaining half onion and add to the meat along with the cloves and bay leaf and a little thyme. After the meat and onions are well browned add the wine and water. Cover and simmer for 1 hour and let the sauce reduce to a thick gravy. Serve hot. Serves 4.

OYSTERS HOLIDAY

1 lb bacon, diced
½ c parsley, minced
2 tbsp garlic, minced
3 bell peppers, minced
worcestershire sauce
3 doz lge oysters, in half shell
1 c turtle soup au sherry*, strained
1 no 3 can pimentos
3 bu shallots, minced
tabasco sauce
salt and pepper to taste
rock salt

Fry the bacon but do not drain. Add the garlic, shallots and bell peppers to the bacon and sauté for a few minutes. Add the soup, parsley and pimentos and season to taste with the tabasco and worcestershire sauces, salt and pepper. Bake the oysters with 6 to a serving on a base of rock salt until the edges begin to curl. Remove the platters of oysters from the oven and cover each with the sauce and serve immediately. Serves 6.

ROCKY MOUNTAIN SHISH KEBAB

2-3 lbs turtle meat, cubed
3-4 bell peppers, in wedges
3-4 tomatoes, in chunks
1 sm bottle Teriyaki sauce
2-3 med onions, in wedges
½ pineapple, peeled, in chunks
4-5 potatoes, parboiled
salt and pepper

Marinate the turtle in the Teriyaki sauce for 2-3 hours. Drain and salt and pepper the turtle lightly. Place on skewers any combination of the vegetables and fruit desired and refrigerate until the coals are ready. Roast slowly, turning often, until the meat is thoroughly cooked. Serves 5-6.

TURTLE KEBAB

turtle steak, cubed
olive oil
sherry peppers
bell peppers, cubed
rum
rice, cooked
red wine
1 clove garlic, crushed
onions, cubed
tomatoes, cubed
seasonings

Marinate the turtle in well-seasoned wine, olive oil, garlic and a few sherry peppers. Place the meat on a skewer alternately with blanched onion cubes, bell peppers and tomatoes. Grill for about 10 minutes turning often to cook all sides and serve flambéed with rum on a bed of hot rice.

TURTLE DRUMSTICKS

turtle meat, cubed
plain flour
oil
veal meat, cubed
1 egg, beaten
water

Place alternate cubes of meat on skewers. Combine flour and egg to make a batter and dip the skewers in. Drop in deep hot oil and brown rapidly. Pour off the oil and add about ½ inch of water to the frying pan and roast in the oven at 350°F until tender. Serve hot from a covered dish.

TORTOISE IN JELLY

tortoise* meat, braised
seaweed
hard boiled eggs, sliced
gherkins, pickled,
young carrots (for garnish)
bouillon
salt and pepper
lettuce leaves
young radishes (for garnish)
sour sauce (your choice)

Make a rich jelly of the seaweed boiled in a strong bouillon. Grease a mold and then put a coating of the seaweed jelly in it and then arrange the sliced eggs in the coating when it is almost stiff. Place the braised meat seasoned with salt and pepper in it along with the gherkins sliced lengthwise. Pour the remainder of the jelly into the mold and place in a cool place until firm. To serve, turn it out on a bed of lettuce leaves and garnish with the radishes and carrots. A sour sauce of your choice is served with it.

TURTLE CURRY

1 lb turtle meat, cubed
1 med potato, diced
1 sprig parsley, chopped
1 tsp salt
butter
2 med onions, diced
1 carrot, diced
½ tsp black pepper
½ tsp curry powder
boiled rice

Brown meat in butter with the onions and then place into the pot the remaining ingredients and let simmer until meat is tender. Make molds by hollowing out cups of boiled rice and serve in the molds. It tastes like chicken or veal curry.

CREAM-STYLE TURTLE

2 lbs turtle meat, cubed
1 med onion, chopped
3 tbsp plain flour
6-8 allspice
black pepper to taste
3 tbsp butter, melted
1 med bell pepper, chopped
½ c milk
salt

Cook the turtle in salted water, drain, and set aside. Sauté the onions and bell pepper in the butter until soft. Add the allspice and turtle and season to your taste. Simmer. Combine the milk and flour and mix until smooth and then add to turtle mixture and stir. Add enough milk to make a thick cream gravy. When thickened remove from the heat. Serve warm on plain or toasted bread.

DELICIOUS TURTLE

2¼ lbs snapping turtle meat, cut in 1 in pieces
1½ tsp paprika
¼ c + 2 tbsp butter
1 12 oz can evaporated milk
1 10 oz can cream of chicken soup
2 tbsp snipped fresh parsley
1 c all-purpose flour
¾ tsp salt
¼ tsp black pepper
1 10 oz can cream of mushroom soup
1 c water

Heat oven to 325°F. In a medium bowl combine the flour, paprika, salt and pepper. Dredge the turtle meat in the flour mixture to coat. In a 12 inch skillet melt the butter over a medium-low heat. Add the turtle pieces and cook for 3-4 minutes or until browned. Remove. Spray a 3 qt casserole with a nonstick vegetable cooking spray. Spoon the meat into this casserole and set aside. In a medium mixing bowl combine the milk, soups, water and parsley. Pour over the turtle. Stir to coat and bake covered for 2-3 hours or until turtle is tender, stirring once or twice. Serves 6-8.

TURTLE A LA KING

2 c turtle meat, cooked, chopped
2 tbsp butter
dash salt, pepper, nutmeg
6 hard boiled egg yolks
2 c cream, scalded

Mash the yolks through a sieve and cream them with the butter. Add the cream seasonings and turtle, mixing well and cover the pot and simmer until well heated, about 5-10 minutes. Serve garnished to taste.

TURTLE SCALLOPINI

½ lb turtle meat
white wine
garlic, minced
salt and pepper to taste
onions, diced
red and green bell peppers, diced
plain flour
butter

Slice the turtle meat very thin and salt and pepper each side and dredge in flour. Sauté the meat in the butter with the bell peppers, onions and garlic. When ready to serve add the wine. Serves 2.

TURTLE CACCIATORE

1 lb turtle meat, bite size pieces
½ tsp black pepper, freshly ground
4 tbsp margarine
½ lb chicken livers, chopped
½ tsp garlic salt
1 tsp dried rosemary
¼ c olive oil
1 tsp parsley, chopped
2 ozs salt pork, diced
½ tsp salt
4 tomatoes, chopped
1 tbsp tomato paste
hot cooked spaghetti (optional)

Heat olive oil, salt pork and margarine together. Add turtle and chicken livers and cook 10 minutes until browned. Add remaining ingredients (except spaghetti) and simmer 30 minutes until turtle is tender. Serve over the spaghetti.

CREAMED TORTOISE

1 tortoise*
ginger, powdered
1 blade mace
1 c cream, boiling, thickened
white wine
white pepper
salt
bread crumbs moistened with milk (optional)

Dress the tortoise carefully and cut meat into small pieces removing all the gristle. Then simmer in the wine with a little pepper, ginger and the mace. When tender add a little salt and the boiling, slightly thickened cream just before serving. A handful of white crumbs moistened with milk can be added if desired.

TURTLE WITH CREAM SAUCE

4 turtle steaks
plain flour
salt and pepper to taste
4 tbsp water
1 tbsp capers
½ tsp paprika
1 tsp prepared mustard
2 ozs butter
¼ pt cream
juice of 1 lime

Beat the steaks and season with salt and pepper and dredge in the flour. Fry in the butter until lightly browned. Remove and keep hot. Chop the capers and then pour the water into the pan together with the paprika, capers and other seasonings and simmer. When hot remove from the fire and gradually stir in the cream. Return to the fire and add the lime juice. When hot pour over the turtle and serve immediately. Serves 4.

GREEN TURTLE STEAKS

2 lbs turtle steaks
sliced lemon
currant jelly sauce or sauce poivrade
2 tbsp butter
salt and pepper to taste
parsley

Select the female turtle, if possible, as the meat is best. Turtle meat is very irregular, therefore cut the meat into thick slices or steaks, about the size of a fillet of beef, and batter down with the hands to make smooth and regular. Then fry in butter. Season with salt and pepper and garnish with parsley and lemon slices and serve with currant jelly sauce or the delightful sauce poivrade.

CURRANT JELLY SAUCE

½ tumbler currant jelly
4 tbsp butter
4 oz water
4 ozs port or madeira wine
salt
sugar

Melt the butter and add the jelly, blending well, and then add the wine and water. Add a little salt and sugar to taste. The sauce is much finer when made of wine without water, but this is a question of taste. If the wine only is used double the amount.

SAUCE POIVRADE

1 carrot, minced
1 wine glass of madeira
dash of cayenne pepper
1 bay leaf
1 tbsp plain flour
2 sprigs, parsley, minced
1 small rib celery, minced
2 sprigs, thyme, minced
½ lemon, grated
½ pint consomme
1 tbsp butter
1 onion, minced
salt and pepper to taste

Put the butter in the saucepan and as it melts, add the flour. Let it brown slowly and then add the consomme. Let it simmer, add the minced herbs and vegetables and the zest of half a grated lemon. Let all simmer slowly for 1½ hours. Add the wine and season with salt, pepper and a dash of cayenne. Let it simmer for 10 minutes longer. Take off the stove and strain. Serve.

TURTLE STEAK GRAND CAYMAN

6 turtle steaks
seasoned bread crumbs*
¼ + c butter
1 c Bordeaux
dash nutmeg
½ c sherry
watercress (for garnish)
vinegar
eggs, beaten
1 tbsp scallions, chopped
salt and pepper to taste
½ c rich beef stock
1 c mushrooms, sliced thin

Rub the steaks with a damp cloth dipped in vinegar. Dredge each steak in seasoned bread crumbs, then eggs, and again in bread crumbs. Heat the ¼ c butter and stir in the scallions. Cook the steaks in this to a delicate brown on both sides. Pour the Bordeaux over the steaks and season with salt, pepper and nutmeg. Cover the pan and simmer gently 15-20 minutes. Remove steaks and keep on a hot platter. Reduce the sauce over a hot fire to almost nothing. Stir in the beef stock, sherry and mushrooms which have been cooked in a little butter. Taste for seasonings. Pour a little sauce over each steak and serve remainder in a sauceboat. Garnish with watercress. Serves 6.

CAYMAN TURTLE STEAK MIRZA

2 lbs turtle steaks
½ lb white grapes
½ lb red grapes
2 bananas, peeled
4 pear halves
½ lb butter
½ c plain flour
salt
4 peach halves
parsley (for garnish)

Cut turtle steaks into thin escalopes and season with salt and flour lightly. Melt part of the butter in a frying pan and sauté the steaks. In another frying pan melt the remaining butter and add the grapes, with the bananas cut in half and sauté. In a small saucepan warm the peaches and pears. When the steaks are cooked add the bananas and grapes over them, and garnish with the parsley. Serve on a hot plate with the peaches and pears. Serves 4.

TURTLE STEAKS 1

- 1 lb turtle steaks, ½ in thick
- ¼ c sherry
- ½ c plain flour
- 2 oz butter
- ¼ c olive oil
- 1 tsp salt
- ½ tsp granulated garlic

Heat butter and oil in skillet with garlic. Salt turtle steaks and roll in flour. Brown on both sides about 10-15 minutes. Pour off drippings. Add sherry and steam for 5 minutes. Serves 2.

TURTLE STEAKS 2

- 1½ lbs turtle steaks
- salt and pepper to taste
- ½–⅓ c sherry
- 2 eggs, beaten
- plain flour
- 2 c plain bread crumbs

Wipe turtle meat well with a moist cloth. Flour thoroughly and dip in beaten eggs and then in seasoned bread crumbs. Quickly brown in butter on both sides. Reduce heat and sauté about 15 minutes or until meat is tender, turning occasionally. Add ½ cup sherry, cover, and simmer until wine is reduced by half. Pour remaining wine over turtle meat and serve. Serves 6.

TURTLE STEAKS 3

- 2¼ lbs turtle steaks
- ¼ tsp nutmeg
- 1½ c plain flour
- 4 tsp lemon juice
- ½ c milk
- 4 ozs butter
- ½ tsp black pepper
- 1 tbsp salt
- 2 eggs
- ½ c olive oil

Pound steaks until about ½ inch thick. Beat the eggs and milk together. Combine the flour, salt, pepper and nutmeg and mix thoroughly. Dip the steaks in the egg mixture, then roll in the seasoned flour and coat evenly. Place on a platter to dry about 5 minutes before sautéing. In a heavy skillet add the olive oil and butter and heat until fairly hot. Fry the steaks until golden brown and crisp, about 5 minutes on each side. Remove to an oven platter. Heat oven to 425°F and place the platter in the oven 12-15 minutes. Remove from oven and serve on heated plates and sprinkle each steak with 2 tsp lemon juice. Serve immediately.

GREEN TURTLE STEAKS 2

- 1½-2 lbs green turtle steaks
- oil
- lime juice
- ¾-1 c bouillon or water
- salt and pepper to taste
- plain flour

Cut meat in serving sizes and cubed or beat with a meat mallet. Brush with lime juice on both sides and set aside for 3-4 hours in a refrigerator. Roll each steak in well-seasoned flour* and brown on both sides in oil. Place in a casserole and add the bouillon or water. Cover and cook in a moderate oven for 40 minutes. As an alternative finish the steaks in the skillet by adding the bouillon or water and cover tightly and simmer for about 25 minutes or until very tender. Two pounds serves 4.

TURTLE PARMESAN

4 lge turtle steaks, ½ inch thick and pounded
4 tbsp parmesan cheese, grated
¼ tsp oregano
¼ c olive oil
2 c tomato sauce
1 c bread crumb
2 eggs, beaten
¼ tsp salt
⅛ tsp black pepper
4 slices mozzarella cheese

Add the salt and pepper to the eggs. Mix well the bread crumbs, parmesan cheese and oregano. Dip the steaks into the eggs and then roll in the bread crumbs. Fry in olive oil until golden brown on both sides. Remove browned steaks from pan and place them into a greased baking dish. Pour the tomato sauce over them. Place a thin slice of mozzarella cheese on top of each. Bake in a 375°F oven for 10 minutes. Serves 4.

TURTLE STEAKS A LA LOBSTER POT

4 turtle steaks (6 oz)
3 ozs mushrooms, chopped
2 ozs onions, chopped
1 bell pepper, sliced
½ pt rich brown gravy
1 tsp Pickappeppa sauce
2 ozs butter
1 tbsp brandy

Heat butter in a frying pan and sear steaks on both sides and remove. Brown the onions and add the gravy and other ingredients except brandy and heat gently for 5 minutes. Return the steaks to the pan and simmer gently for an additional 5 minutes. Pour the brandy over them and flambé. Serve at once on hot plates with your choice of vegetables. Serves 4.

BAKED TURTLE

turtle meat
plain flour
water
salt and pepper to taste
1 cube butter

Dip the turtle in seasoned flour* and put in a baking dish and cover with water. Melt the cube of butter and put over the turtle and bake in a 350°F oven for 3-4 hours.

ROASTED TURTLE

turtle meat
2 tbsp salt
3 tbsp shortening
1 sm onion, chopped
½ c vinegar
1 qt water
black pepper to taste
½ tsp caraway seeds

Marinate the turtle for 2 hours with the salt, vinegar and water. Drain well and put in a roaster. Add the shortening, pepper, onion and caraway seeds and roast in a 375°F oven for about 2½ hours or until meat is tender and brown.

CROCK POT TURTLE

1 snapping turtle, dressed
water
oil
black pepper
salt
plain flour
1 10¾ oz can golden mushroom soup

Soak the meat in salted water overnight. Drain, salt and pepper and then dredge in flour and fry in a cast iron skillet or dutch oven and brown on all sides. Put in a crock pot and add the soup and cook on low for 5-6 hours until tender.

BRAISED TURTLE STEAKS

4 turtle steaks, 4 in long, 1 in thick
4 red hot pickled peppers
¼ tsp cloves, ground
salt and pepper to taste
steamed rice
1½ c hot water
4 tbsp lard
1 lb can tomatoes, chopped
¼ tsp thyme
2 tbsp flour
1 c + water

Melt 2 tbsp lard in a heavy skillet and put the steaks in and lower the heat and cook slowly. Brown on both sides. Add the peppers, tomatoes, cloves and thyme. Add ½ c water and cover the skillet and let the steaks braise until the water is nearly gone. Scrape the bottom of the pan and turn the steaks over and add remaining water and continue to cook slowly turning now and then. Add more water if necessary to keep steaks from drying out or burning. Cook for 2 hours or until the steaks are tender. About 20 minutes before the end cook your rice. Near the end sprinkle with salt and pepper to taste. Place them on a warm serving plate and keep them warm. Add remaining lard and flour to the braising residue in the skillet and blend well. Add 1½ cups of hot water and raise the heat and blend well scraping the bottom of the skillet. Season to taste. Serve in a gravy boat to be spooned over the steaks and rice. Serves 4.

TURTLE STEAK WITH SOUR CREAM

2 lbs turtle steaks
1 clove garlic, mashed
3 anchovies, mashed
½ tsp thyme
3 tbsp butter
1 c dry white wine
½ tsp black pepper
1 tbsp plain flour
⅓ c sherry
½ c sour cream
1 tbsp capers
1 tsp salt

Slice the steaks thin and pound with a meat hammer. Cream together the garlic, anchovies, thyme and half the butter. Spread over the meat and roll and tie in several places. Place the meat in a glass bowl and pour the wine over it and marinate for 3 hours or more, turning and basting occasionally. Drain and dry the meat and reserve the marinade. Rub the meat with the salt and pepper. Heat remaining butter in a dutch oven and brown the meat in it on all sides. Slip a rack under the meat and add the reserved marinade. Cover and cook over a low heat for 2 hours. Transfer the meat to a hot serving platter. Blend the flour with the sherry and stir into the gravy until thickened. Then cook 2 minutes longer. Blend in the sour cream and capers and taste for seasoning and heat but do not boil. Slice the meat, pour some of the sauce over it, and serve the rest separately. Serves 6.

TURTLE STEAK FLORIDIAN

1½ lbs turtle steaks
1 c sour cream
½ c wine
6 tbsp butter
2 tbsp parsley, chopped
plain flour
1 tbsp paprika
salt and pepper to taste

Have turtle steaks cut paper thin and pound with the edge of a plate. Dip them in flour. Melt the butter in a skillet that has a cover and brown the pieces of turtle very quickly. Salt and pepper to taste and add 1 tsp paprika. Pour wine over them, cover, and simmer for 1 hour. Remove the steaks to a hot platter. Add sour cream to the drippings and stir well until it is heated through and blended. Pour the sauce over the turtle steaks and sprinkle with remaining paprika and parsley.

TURTLE STEAK FLORIDA KEYS

turtle steaks
lime juice
garlic powder
egg, beaten
seasoned bread crumbs*
oil

Slice the steaks into thin slices and place in a layer in a flat pan and pour lime juice over them. Sprinkle with a little garlic powder on top of the lime juice. Keep adding layers until all the turtle is used up. Let marinate 4-6 hours. Drain and dip in egg and then into bread crumbs. Fry in hot oil until just light brown.

TURTLE STEAK ST. THOMAS

2 lbs turtle steaks, cut 1½ in
by ½ in strips
2 ribs, celery, chopped
1 onion, minced
1 tomato, minced
3 lge mushrooms, chopped
1 clove garlic, minced
1 sm can tomato puree
½ pt dry white wine
dash brandy
1 tbsp plain flour
1 tsp salt
pinch sweet basil
pinch thyme
pinch marjoram
2 sprigs parsley, chopped

Combine and mix well the celery, onion, tomato, mushrooms, garlic, tomato puree, parsley and wine. Mix the flour, salt and herbs and stir into the mixture and bring to a boil stirring from time to time. Add the turtle strips and blend. Transfer to a casserole. Bake for 1½ hours in a 350°F oven. Just before removing from the oven, stir in the brandy. Serves 4-6.

TURTLE CUTLET

turtle meat, lean
cracker crumbs
1 egg, beaten
butter

Take the lean meat and pound until it is like a hamburger steak. Dip it into the egg and roll in the cracker crumbs and fry in butter. Tastes just like a veal cutlet.

LAGLER'S FRIED TURTLE

turtle meat
butter
1 qt + water
a little chopped onion
beef tallow
1 tbsp vinegar
salt and pepper to taste

Soak meat overnight in the vinegar and quart of water. Rinse and brown the meat with the beef tallow and butter. Pour off the grease and season and add a few bits of onions and enough water to cover and simmer until the flesh begins to fall from the bones. Serve hot or cold.

TURTLE STEAK WITH A SAUCE PIQUANT

1 lb turtle steak
oil
salt and pepper
sauce piquant

Season the steak and sprinkle with oil. Broil over a quick fire for about 15 minutes. Put it on a hot platter and pour over it the sauce Piquant.

SAUCE PIQUANT

2 ozs butter
½ pt vinegar
1 pt half glaze*
1 tsp parsley, chopped
2 onions, minced
1 doz sour pickles, chopped
salt and pepper to taste

Put into a saucepan the butter and onions and sauté until golden. Drain off the butter and add vinegar and let reduce almost completely. Add the pickles and moisten with the half glaze and let it boil for 10 minutes skimming well while cooking. Season to taste and add the parsley.

BATABANO

2 lbs turtle steak, ground
2 lge bell peppers, diced
sm jar capers
2 8 oz cans tomato sauce
½ tsp garlic powder
¼ tsp oregano
¼ tsp paprika
1 tbsp Pickappeppa sauce
2 med onions, diced
med jar stuffed olives
7½ ozs raisins
½ tsp salt
½ tsp celery salt
¼ tsp black pepper
4 tbsp olive oil
2 tbsp worcestershire sauce

Sauté the onions and bell peppers in half the olive oil. Sauté the meat in the remaining olive oil and stir in the seasonings, worcestershire and Pickappeppa sauces, half the juice from the olives and the tomato sauce and stir well. Add the sautéed onions and bell peppers and the remaining ingredients except the liquid remaining in the olives. Cook for about 40 minutes over a slow heat, covered. Serve with white or yellow rice.

SAVOURY TURTLE STUFFED BREADFRUIT

2 c turtle steak, ground
1 firm whole breadfruit
1 onion, chopped
oil
water
1 tomato, chopped
gravy to moisten
seasonings to taste
salt

Peel the breadfruit and remove the stalk and core. Parboil in salted water and remove a little more of the fruit from the cavity. Mix together the remaining ingredients to a firm mixture and fill the fruit, brush with a little oil, and bake in a preheated oven for about 45 minutes. Garnish and serve hot.

STUFFED BELL PEPPERS

2 c ground turtle meat, cooked
6-8 lge bell peppers
1½ c bread crumbs
1 sm onion, chopped
seasonings to taste

Remove the stem and seeds from the bell peppers and parboil for 2 minutes. Combine the remaining ingredients and fill the pepper shells. Shake a few breadcrumbs that have been soaked in a little butter on the top of each and bake in a moderate oven for 10-15 minutes. Before serving brown tops under the grill. Goes well with rice and a hot tomato gravy. Serves 6-8.

TURTLE STEAK AND GRAVY MISSISSIPPI STYLE

2 lbs turtle steaks
5 tbsp red wine
5 tbsp capers
3 tbsp butter
plain flour
garlic salt
⅓ c milk
1 clove garlic, chopped
1 10¾ oz can cream of mushroom soup
salt
black pepper

Cut the steaks crosswise of the grain and pound with a mallet and season with salt, pepper and garlic salt. Work into the meat with a fork. Fry and remove from the skillet. Add the butter and chopped garlic and simmer for 10 minutes. Add the soup and wine while stirring. Lower the heat and add the milk which has been heated and stir until smooth. Place the steaks in the gravy and cook on low heat for 10 minutes.

BROILED TURTLE STEAKS

turtle steaks, 1½-2 in thick
salt
parsley, chopped
butter
black pepper, freshly ground
lemon slices

Spread butter generously over the steaks and broil them rapidly until they are tender and golden brown. Season to taste. Arrange steaks on a heated platter and pour the juices from the pan over them. Sprinkle with the parsley and serve with the lemon slices.

SNAPPER WITH MADEIRA

1¼ lbs turtle meat	3 tbsp butter
plain flour	salt and pepper
¾ c madeira	½ c heavy cream
3 drops tabasco sauce	parsley, chopped
paprika	buttered toast points

Cut meat away from the bones and cut into paper-thin pieces and then dredge in flour. Heat the butter in a heavy skillet and brown the meat lightly. Season with salt and pepper. Add ½ cup madeira, cover and simmer until the meat is tender, approximately 1 hour. Add a little water if necessary. Just before serving add the remaining madeira and the cream. Heat to the boiling point and then blend in the tabasco sauce. Serve on buttered toast points topped with a little parsley and a sprinkling of paprika. Serves 4.

TURTLE STEAKS SAUTEED

2 turtle steaks, 1½ in thick	seasoned flour*
¼ c butter	madeira
salt	water

Soak the steaks overnight in salted cold water. Wipe dry and dredge thoroughly in the flour. Melt the butter in a skillet and quickly sear the steaks on both sides. Reduce the heat and continue cooking them until they are tender and brown. Serve with the pan juices blended with a little madeira. Serves 2.

TURTLE STEAKS WITH SHERRY SAUCE

6 turtle steaks	seasoned flour*
vinegar	2 eggs, beaten
bread crumbs	¼ c + butter
1 tbsp shallots, minced	1 c Bordeaux
salt and pepper to taste	dash nutmeg
½ c beef stock	¼ c sherry
1 c mushrooms, thinly sliced	watercress (for garnish)

Rub the steaks with a damp cloth dipped in vinegar. Dredge in the flour and then the eggs and then the bread crumbs. Heat the ½ cup butter in a large skillet and stir in the shallots. Cook the steaks in the shallot butter until they are delicately browned on both sides. Pour the Bordeaux over them and season with salt, pepper and nutmeg. Cover the skillet and simmer gently for 15-20 minutes. Arrange the steaks on a hot platter and keep them hot. Reduce the sauce in the skillet to almost nothing. Stir in the beef stock, sherry, and mushrooms which have been sautéed in a little butter. Adjust seasoning. Pour a little of the sauce over each steak and serve the remainder in a heated sauce boat. Garnish with watercress. Serves 6.

TURTLE MARENGO

2 lbs turtle steaks, cubed	2 slices onion
¼ lb mushroom buttons	1 clove garlic, chopped
¼ pt dry white wine	¼ pt water
seasoned flour*	4 ozs tomato paste
parsley	4 tbsp butter
1 bouquet garni*	fried croutons

Dredge the meat in the seasoned flour. Heat the butter in a pan and add the meat and cook to seal in the juices. Add the garlic and onion and cook until tender. Add a tbsp of the flour and cook until it browns. Gradually stir in the liquid and bring to a boil. Add the tomato paste, seasonings, and herbs. Cover the pan and simmer gently for 1 hour. Add the mushrooms and cook for a further 10 minutes. Serve in a deep dish. Cover with the sauce and garnish with chopped parsley and fried croutons.

BROILED SEA TURTLE

3 lb turtle meat	1 c lime juice
olive oil	allspice, powdered

Soak meat in lime juice several hours turning occasionally. Drain and rub meat first with allspice and then with olive oil. Place under broiler or over coals in aluminum foil for 45 minutes or until tender. Serves 6-8.

FRIED SNAPPING TURTLE

snapping turtle meat	shortening or oil
plain flour	

Roll the meat in flour and brown in the shortening. Place on a rack in a roaster and bake for 2 hours at 350°F.

GENTLEMAN JEFF'S VIRGINIA STYLE SNAPPER

1 10-15 lb snapper	1 18 oz bottle barbecue sauce
I lge onion, sliced	1 lge bell pepper, sliced
salt	black pepper

This is best when cooked slowly in a 3 quart electric crock pot. Cut the dressed turtle meat into manageable pieces leaving the meat on the bones. Place half the onion and bell pepper in the pot and then place half the meat on top and salt to taste and half the barbecue sauce and then repeat with the remaining ingredients. Set on low and cook for 8-10 hours. Black pepper should be added about 1 hour before serving.

ERNIE'S TURTLE SAUCE PIQUANT

15 lbs turtle meat
24 ozs tomato paste
1 20 oz can tomato sauce
4 lbs red onions, chopped
9 lemons, seeded, sliced
1 13 oz jar salad olives
2 gal whole tomatoes
few bay leaves
1 7½ oz can salsa de jalapeno
sherry (optional)
1½ bu celery, chopped, stems only
2 cayenne peppers
3 ozs worcestershire sauce
½ c oil
salt and pepper to taste
3 8 oz cans mushrooms, stems and pieces
5 lge bell peppers, chopped
3 tbsp sweet basil
3 tsp granulated garlic
1 bu scallions, chopped

Drain mushrooms and save liquid and sauté with celery, onions and bell pepper in the oil. When the vegetables are tender pour in 1 gallon of whole tomatoes, drain 1 gallon reserving the liquid and add along with the tomato sauce and paste. Use the water from the mushrooms to wash out the cans and add to mixture. Put in the cayenne peppers whole along with the bay leaves, garlic and sweet basil. Let simmer about 4 hours stirring every so often to keep from sticking. After sauce has been thickened add the turtle meat, (the turtle meat can be fried in a little additional oil if desired as some say this improves it), worcestershire sauce and lemons. Let simmer. Add the black pepper, salt and salsa de jalapeno (use tabasco sauce if salsa not available) to taste. It should be hot to be good. Cook until meat is tender and is falling apart and desired thickness is attained. Add the sugar. If too thick add a little of the reserved tomato liquid. It should take about 8-10 hours. Wash and add the olives and scallions and cook a few minutes longer. Add sherry if desired and serve over rice. Makes about 4 gallons.

TURTLE SAUCE PIQUANT

5-6 lbs turtle meat
4 hard boiled eggs, chopped
salt
½ c plain flour
½ c oil
cayenne pepper
4 lge onions, minced
1 no 2 can tomatoes
1 c sherry
2 bell peppers, minced
2 bell peppers, minced
1 10 oz can Ro-Tel Tomatoes
¾ c scallion tops, chopped
6 cloves garlic, smashed
4 stems celery, minced
¼ c parsley, chopped
1 tsp thyme
½ tsp thyme
½ tsp sweet basil
3 qts hot water

Season turtle meat with salt and cayenne and brown very thoroughly in hot oil. Remove meat and add the onions, stirring until a deep brown and all pan drippings are absorbed. Add the tomatoes and cook down until browned. Add bell pepper, celery, garlic, thyme, sweet basil and hot water. Let simmer about 1 hour until vegetables are done. Add the turtle meat and the boiled eggs and cook slowly until the turtle is done, about 2 hours. Add sherry, parsley and scallions during the last 4 minutes. Serve over rice.

TURTLE STEW PIQUANT

3-5 lbs turtle meat
1 c plain flour
10 lge onions, chopped
3 tbsp garlic, minced
3 bu celery, chopped
2 bu parsley, chopped, separated
¼ tsp cloves, powdered
¼ tsp thyme, powdered
1 14½ oz can tomatoes, chopped
1 6 oz can tomato paste
1 tbsp cayenne pepper
12 eggs, hard boiled
1 stick butter
2 hot green peppers, seeded, minced
paprika
6 bell peppers, chopped
2 bu shallots, chopped, stems
and bottom separate
½ tsp allspice, powdered
2 bay leaves
2 qts hot water
3 tbsp salt
sherry
1 lemon, sliced, seeded

Brown the flour in the butter until it is a reddish-brown color. Add the bell peppers, peppers, onions, garlic, celery, shallot bottoms and parsley stems. Cook until the onions are transparent. Add the cloves, allspice, thyme, bay leaves, tomatoes and tomato paste blending well. Add the hot water and bring to a boil and boil for 1 hour. Season the meat with salt and cayenne pepper and add to the pot and lower the flame and simmer for 2 hours. Add some of the shallot and parsley tops reserving some for garnish. Correct seasoning to taste. Prepare the eggs by separating the yolks from the whites and mash the yolks and season with salt and cayenne pepper and add 15-20 tbsp sherry and mix into a paste and stir into the pot. Slice egg whites into the stew. In 15 minutes dish into soup plates and place lemon slices dusted with paprika in the center of each plate and sprinkle shallots and parsley on top. Serve hot. Serves 15.

RIVER ROAD SAUCE PIQUANT

15-20 lbs turtle meat
1 qt oil
16 cloves garlic, chopped
6 6 oz cans tomato paste
1 fifth burgundy
1 bu shallots, tops, chopped
8 kitchen spoons plain flour
10 lbs onions, chopped
6 bell peppers, chopped
2 c water
1 bu parsley, chopped
3-4 bay leaves

Make a roux* of the flour and oil and then add the onions, garlic and bell peppers and cook until tender. Add the tomato sauce and cook about 35 minutes. Add the meat and water. When the meat is tender add the burgundy (or dry white wine, if preferred) and let cook on a slow heat for about 1 hour. Before serving, add the parsley, scallion tops and bay leaves and let remain on the fire until the parsley curls. Serves 50.

CAJUN TURTLE STEW PIQUANT

5 lbs turtle meat, cubed
1 gal hot water (or more)
2 sticks butter
6 ribs celery, chopped
2 bell peppers, chopped
1 c parsley, chopped
1 6 oz can tomato paste
6 red hot pickled peppers, minced
½ tsp allspice, ground
½ tsp thyme
6 hard boiled eggs, chopped
lemon slices (for garnish)
2 c seasoned flour*
1 c plain flour
3 lge onions, chopped
6 cloves garlic, minced
10 scallions, sliced
1 32 oz can plum tomatoes, chopped
½ lemon, sliced, seeded
2 bay leaves
½ tsp cloves, ground
1 tbsp cayenne pepper
2 c sherry
scallion tops, chopped (for garnish)

Melt the butter in a heavy pot and add the plain flour and cook over a low fire for 20-30 minutes, stirring, until a dark brown. Add the scallions, celery, garlic, bell pepper, onions and parsley and cook until the vegetables are soft. Add a little water if it gets too dry. Add the tomatoes and tomato paste. Add enough water to cover all the materials in the pot by 2 inches. Add the sliced lemon, pickled peppers bay leaves, allspice, thyme, cloves and cayenne pepper. Bring to a boil and then reduce the heat to a slow simmer, stirring frequently. Cook for 1 hour. After the vegetables have cooked for an hour add the turtle meat which has been dredged in the seasoned flour (salt, black pepper, cayenne pepper). The liquid should be about 1 inch above the materials in the pot. If not add more water or beef broth. Bring to a boil and then lower the heat and simmer for 1 more hour, stirring frequently to keep from burning. At ½ hour before it is finished add the sherry and eggs. Serve in preheated soup bowls and sprinkle the top with scallions and add a slice of lemon on top. Can be served over rice if desired.

SIMPLE TURTLE SAUCE PIQUANT

5 lbs turtle meat, in chunks
2 6 oz cans tomato paste
2 tbsp oil
black pepper
3-4 banana peppers, chopped
1 lge onion, chopped
Louisiana Hot Sauce
cayenne pepper

In a deep skillet brown the turtle in the oil. Add the tomato paste and simmer until the grease rises to the top. Skim off. Add peppers and onion and cook until soft. To this add some Louisiana Hot Sauce, black and cayenne pepper to taste. Simmer for 2 hours, stirring occasionally until the meat is tender. If the sauce is a too thick add a little water.

FOLSE'S TURTLE SAUCE PIQUANT

20 lbs turtle meat
2 bu shallots, diced
1 garlic, whole, chopped
1 sm bu parsley, minced
4 10¾ oz cans cream of mushroom soup
12 tbsp plain flour
3 ozs worcestershire sauce
salt and pepper
5 lbs onions, diced
4 bell peppers, diced
⅓ bu celery, diced
2 8 oz cans mushrooms
1 pt oil
2 10 oz cans Ro-Tel tomatoes
3 tbsp french mustard

Brown the turtle meat and then add the vegetables and soup and cook until the onions and seasonings cook to juice. Brown the flour and oil until dark brown and add to the gravy and let simmer for 4-5 hours. Add salt and pepper to taste. Add the worcestershire sauce and mustard about 20 minutes before serving. Serve over rice. Serves 50.

TURTLE SAUCE PIQUANT 2

2 lbs turtle meat, seasoned
oil
1½ onions, chopped
3 ribs, celery, chopped
1 bell pepper, chopped
1 10 oz can Ro-Tel Tomatoes
1 lb can whole tomatoes
1 10¾ oz can golden mushroom soup
4 cloves garlic, pressed
1 soup can dry vermouth
worcestershire sauce
bitters
1 tsp rosemary
1 tsp thyme

Add enough oil to just cover the bottom of a pot and add the turtle meat and brown. Discard the excess oil. In the same pot sauté the chopped vegetables. Add the rest of the ingredients and bring to a boil and then lower the heat to a simmer. Simmer for 1½-2 hours or until the meat is tender. Serves 6-8.

TURTLE SAUCE PIQUANT 3

5 lbs turtle meat
1 bu shallots, chopped
4 cloves garlic, chopped
2 tbsp strawberry preserves
2 tbsp bourbon
1 8 oz cans mushrooms, stems and pieces
1 28 oz can whole tomatoes
2 14½ oz cans stewed tomatoes
1 10 oz can Ro-Tel Tomatoes
plain flour
2 8 oz cans tomato sauce
1 10¾ oz can cream of mushroom soup
½ bu parsley, chopped
2 lbs onions, chopped
1 rib celery, chopped
1 bell pepper, chopped
worcestershire sauce
white port wine
oil
salt and pepper to taste

Make a roux* with the oil and flour as desired. Add all chopped vegetables and sauté until wilted. Add all the tomatoes and cook approximately 1 hour. Add the tomato sauce and soup and cook for another hour. Add the preserves, bourbon and mushrooms and cook for another hour. Cook the meat separately in another pot with worcestershire sauce and wine. Liquid should completely cover the meat. Salt and pepper to taste. Cook until all liquid is evaporated or meat is tender. Add drained cooked meat to sauce and cook together approximately 30 minutes.

TURTLE SAUCE PIQUANT 4

10 lbs turtle meat
1 8 oz can mushrooms, drained
sm bottle Louisiana Hot Sauce
1 tbsp garlic powder
salt
water
2 pkgs Chef's Frozen Seasoning
1 med jar stuffed olives, drained, chopped ½
1 46 oz can V-8 Juice
2 tsp cayenne pepper (optional)
oil

Boil the turtle meat in just enough water to cover for about 2 hours or until the meat begins to fall away from the bones. In a separate pot pour enough oil to cover the bottom well. Add Chef's Seasoning, mushrooms, olives and garlic powder. Cook until the seasonings are soft in about 20 minutes. Add the V-8 Juice and boil lightly for 1 hour. When the meat begins to fall from the bones remove the bones and add the seasonings, cover and boil about 1 hour. Add the hot sauce, cayenne and salt to taste. Serve over rice and with garlic bread. Serves 15.

TURTLE ON THE BAYOU

4 lbs turtle meat, boneless
6 cloves garlic, chopped
salt and pepper to taste
water
1 large onion, chopped
oil
plain flour

Add a small amount of oil in a dutch oven. Dredge turtle in flour and fry well. Add onions, garlic and small amount of cold water as needed. Watch the pot closely and keep stirring with just enough heat to simmer. Cook about 2 hours until the meat is tender. Serve over rice. Serves 6.

SMOTHERED TURTLE

3 lbs turtle meat
cayenne pepper
2 c water
¾ oil
salt and pepper to taste
garlic salt
3 onions, chopped

Season the meat to taste and cook with the oil and 1 cup of water for about 30 minutes. Add the onions and simmer with the remaining water for another 35 minutes or until meat is tender.

TURTLE BIRDS

2 lbs turtle steaks
salt and pepper to taste
½ pkg stuffing mix made to pkg directions
1 c dry white wine
½ lb mushrooms
2 tbsp parsley, minced
4 tbsp butter
1 c chicken broth
3 tbsp onion, minced

Pound the steaks as thin as possible and then cut into 6 pieces. Salt and pepper each piece lightly. Spread the stuffing mix on the turtle pieces and roll them up. Tie securely with thread or toothpicks. Melt the butter in a deep skillet and brown the turtle rolls in it. Add the wine, onion, parsley, broth and mushrooms. Cover and bake at 375°F for 45 minutes or until tender. Serves 6.

TURTLE RIVOLA

½ lb turtle meat, chili ground
¼ lb cheese, shredded
1 c plain flour
butter
2 onions, diced
salt and pepper to taste
¼ tsp salt
1 egg

Cook the turtle and onions until tender in a little water. Drain, reserving the liquid, and add the cheese, and salt and pepper to taste. Take the flour, egg, ½ tsp salt and make a thick batter and roll out thin into a sheet of noodle dough and cut into 2 inch squares. Take 1 spoon of the turtle mixture and lay on the noodle squares and fold the corners to enclose the meat. Cook like noodles in the water that the turtle meat was cooked in and add a spoonful of butter.

TURTLE STROGANOFF

1½ lbs snapping turtle meat, 1 inch cube
¼ c vegetable oil, divided
1 med onion, cut in half and thinly sliced
¼ c red wine
1 tbsp ground ginger
1 bay leaf
1 c all-purpose flour
2 tbsp butter
1 10½ oz can condensed beef broth
2 tbsp soy sauce
1 tbsp worcestershire sauce

MARINADE

2 tbsp soy sauce
1 tbsp vegetable oil
1 tsp instant minced onion
pinch sugar
1 10 oz pkg uncooked wide noodles
1 tbsp butter, melted
1 tsp ground ginger
1 tsp worcestershire sauce
1 lb fresh mushrooms, thinly sliced
snipped fresh parsley

Heat oven to 325°F. Place the flour in a shallow dish and dredge the turtle to coat. In a 4 qt dutch oven heat 3 tbsp oil over a medium heat. Add the turtle and cook for 3-4 minutes or until browned. Remove the meat from the dutch oven and set aside. Spray a 3 qt casserole with nonstick vegetable cooking spray and set aside. In the same dutch oven melt 2 tbsp butter over medium heat. Add onion and cook for 2-3 minutes or until tender, stirring frequently. Stir in the broth, wine, soy sauce, ginger, worcestershire sauce and bay leaf. Bring to a boil. Add the turtle meat and remove from the heat. Transfer the mixture to the casserole and bake covered for 45-50 minutes or until meat is tender. Remove the cover during the last 15 minutes. Reduce the oven temperature to 175°F. Cover the casserole and keep warm in the oven. In a small to medium mixing bowl combine the marinade ingredients. Add mushrooms and toss to coat. Set aside. Prepare noodles as directed on the package. Drain and cover to keep warm and set aside. In a 12 inch skillet heat the remaining 1 tbsp oil over a medium heat. Add the mushroom mixture and cook for 4-5 minutes or until the mushrooms are tender, stirring frequently. Add mushroom mixture to the turtle mixture and mix well. Place noodles on a large serving platter and spoon the turtle mixture evenly over the noodles. Garnish with the snipped parsley. Serves 4-6.

BARBECUED TURTLE IN PRESSURE COOKER

2 lbs turtle meat
paprika
2 tbsp vinegar
salt and pepper to taste
½ c chili sauce
1 onion, minced
½ c water
cornstarch

Sprinkle the turtle with the paprika and place into the pressure cooker. Combine the remaining ingredients except cornstarch and pour over the meat. Close cover securely and cook at 15 lbs pressure for 15 minutes or until tender. Cool cooker and reduce pressure at once. Remove the meat from the bones and cut into bite size pieces. Return the meat to the sauce and thicken with a little cornstarch. Serve barbecued turtle over buns or hot rice. Serves 4.

BARBECUED TURTLE STEAK

turtle steaks
barbecue sauce (your choice)
garlic salt

Pound the steaks until fairly thin and then sprinkle both sides with garlic salt. Cook over a barbecue grill basting with your favorite sauce until done. (About ½ hour.)

TURTLE TOMATO SUPREME

1 turtle, dressed, cut up
1 8 oz can tomato sauce
¼ c oil
¼ c shallots, chopped
6 cloves garlic, minced
salt and pepper to taste
water to make sauce
1 14½ oz can stewed tomatoes
1 c onions, chopped
¼ c celery, chopped
½ c parsley, chopped
4 drops liquid crab boil*
¼ c bell pepper, chopped

In a large iron pot brown the onions in the oil and when done add the stewed tomatoes and tomato sauce and let simmer for about 30 minutes. Then add remaining vegetables and simmer for another 30 minutes. Add the turtle meat and smother in sauce until the water in the turtle is cooked out and smother a while. Then add water to make sauce and cook until meat is tender. If turtle has eggs, add them about 10 minutes before turtle is done and cook in the sauce. Serve over rice.

TURTLE SAUTEED CREOLE STYLE

turtle meat
salt and pepper to taste
bell peppers, sliced
mushrooms, julienne cut
tomatoes, fresh, peeled
parsley, chopped
butter
onions, sliced
ham, diced
1 clove garlic, crushed
beef stock
cooked rice

Cut the meat into bite size pieces and season with salt and pepper and fry in butter to a nice color. Add some onions, bell peppers, ham, mushrooms and the garlic. Also add some fresh tomatoes cut into small squares and moisten all with some stock and cook until the meat is tender. When done serve over rice and sprinkle with parsley.

FRICANDEAU OF TURTLE, COUNTESS STYLE

turtle steaks
carrots, sliced
celery, sliced
1 bay leaf
8 peppercorns
salt and pepper to taste
beef or turtle stock
sherry
larding pork
onions, sliced
parsley, chopped
3 cloves, whole
1 clove garlic, minced
butter, melted
half glaze*

Lard the steaks with some thin slices of larding pork on its smoothest side. Cover the bottom of a saucepan with sliced carrots, onions, celery, and parsley, bay leaf, cloves, peppercorns and garlic. Lay the meat on top and season with salt and pepper. Baste with the butter and let it cook over a moderate fire until a nice color. Moisten with some stock and let it fall slowly to a glaze and then moisten without covering the meat and let it come to a boil. Then cover the saucepan and put it into a hot oven and finish cooking the meat, basting frequently. When done lay the steaks on a hot platter and strain the gravy and take off all the fat and reduce it properly adding a little half glaze and a little sherry. Pour the sauce over the meat. Garnish and serve.

TURTLE CHOWDER DELUXE

⅓ slice bacon, diced
1 med onion, chopped
1 10¾ oz can cream of potato soup
4 carrots, diced and cooked
1 12 oz can evaporated milk
1 can + water, as needed
2 lbs turtle meat, diced
1 10¾ oz can cream of celery soup
4 sm potatoes, diced and cooked
1 sm can mixed vegetables
½ bell pepper, chopped

Sauté the bacon, bell pepper and onion. Drain off the bacon grease and add the celery and potato soup and 1 can of water. Add the carrots and potatoes and add more water if the chowder is too thick. Simmer and then add the vegetables with the liquid and add the turtle and cook for a short time or until the turtle is tender. Add the milk just before serving. Do not boil after adding the milk. Serves 4.

TURTLE CHOWDER

½ lb turtle meat, cubed
3 onions, diced
parsley, chopped
¼ lb salt pork, diced
1 tsp butter
½ bell pepper, chopped
2 med potatoes, diced
3 carrots, diced
2 scallions, diced
1 tsp black pepper
1 stem celery, chopped

Combine all ingredients well and simmer over a low fire about 2 hours.

CONCH OR TURTLE CHOWDER

3 conches, ground (or substitute turtle)
1 clove garlic, minced
1 8 oz can tomato sauce
water
1 lge onion, chopped
1 bell pepper, chopped
1 lb potatoes, peeled, diced

Cover meat with water and boil for 1½ hours. Add the onions, garlic, bell pepper and tomato sauce and cook for about 20 minutes. Then add the potatoes and a couple cups of water. Salt and pepper to taste. Simmer until potatoes are cooked.

RAGOUT OF TURTLE

3 c turtle meat, diced
1 tbsp parsley, minced
3 tbsp butter
pinch of thyme
2 tbsp plain flour
1 bay leaf
3 c chicken stock
12 button mushrooms
¾ tsp salt
1 slice bacon, finely chopped
dash of cayenne pepper
½ onion, sliced and parboiled

Fry turtle meat in butter until golden brown. Add flour and brown this also. Gradually pour in stock and bring to boiling point, stirring constantly. Add seasonings, then onions, bacon and mushrooms. Cover and let simmer until the turtle is tender, about 45 minutes. Serves 4.

TURTLE RAGOUT "PRINTAINERE"

2 lbs turtle meat, diced
1 turtle flipper
1 onion, diced
½ lb carrots, diced
½ lb potatoes, diced
½ lb celery, diced
2 cloves garlic, diced
water
2 tomatoes, diced
salt and pepper to taste
paprika
2 bay leaves
2 cloves, whole
thyme
butter
noodles, cooked

Cover the flipper with enough hot water to cover and let it boil for stock. Season the meat and brown in the butter. When the meat is browned add the diced vegetables and brown with the meat. Then add the turtle stock and simmer for 1½ hours. Serve on a bed of noodles.

TURTLE RAGOUT A LA APPALACHIAN

2 lbs turtle meat, diced
1 onion, chopped
2 tbsp butter
1 tbsp plain flour
1 bay leaf
1 clove garlic, chopped
1 c water
1 c wine

Cook the onion in the butter and then add in the flour stirring well. Add the remaining ingredients and simmer for 30 minutes.

SNAPPING TURTLE IN A POT

1-2 lbs turtle meat
2 tsp instant minced onion
⅛ tsp basil, dried
2 c water
8 sm red potatoes, halved, unpeeled
¼ c sherry (optional)
2 carrots. sliced
salt
2 ribs celery, cut in pieces

Salt turtle meat well and place in your slow cooker. Add all other ingredients mixing well and then cover and cook on low heat for 6-7 hours or until the turtle meat is tender. Remove the meat from the pot and cut into bite size pieces and return to pot, cover, and continue to cook on low heat for an additional 2 hours or until vegetables are done.

TURTLE JAMBALAYA

2 c turtle meat, diced
water
1 c bell pepper, minced
1 c ham, diced
2½ c chicken broth
1 tbsp parsley, chopped
1½ tsp salt
2¼ c tomatoes, canned
2 tbsp butter
1 c onion, minced
2 cloves garlic, minced
12 pork sausages, cut in pieces
½ tsp thyme
¼ tsp chili powder
¼ tsp black pepper
1 c raw white rice

Simmer the turtle meat in water until tender. Melt the butter in a skillet and add the onion, bell pepper and garlic and cook slowly, stirring often, until the vegetables are tender. Add the ham, sausage (breakfast) and drained turtle meat (reserve the liquid for another dish*) and cook for 5 minutes. Add the rice, tomatoes, chicken broth, thyme, parsley, chili powder, salt and black pepper. Mix well and place into a large casserole and bake covered for 1½ hours or until the rice is tender.

* or use in place of chicken broth.

TURTLE FRICASSEE

2 lbs turtle meat
1 sprig thyme, chopped
1 tbsp oil
½ c tomatoes
1 tbsp plain flour
water
1 cayenne pepper
1 c sherry
salt and pepper to taste
2 hard boiled eggs, sliced
4 cloves garlic
parsley

Put turtle meat, cut in 3 inch pieces, into pan with oil and fry. When brown, sprinkle in the flour, then put in cayenne, salt, pepper, garlic, thyme and tomatoes. Cover and boil gently for 4 hours or until tender, adding a little water from time to time. A ½ hour before it is done put in the sherry. After putting it on a platter, place sliced eggs on top and garnish with curled parsley. Serve with rice.

TURTLE FRICASSEE 2

turtle meat, cubed
salt and pepper to taste
water
1 tbsp plain flour
butter
1 c onion, diced
cream
rice, cooked

Wash the meat in salted water and then brown in a heavy skillet with plenty of butter. When brown add onion and just enough water to cover the meat. Cover and simmer until the meat is tender. Remove meat to a separate platter and add flour and a little cream to the pan juices. Stir briskly until thickened and then add the turtle. Pour the gravy and meat over a bed of rice and serve.

SIMMERED TURTLE

1 lb turtle meat, cubed
1 chili pepper
2 med onions, diced
butter

Brown the meat in butter with the onions and simmer covered until tender and add the chili pepper while simmering. To serve pour over boiled rice.

STEAMED TURTLE

turtle meat
black pepper
butter
worcestershire sauce

Season meat heavily with the black pepper and add some bits of butter and steam until the flesh separates from the bones and then add the worcestershire sauce. (Black sauce which is the soy bean sauce found in chinese restaurants can be substituted for the worcestershire sauce if desired).

TURTLE ETOUFFÉ

2 lbs turtle meat, cubed
8 c water
1 lge onion, quartered
1 stem celery, cut into 4 pieces
1 tsp + 2 tsp salt
2 tsp + 1 tsp black pepper
½ bell pepper, cut in 4 pieces
½ c oil, divided
½ c plain flour
⅔ c bell pepper, chopped
½ c celery, chopped
⅓ c garlic, minced
2 8 oz cans tomato sauce
4 c turtle stock
1 14½ oz can chicken broth
1 tsp soy sauce
1 tbsp lemon juice
1 tsp paprika
2 tsp hot pepper sauce
2 c onions, chopped
4 bay leaves
1 oz pkg processed cheese

In a large pot combine the water, turtle meat, quartered onion, large pieces of celery, large pieces of bell pepper, 1 tsp salt and 1 tsp black pepper and boil for 45 minutes. Remove the turtle and reserve 4 cups of the stock. In a medium skillet combine ½ cup of the oil and the flour and make a roux until it is peanut butter colored. Set it aside. In a large 2 gallon pot put the remaining oil and heat. Add the remaining onions, bell pepper, celery, garlic and turtle meat and sauté until the vegetables are tender. Add the tomato sauce and stir to combine. Add the roux and mix thoroughly. Add the reserved stock, chicken broth, soy sauce, lemon juice, pepper sauce, remaining black pepper and salt, paprika, and bay leaves. Cook for 1 hour and in the last 5 minutes add the cheese and let it melt. Serve over noodles or rice. Serves 12.

TURTLE ETOUFFÉ 2

10 lbs turtle meat
1 c parsley, chopped
3 bell peppers, chopped
8 large onions, chopped
1 tbsp worcestershire sauce
1 clove garlic, chopped
juice of ½ lemon
olive oil
salt and pepper to taste

Wash and drain turtle meat. Salt, pepper and brown in olive oil. Put in heavy pot and add all other ingredients. Cook on low, low heat for 6-8 hours until turtle meat is tender. Serve with rice.

TURTLE RAGOUTFAIM

1 lb turtle meat, cubed
salt and pepper to taste
onions
garlic
noodles, cooked
dill pickles
plain flour
vinegar or lime juice
white wine
butter

Boil the meat with salt, pepper, onions and garlic until tender in water to cover. Remove meat and then sauté in butter with some diced onions. Thicken with flour and use the turtle stock for the sauce. Add dill pickles, vinegar or lime juice and white wine to the sauce. Simmer the meat and sauce for about 15 minutes. Serve on a bed of noodles.

TURTLE AU GRATIN

4 lbs turtle meat, boneless
1 tbsp plain flour
1 clove garlic, chopped
2 doz stoned olives
1 hard boiled egg, sliced
toasted bread crumbs
2 sprigs parsley, chopped
½ c butter
1 onion, chopped
1 c salad olives, chopped
1 wine glass sherry
1 lemon, sliced
salt and pepper to taste

Boil turtle meat until tender in just enough water to cover. Dice. Sprinkle flour in hot butter letting it brown thoroughly. Add the onion, parsley and garlic. Now put in the diced turtle meat, cover with the stock, add salt and pepper to taste, and let all simmer gently until tender and the stock will be reduced to a thick gravy. Soak the olives for chopping in water, drain, and add them to the simmering turtle meat with the sherry and pour all into a shallow pyrex baking dish. Cover the top with toasted bread crumbs and decorate with the stoned olives, the sliced hard boiled egg and sliced lemon. Serve very hot.

TURTLE CREOLE

turtle meat
1 onion, minced
salt
1 onion, sliced
1 8 oz can tomato sauce
3 cloves garlic, chopped
mashed garlic
vinegar
butter

After cleaning turtle, disjoint and soak meat overnight in vinegar, sliced onion and mashed garlic. Then cook in boiling water to which a little salt has been added. When meat is tender, remove (reserve stock for some other turtle recipe) and fry in butter until brown. Add minced onion and chopped garlic. When vegetables are tender add the tomato sauce and let all simmer together 15 minutes.

TURTLE BALTIMORE

turtle meat
plain flour
1 egg, beaten
plain bread crumbs
butter
salt and pepper to taste

Select a nice cut of lean meat from under the carapace, cut into thin slices like veal, flatten and season lightly. Dredge the slices in flour, dip in egg and roll into fresh bread crumbs. Fry the pieces in clarified butter, turning once. Serve with port wine sauce.

PORT WINE SAUCE

½ pt espagnole sauce
2 glasses port wine

To espagnole sauce add the port wine.

ESPAGNOLE SAUCE

3 tbsp butter
1 pt beef consomme
salt and pepper to taste
3 tbsp plain flour

Melt butter and let it brown slightly. Add flour and let simmer, stirring, until it becomes nicely browned. Have ready consomme and add this, ½ pint at a time, to the browned butter and flour stirring constantly until thick and creamy. Each ½ pint consomme must be well incorporated before the next ½ pint is added. Simmer the sauce until it is reduced to ½ of its original volume, taste for seasoning, and it is now ready as the base for many other sauces. The flavor of the finished sauce will greatly depend on the richness of the consomme.

PREPARED TERRAPIN

female terrapins
salted water

Keep terrapins for at least 1 week without feeding before killing by plunging into boiling salted water. Remove and strip off outer layer of skin. Return. Parboil for 1½ hours until feet fall off and the shell cracks. Remove the turtle and place on its back until cool enough to handle. Discard the heavy part of the intestines, gall and sand bags. Save the liver, heart, and eggs along with the meat. Pick to pieces and dice. The meat is then ready for various recipes.

PREPARED TERRAPIN 2

terrapin meat, cut up
cold water to cover
1 tsp salt
1 med onion, sliced
2 carrots, sliced
½ c celery, chopped
1 bay leaf
2 whole cloves

Cut the meat into small pieces with the bone in. Cover with the water and add all the other ingredients and cover and simmer for 30 minutes. Remove the turtle meat and bone it. It is now ready for various recipes.

TERRAPIN BUTTER

12 hard boiled turtle egg yolks
1 lb butter, fresh
salt
cayenne pepper
nutmeg, grated

Pound the egg yolks with the butter and add the seasonings. Press through a sieve and then use as directed.

SHE TURTLE SOUP OR TERRAPIN TUREEN

1 lb terrapin meat, boned
½ gal milk
1 10¾ oz can cream of mushroom soup
1 lb mushrooms, chopped
2 tbsp dill, chopped
2 tbsp white pepper
3 sticks butter
4 med potatoes, chopped
3 c onions, chopped
6 stalks celery, chopped
24 ozs sour cream
2 tbsp basil, chopped
2 tbsp Beau Monde Seasoning

Sauté the onions, potatoes, celery and mushrooms in the butter until tender. Let cool completely and then puree. Place in a large pot and add the remaining ingredients and cook over a medium heat. Watch carefully and stir often.

TERRAPIN SOUP

1 female terrapin, dressed, meat and eggs
nutmeg, ground
½ c currant jelly
salt to taste
3 tbsp butter
1 egg yolk, beaten
1 pinch cayenne pepper
mace

Put the meat and eggs in a stew pot with 2 tbsp butter and let simmer until quite hot throughout and well-covered. Serve with a sauce made with the rest of the ingredients.

TERRAPIN SOUP 2

1 terrapin, dressed
1 carrot, minced
pinch of basil
pinch of thyme
½ c butter
1 c turtle stock
salt and pepper to taste
4 hard boiled egg yolks, whole
1 c heavy cream
1 rib celery, chopped
1 bay leaf
pinch of lemon peel
pinch of marjoram
2 tbsp plain flour
3 c milk
2 tsp parsley, chopped
4 tbsp lemon juice
1-2 ozs sherry

Each terrapin should have 4 large chunks of meat. Boil it in salted water just to cover seasoned with the celery, carrot, bay leaf, basil, lemon peel, thyme and marjoram. Remove the meat and reserve the stock. Bone and dice the terrapin meat. In a skillet melt the butter and add the flour and vigorously add 1 cup of stock, milk, salt, and pepper. When this is hot add the meat, the remaining stock minus the bay leaf, parsley, lemon juice, thick cream and the egg yolks being careful not to break. Heat over a low fire, do not boil, stirring gently. When piping hot add the sherry and serve immediately making sure an egg yolk is in each soup dish. Serves 4.

TERRAPIN WASHINGTON STYLE

3 c prepared terrapin*
dash of cayenne pepper
2 tbsp plain flour
3 tbsp butter
1 c mushrooms, sliced
2 tbsp white wine
¼ tsp pepper
⅔ tsp salt
toast
1 c light cream

Sauté mushrooms in butter for 5 minutes. Add flour and when smoothly blended add the seasonings and cream. Bring slowly to boiling point, stirring constantly, add diced terrapin and simmer 10 minutes. Put in the wine at the last moment and serve on toast. Serves 4.

TERRAPIN PHILADELPHIA STYLE

1 prepared terrapin*
2 tbsp butter
turtle stock
½ c cream
2 egg yolks
1 c mushrooms
sherry
1 c velouté

Add terrapin and strained stock to cover and simmer for 25 minutes. Make 1 cup of velouté. Add this to the terrapin with mushrooms sautéed in the butter. Beat egg yolks with cream until smooth and stir this into the mixture slowly. Do not let it boil. Add sherry just before serving.

VELOUTÉ

1 tbsp plain flour
salt
1 tbsp butter
½ c turtle stock
freshly ground black pepper

Combine the flour and butter and cook together until they are slightly browned. Gradually stir in the strained turtle stock and stir until it thickens. Simmer for 10 minutes and season to taste. Makes 1 cup.

TERRAPIN A LA MARYLAND

3 cups prepared terrapin*
1 c cream
1 tsp lemon juice
1 wine glass white wine
toast or puff pastry triangles
2 tbsp butter
cayenne pepper
salt
3 hard boiled egg yolks

In the upper part of a double boiler, combine terrapin, butter, lemon juice, wine and seasonings and bring to just to the boiling point. Add cream and crushed egg yolks and make all very hot without actually boiling. Serve garnished with toast or triangles of puff pastry. Serves 4.

MARYLAND TERRAPIN

2 prepared terrapins*
1 c heavy cream
½ tsp salt
2 truffles, chopped
2 tbsp madeira
¼ c butter
dash mace
dash cayenne pepper
2 egg yolks, beaten

Melt butter and add terrapin, truffles and seasonings and stir until thoroughly heated. Mix cream and beaten egg yolks and add slowly, stirring constantly. Do not allow to boil. Stir in wine just before serving. Serves 3-4.

TERRAPIN MARYLAND

1 c prepared terrapin*
little nutmeg, grated
½ glass + dry sherry
2 egg yolks
toast
salt and white pepper to taste
taste
1½ c heavy cream
1 oz butter

Put the terrapin in a chafing dish and add a little nutmeg, salt, pepper and ½ glass sherry and boil until reduced by half and then add 1 cup cream. Boil. Combine egg yolks, ½ c cream and butter beaten together. Combine with the terrapin and heat but do not boil. Serve in chafing dish with sherry and toast on the side.

TERRAPIN MARYLAND STYLE

1 pt prepared terrapin*
plain flour
salt
sherry
½ pt + brandy
3 oz terrapin butter*
cayenne pepper

Take the terrapin meat and place it in a saucepan with the ½ pint of brandy and let it boil for a few minutes. Thicken with a little flour and the terrapin butter and cook together about 5 minutes. Season with salt and cayenne pepper. Before serving add a little more brandy and a little sherry. Serve in a chafing dish.

TERRAPIN CARDINAL

Prepare the same as for terrapin Maryland style except instead of using terrapin butter* use 2 ounces of lobster butter.

LOBSTER BUTTER

Prepare the same as crawfish butter except use lobster shells instead of crawfish shells (1 pound of shells to 2 pounds of butter).

CRAWFISH BUTTER

5 doz cooked crawfish shells	2 lbs fresh butter

Pound the shells from the cooked crawfish and then add the butter and put it into a saucepan and cook slowly until the butter is thoroughly clarified. Then strain it through a chinese cap into a bowl and beat with a wooden spoon until it becomes thick. If the butter is not red enough add a little red carmine. Set the butter aside and use when needed.

TERRAPIN BALTIMORE

2 prepared terrapins*	½ tsp beef extract
¼ c madeira	salt and pepper to taste
1 tbsp brandy	terrapin eggs, if any
2 c chicken consomme, condensed	½ c butter

Boil terrapin meat in just enough consomme to cover for 20 minutes. Remove terrapin and reduce consomme to ⅓. Remove from heat and add terrapin eggs and beef extract and slowly stir in the butter. Season and then add brandy and wine. Reheat but do not allow to boil. Pour over terrapin. Serves 3-4.

TERRAPIN CLUB STYLE

2 prepared terrapins*	1 c cream
3 tbsp butter	3 egg yolks, beaten
½ tsp salt	2 tbsp dry sherry

Melt butter and add terrapin and salt and heat thoroughly. Add cream and let simmer gently while stirring for 2 minutes. Remove from heat and slowly stir in the egg yolks. Add sherry just before serving. Serves 3-4.

TERRAPIN STEW BALTIMORE

1 prepared terrapin*	terrapin stock
terrapin eggs, (if any)	¼ c butter
2 egg yolks, beaten	pinch cayenne pepper
pinch salt	pinch black pepper
sherry to taste	

Cook the meat of the prepared terrapin in the stock to cover in the top of a double boiler over boiling water for 1 hour. The stock will be reduced by half. The stew should be finished in a chafing dish at the table. Put the meat and the reserved eggs in the chafing dish with the butter and heat together. Add a little of the stock to the egg yolks and stir the yolks into the pan. Add the seasonings and sherry just before serving. Serves 3-4.

TERRAPIN, BALTIMORE STYLE

1 pt prepared terrapin
sherry
brandy
2 ozs + fresh butter
half glaze*
salt and pepper to taste

Drain the broth from the terrapin meat and reserve. Cook the 2 ounces of butter to a hazelnut color and then add the meat and fry for a few minutes. Then add the broth from the terrapin and a little sherry. Thicken with the same quantity of half glaze as broth and let boil for 5 minutes. Finish with a piece of fresh butter and a little brandy. Season to taste and serve in a chafing dish.

TERRAPIN JOCKEY CLUB

The same as terrapin Baltimore style except that sliced fresh mushrooms fried in butter, sliced truffles and olives are added.

TERRAPIN CHESAPEAKE BAY

2 prepared terrapins*
1 c bouillon
2 c sherry
juice of ½ lemon
milk
grated rind of ½ lemon
¼ c celery, chopped
cayenne pepper
2 tbsp plain flour
½ c cream
1 tsp nutmeg
2 tsp worcestershire sauce
4 hard boiled eggs
1 sm onion, minced
salt
toast

Mash egg yolks with lemon juice, grated rind, nutmeg and flour. Blend in bouillon and add onion, celery, terrapin, sliced terrapin liver and eggs, if any. Add enough milk to cover and cook in double boiler until meat separates from bones (if any). Remove bones. Add chopped egg whites, worcestershire sauce, cream and sherry. Season to taste, reheat, and serve on toast. Serves 6.

TURTLE SCOTCH BROTH

1 lb turtle meat
3 turnips, diced
⅓ c pearled barley
1 tbsp parsley flakes
21 ozs water
3 carrots, diced
1 onion, diced
1 tsp salt
1 10½ oz can beef broth

Place the vegetables and barley in a crock pot and sprinkle with the salt and parsley. Place the meat on the top and add the broth and water. Cover and cook on low 8-10 hours. Lift out the meat (remove bones, if any) and cut into small pieces. Return the meat to the crock pot and correct the seasoning. Serves 4.

STEWED TERRAPIN WITH CREAM

1 pt prepared terrapin*
1 tbsp rice flour
1 tbsp salt
¼ tsp nutmeg, grated
4 egg yolks, well beaten
2 tbsp butter
1 pt thin cream
½ tbsp white pepper
pinch cayenne pepper
1 tbsp lemon juice

Place in a saucepan the butter and flour and stir over a fire until it bubbles and then stir in the cream slowly, the salt, the peppers, nutmeg and turtle meat. Stir until scalding hot. Remove from fire and stir in the egg yolks. Do not boil again but pour immediately into a tureen containing the lemon juice. Serve hot.

CREAMED TERRAPIN

1 prepared terrapin*
½ c cream
1 tsp salt
1 tsp lemon juice
1 c white sauce*
1 c mushrooms, chopped
¼ tsp cayenne pepper
2 eggs, beaten

Heat sauce and add cream, eggs, seasonings and lemon juice. Cook over boiling water for 5 minutes. Add the meat and mushrooms and cook until thoroughly heated. Serves 6.

TERRAPIN IN CREAM

2 c prepared terrapin*, diced
8 tbsp butter
½ tsp salt
allspice to taste
½ c sherry
6 hard boiled eggs
2 c heavy cream
white pepper to taste
nutmeg to taste

Sieve the egg yolks and then cream them in the butter. Scald the cream over hot water and add the seasonings and beat in the egg mixture. Add the terrapin and sherry and heat thoroughly but do not boil. Serves 4.

TERRAPIN NEWBERG

1 pt prepared terrapin*
salt
4 egg yolks, beaten
1 oz butter
¼ pt + sherry
cayenne pepper
½ pt heavy cream

Put the terrapin in a saucepan and reduce the broth almost completely and then add a little sherry. Season with salt and cayenne pepper and thicken with the egg yolks diluted with the cream. Stir on the fire without boiling and add the butter. Before serving add the ¼ pint of sherry and adjust the seasoning. Serve in a chafing dish with toast on the side.

TERRAPIN, INDIAN STYLE

As for terrapin Newberg only add a tbsp of curry powder diluted with a little water and garnished with cooked rice.

BOOKBINDERS TURTLE SOUP

2 prepared terrapins*
2 tbsp plain flour
pepper to taste
1 tsp parsley, chopped
terrapin eggs (if any)
2 tbsp butter
celery salt to taste
1 qt boiling water
sherry

Put the meat and the eggs (if any) into boiling water. Add chopped parsley, pepper and celery salt to taste. Boil for 20 minutes. Remove a cup of the stock and place it in the freezer. Take a frying pan and melt the butter and add flour until smooth and creamy. Add ¼ cup of chilled turtle stock from the freezer. Never use hot turtle stock or it will make the mixture lumpy. Put on the stove over a medium heat and stir and let cook until the mixture thickens. Add enough turtle stock until the mixture is about as thin as half and half cream and milk, no thicker. It is better to have the sauce too thin than too thick. Cut up the terrapin meat into small pieces and add with the eggs. Season with celery salt and pepper again if necessary. Serve with sherry.

SNAPPER SOUP MORE OR LESS LIKE BOOKBINDERS

4 c turtle meat, cooked with skin
4 tbsp butter
3 c onions, chopped
1 c celery, chopped
½ tsp marjoram, dried
1 bay leaf
1 qt canned tomatoes
tabasco sauce to taste
2 hard boiled eggs, chopped
sherry
salt and pepper to taste
1 2½-3 lbs veal shank
2 c carrots, chopped
½ tsp thyme, dried
3 cloves, whole
1 c plain flour
4 qts turtle stock
2-6 cloves, crushed
3 lemon slices, seeded

Preheat the oven to 350°F. On top of the stove heat the butter in a small heavy roasting pan. Sprinkle the veal shank with salt and pepper and add. Cook the shank turning to brown it lightly all over. Scatter the onions, carrots, celery, thyme, marjoram, whole cloves and bay leaf around the shank. Cook about 5 minutes, occasionally turning the shank and stirring the vegetables. Place the pan in the oven and roast for 30 minutes, stirring occasionally to prevent sticking. Sprinkle the vegetables with the flour and stir until well coated. Bake 30 minutes longer. Scrape all this into a large soup kettle and add the tomatoes and turtle stock, stirring constantly. Bring to a boil and simmer about 2 hours, stirring frequently to prevent sticking or burning. Add the turtle, crushed cloves, tabasco sauce and lemon slices. Simmer about 15 minutes longer. Add salt and pepper. Stir in the eggs. Serve with sherry on the side. Makes about 1 gallon.

MOCK TURTLE SOUP

½ lb beef, boneless
½ lb lean pork, boneless
1 lb chicken, boneless
1 c onions, ground
salt and pepper to taste
3 tbsp scallions, chopped
4 hard boiled eggs, chopped
2-3 c sherry
¼ c roux, brown*
2 lemons, seeded and chopped

Boil all meats together until tender and cut into small pieces. Fry onions in the roux for 3 minutes. Add the broth and meats and remaining ingredients, except sherry, and let cook ½ hour. Just before serving add the sherry.

MOCK TURTLE SOUP 2

1 calf's head
6 cloves
2 carrots, peeled
12 peppercorns
salt
½ c plain flour
1 c stewed tomatoes, drained
juice ½ lemon
cognac or madeira
2 qts + water
2 onions, whole
handful celery tops
1 lge leek
3 sprigs parsley
¼ c butter
2 c brown stock*
1 c veal stock

Have the butcher cut, clean, and bone a calf's head. Soak in cold water to cover for about 2 hours and drain well. Boil for 30 minutes the 2 quarts water, onions studded with 3 cloves each, celery, carrots, leek, peppercorns, parsley and salt to taste. Simmer the head and the bones in this mixture 2 hours or until the meat is tender. Remove the head to a platter. Boil the stock rapidly until reduced to 1 quart and then strain. Melt butter and add flour and stir over low heat until lightly browned. Add brown stock and bring mixture to a boil, stirring constantly. Add strained court bouillon and tomatoes, cup of diced veal meat and then lemon juice. Heat well and then at last add the cognac or madeira.

MOCK TURTLE SOUP 3

1 calf's head
1 bu parsley, chopped
2 leeks, chopped
1 tsp black pepper
1 pt madeira
cayenne pepper to taste
1 sprig parsley, minced
water
1 slice fat ham
1 sprig thyme, chopped
6 cloves, whole
3 ozs + 1 tbsp butter
1 lemon, sliced thin, seeded
salt to taste
1 tbsp wheat flour
1 egg, well-beaten

Clean and wash a calf's heat and split in two saving the brains. Boil the head in plenty of water until tender. Put the ham, parsley, thyme, leeks cloves, black pepper and the 2 ounces of butter into a stew pan. Fry to a nice brown. Then add the water the head was boiled in and cut the meat from the head and the ham into small pieces and add to the soup. Add the madeira and lemon along with cayenne pepper and salt to taste. Simmer gently for 2 hours. Make a forcemeat of the brains by putting them in a stew pot and pour hot water over them and set over the fire for a few minutes. Remove and chop them fine. With the sprig of parsley, salt and pepper to taste, wheat flour, tbsp butter and egg make into small balls. Drop them in the soup 15 minutes before it is time to take off the fire. In making the balls a little more flour may be necessary. Skim the soup clear and serve. The brain balls may be left out if desired.

MOCK TURTLE SOUP 4

2 lbs boned chicken breasts
2 tbsp oil
1 tbsp plain flour
1 c celery, minced
1 c onions, minced
2 tbsp tomato paste
½ c scallions, minced
2 cloves garlic, minced
2 hard boiled eggs, sliced
2 qts veal stock
2 sprigs basil, chopped
2 sprigs rosemary, chopped
salt and pepper to taste
½ lemon, sliced thin, seeded
1 c sherry
¼ c basil, minced
dash tabasco sauce

Cut the chicken into bite size pieces and lightly brown in hot oil. Remove to a warm plate. Add the flour and cook until it turns brown and then add the celery and onions. Sauté until soft and then add the tomato paste, scallions and garlic. Cook for a few minutes stirring constantly so the mixture does not burn. Return the chicken to the pot and add the veal stock, salt, pepper and lemon slices. Simmer for about ½ hour, stirring occasionally. When ready to serve remove the lemon slices and add sherry, basil and tabasco sauce and allow to simmer 5 minutes longer and then serve in warm soup bowls. Garnish with egg slices. Serves 6.

MOCK TURTLE SOUP 5

1 lb stew meat
1 lb sausage, loose
½ lb salt meat
3 ribs celery, diced
1 8 oz can mushrooms
2 14½ oz cans stewed tomatoes
1 tbsp Accent
salt and pepper to taste
2½ gals water
½ c scallions, chopped
¼ lb spaghetti, broken into pieces
1 sm chicken
2 c ham, chopped
2 lge onions, chopped
4 cloves garlic, diced
1 10 oz can Ro-Tel Tomatoes
1 tbsp tabasco sauce
¼ tsp allspice
¼ tsp marjoram
¼ tsp basil
¼ tsp thyme
¼ tsp rosemary

Clean, cut and season stew meat and chicken with salt, lemon, pepper and Accent and cut salt meat into small dice and break up the sausage meat. Place meats and all ingredients except scallions, parsley, spaghetti and ham into a large soup pot. Bring to a boil and simmer for 1-1½ hours. Add the remaining ingredients during the last 15 minutes of cooking. Serves 14-16.

DELUXE MOCK TURTLE SOUP

2 cans condensed mock turtle soup
1½ cans water
parsley, chopped
½ c sherry
1 hard boiled egg, sliced

Simmer the soup and water for 15 minutes and put through a strainer. Add sherry and serve with a slice of boiled egg and a topping of parsley. Serves 4.

MOCK TURTLE SOUP FAMILY STYLE

1 c red kidney beans
grated peel of 1 lemon
water
salt
1 tbsp butter
dash of tabasco sauce
1 med onion, chopped
sprinkle of cayenne pepper
1 clove garlic, crushed
2 hard boiled eggs, chopped
1 tbsp plain flour
1 c sherry

Soak beans overnight in cold water. Drain and cook, covered, in large amount of boiling water 2-3 hours until tender. Drain and measure water. There should be 3 pints. If too much, reduce by boiling. If not enough add boiling water to make up quantity. Mash beans through a sieve. Brown onion, garlic and flour in butter. Gradually stir in water and add beans and grated lemon peel and season quite highly. Simmer gently 1 hour. Add sherry and eggs just before serving. Serves 6.

MOCK CREOLE TURTLE SOUP

1 stick butter
½ c plain flour
2 lbs lean beef stew meat
½ c tasso*, minced
1½ c onion, chopped
¾ c celery, chopped
½ c carrot, minced
3 cloves garlic, minced
1½ c tomatoes, seeded, chopped, peeled
2 tsp tomato paste
tabasco sauce (optional)
1 lemon, chopped fine
1 tbsp parsley, chopped, fresh
1 tsp thyme leaf, dried
1½ qts beef stock
1 tsp white pepper
1¼ tsp cayenne pepper
3 bay leaves
3 hard boiled eggs, chopped
scallion, chopped
1 tbsp worcestershire sauce

Cut beef into ½ inch cubes. Heat the butter in a heavy saucepan and brown the beef and remove. Add the flour and cook while stirring until roux* is medium brown. Add the onions, celery, carrots and garlic and cook for 5 minutes until soft. Add the meat and all remaining ingredients except eggs and shallots. Stir to blend and simmer until the meat is very tender in about 1½ hours and add a little water if needed during cooking. Taste for salt and pepper. If desired add tabasco sauce. Add the eggs and cook for 5 minutes. Garnish with the shallots. Serves 8-10.

SOUPE FAUSSE TORTUE

2 calf's feet, split	1 lb veal shank
2-2½ lbs beef shank, 2 in pieces	2 tbsp butter
1 onion	2 carrots, sliced
¼ c water	4 qts broth or water
2 whole cloves	½ tsp peppercorns
1 bay leaf	1 tbsp plain flour
¼ c tomato, pureed	1 shallot, minced
1 sprig thyme	1 sprig sage
1 sprig savory	1 sprig rosemary
1 sprig basil	1 sprig marjoram
½ c madeira	⅔ c mushrooms, sliced, sautéed in butter
cayenne pepper to taste	6 hard boiled egg yolks, chopped

Melt half the butter in a large heavy saucepan and add the onion, carrot, veal and beef shanks and the calf's feet (which have been blanched in boiling water for 10 minutes and rinsed). Add the ¼ cup of water and cover and cook over a low heat until the water has completely evaporated and the meats are lightly colored, about 30-40 minutes. Pour in the broth or water and add the cloves, peppercorns and bay leaf. Bring to a boil and put the lid on slightly ajar and cook the meats very slowly until they are tender in about 3 hours. Take out the calf's feet, bone them and put their meat on a plate with a weight on top to press as it cools. Strain and degrease the broth, discarding the shanks. In a large saucepan melt the remaining butter and stir in the flour and add the broth and the pureed tomato. Simmer over very low heat about 30 minutes removing the skin as it forms on the surface of the soup. Put the shallots and herbs in a small pan with the madeira and simmer, covered, for 5 minutes. Cut the meat from the calf's feet into small dice and put it into a saucepan. Strain in the madeira and herb infusion and add the mushrooms and cover to keep the meat warm. Remove any traces of fat from the soup and season with cayenne pepper. Pour the soup through a strainer into a warmed tureen, add the herb infusion with the meat and mushrooms and serve the soup garnished with the chopped egg yolks.

MOCK COOTER* SOUP

3 lbs lean beef	1 pt sweet milk
1 tbsp plain flour	½ tsp mace
salt and pepper to taste	½ tsp dry mustard
1½ qts water	¼ lb butter
½ pt cream	2 hard boiled eggs, cut up

Boil meat and water until 1 quart of liquid remains. Add the milk and butter. Remove the meat and allow to cool and return to stock. Let it cook down a little more. Add the flour dissolved in the cream and add the seasonings.

LOUISIANA MOCK TURTLE SOUP

2 lbs beef brisket, cubed
2 onions, sliced
1 16 oz cans whole tomatoes
3 bay leaves
¼ tsp thyme
1 c sherry
1 lemon, sliced thin, seeded
2 tbsp peanut oil
½ c plain flour
2 qts water
6 whole cloves
salt and pepper to taste
3 hard boiled eggs, sliced

Brown the meat in peanut oil. Add the onions and cook until tender. Remove the meat and onions and add the flour and make a dark roux*. Add meat and onions, stir in tomatoes, water, spices, salt and pepper. Simmer for 3 hours. Remove the bay leaves and cloves. Before serving add sherry and garnish with egg and lemon slices. Serves 8.

KOSHER MOCK TURTLE SOUP

1½ lbs red snapper fillets
1 rib celery, chopped
2 shallots, chopped
1 onion, minced
3 cloves garlic, chopped
1½ sticks pavere margarine*
3 tbsp plain flour
salt and pepper to taste
8 tbsp sherry
1 tbsp worcestershire sauce
1 lemon, sliced thin, seeded
1 sprig parsley, chopped
1 hard boiled egg, grated
2 cloves garlic minced
1 8 oz can stewed tomatoes
water

Place whole red snapper in salted water and boil until tender. Set aside to cool and then cut into bite size pieces. Continue cooking the broth and skimming off the fat when necessary. Remove from heat and add celery, shallots, chopped garlic and parsley. Make a roux* with the flour and margarine and then add the onion, tomatoes, minced garlic, worcestershire sauce, salt and pepper to taste. Combine both and cook 15 minutes. Add the fish, sherry and egg at the last minute.

BAIGAN SOUP (MOCK TURTLE SOUP)

2 lge eggplants
3 c beef broth
2 tbsp water
2 tbsp butter
¼ tsp black pepper, freshly ground
1 qt milk
2 tbsp plain flour
1 tbsp anchovy paste
½ tsp salt
3 tbsp parsley, chopped

Peel the eggplants and cut into 1 inch cubes. Combine in a saucepan with the milk and broth and bring to a boil. Cook over low heat for 45 minutes or until the eggplant is very soft. Mix the flour and water to a smooth paste and add it to the eggplant mixture, stirring constantly. Cook for 5 minutes, stirring frequently. Puree the mixture in an electric blender or force through a sieve. Add the anchovy paste, butter, salt and pepper. Correct the seasoning if necessary. Heat again but do not allow to boil. Serve sprinkled with the chopped parsley. Serves 8-10.

CANNED TURTLE

turtle	pt or qt jars
water	salt

Dress out the turtle and place in a little water and salt in a pressure cooker and cook until the meat separates from the bones. Remove the bones and place the meat with the broth in sterilized jars, adding water if necessary, and then pressure cook according to your pressure cooker instructions. This can now be stored until needed for some other delicious dish.

CANNED TURTLE SOUP

1 pt turtle or more	potatoes, diced
green beans	carrots, diced
tomatoes	celery, diced
peas	red cabbage, minced
water	seasonings to taste

Combine all ingredients with a little water to cover and simmer until the vegetables are tender. Season to taste.

CANNED TURTLE SOUP 2

1 10½ oz can green turtle soup	1½ tbsp oil
2 tbsp plain flour	1 lge onion, chopped
2 cloves garlic, chopped	2 carrots, grated
1 8 oz can tomato sauce	2 c water
¼ c bell pepper, chopped	2 hard boiled eggs, sliced
1-2 tbsp sherry	dash worcestershire sauce
salt and pepper to taste	

Mix the oil and flour and cook slowly until dark brown. Add onion, garlic and bell pepper and sauté until tender. Add the tomato sauce, carrots and water and cook for 15 minutes, stirring occasionally. Add the can of soup and worcestershire sauce and bring to a boil, lower the heat, and simmer for 20-30 minutes. Season to taste. Add sherry and slices of egg just before serving.

TURTLE SOUP WITH MEATBALLS

2 cans turtle soup	salt and pepper to taste
¼ lb turtle steak, ground	nutmeg, grated
plain flour	2 ozs fine vermicelli

Season the meat with salt, pepper and nutmeg and shape into small balls and roll them in flour. Add them to the soup which has been heated and break in the vermicelli. Cover and simmer for 20 minutes. Serve at once.

TURTLE SOUP WITH MADEIRA

1 10 oz can turtle meat
1 tbsp madeira for each bowl soup
3 10½ oz cans turtle soup

Cut the turtle meat into small pieces and add the soup. Simmer about 40 minutes uncovered. Add madeira when serving. Serve with toast. Serves 6.

SOPA ALETUS

1 10 oz can turtle meat
½ c onions, chopped
1 8 oz can spanish style tomato sauce
¼-½ tsp oregano
1 c mixed vegetables, cooked, canned or frozen
¼ c oil
3-4 c beef stock
1 clove garlic, minced
⅛ tsp black pepper
salt to taste

If canned turtle meat not available use 1-1½ cups of cooked fresh turtle meat. Drain canned turtle meat and reserve the liquid and cut meat into bite size pieces. If using fresh or frozen turtle simmer until tender in salted water and then cut into bite size pieces and reserve the broth. In a soup pot heat the oil and add onions and cook until soft and yellow. Add the beef stock, tomato sauce and the reserved turtle broth. When the soup is simmering add the garlic, oregano, turtle meat, pepper, and vegetables and simmer for about 30 minutes. Add salt to taste if needed. Serves 6 as an opener or 2-3 as a main course.

PENDENNIS TURTLE SOUP

1 10 oz can turtle meat
1½-2 lbs veal or beef bones
2 med onions, sliced
2 qts water
1 no. 1 can tomato puree
¼ tsp thyme, crushed
½ c sherry
1 slice grapefruit, thin
2 tbsp butter
2 med carrots, sliced
3 tbsp plain flour
2 tsp salt
3-4 cloves, whole
½ tsp black pepper
1 thin slice lemon
2 hard boiled eggs, chopped

Melt the butter carefully in a heavy kettle and add the bones and brown slowly. Add the carrots and onions to the bones and brown. Stir in the flour and brown. Do not let it burn. Add the water and salt and simmer slowly for 2½ hours. Cool this stock slightly and strain. To the strained stock add the tomato puree, cloves and thyme and stir well. Simmer for about 3 minutes. Add the turtle meat which has been diced and its stock and simmer until thoroughly heated. Add additional salt if needed and pepper. Just before serving remove the cloves and add the sherry. Cut the slices of lemon and grapefruit into 6-8 small pieces. Ladle soup into cups and garnish each with a piece of lemon and grapefruit and eggs. Serves 8. (If fresh turtle is used cover with water and add 1 tsp salt for each pound of meat and simmer for 2 hours. Remove meat from bones and dice.)

IRISH TURTLE SOUP

- 3-4 c irish clear soup or 2 10½-oz cans condensed beef consomme diluted with 2 cans water
- 1 10 oz can turtle meat
- salt to taste
- ¼ c light rum or irish whiskey

Dice the turtle meat into bite size pieces and add to the clear soup (or consomme) along with its broth in a kettle and heat to simmering. Add additional salt if needed. Just before serving stir in the rum or whiskey. When the soup simmers again pour into bouillon cups and serve. Serves 6.

IRISH CLEAR SOUP

- 3-4 lbs beef shank
- 4 qts water
- 2 carrots, whole, unscraped
- 1 leek, whole
- 6 peppercorns, whole
- 1 sprig parsley
- ¼ tsp thyme, crushed
- ¼ c irish whiskey
- veal bones
- 2 lge onions, halved
- 2 ribs celery and leaves
- 6 cloves, whole
- 1 bay leaf
- 1 tbsp salt
- 2 egg whites

Combine the beef and veal bones with the water in a large kettle and bring to a boil. Skim off any scum that forms. Turn heat down to a simmer. Add the vegetable along with the cloves, peppercorns, bay leaf, parsley, salt and thyme. Simmer slowly for 5-6 hours, skimming occasionally if needed. Cool slightly. Wet a piece of muslin or cheesecloth and place it in a wire sieve. Strain the stock through the sieve. Chill until the fat hardens on top and remove it. Clarify the stock by bringing to a boil and beat in the egg whites with a wire whip. Turn the heat down and simmer for about 30 minutes. Strain it again through the cloth and sieve. To serve add the irish whiskey.

PURE CLEAR GREEN TURTLE SOUP

- 1 10½ oz can clear green turtle soup
- 1 qt soup stock, well seasoned
- salt and pepper

To the soup stock add the can of turtle soup and season. Simmer for 15 minutes.

SOUP STOCK

To make a soup stock simmer in water for an hour or more any kind of meat and bones along with any combination of several of the following: onion, leek, turnip, carrot, celery, thyme, bay leaf, sage, clove and pepper. Strain.

TURTLE (CANNED) WITH WHITE SAUCE

1 10 oz can turtle meat, diced
½ c mushrooms, chopped
½ c sherry
parsley, minced
2 c white sauce*, thick
butter
toast

Sauté mushrooms in a little butter and add to the white sauce. Add the sherry and the turtle meat, drained and mix together and heat through. Serve on toast and sprinkle with parsley.

A TASTE OF TURTLE

1 10½ oz can clear turtle soup
1 tbsp unflavored gelatin
⅔ c unflavored yogurt
¼ tsp curry, powdered
1¼ lb jar lumpfish roe (garnish)
lemon slices (garnish)
⅔ c water
½ c yogurt curd cheese or cream cheese
1 tsp garlic salt
½ tsp tabasco sauce
cucumber, sliced thin (garnish)

Place the water in a small saucepan and add the gelatin. When it has swollen, heat the mixture to dissolve the gelatin. Combine with the remaining ingredients and puree quickly in a blender or whisk until smooth. Divide in 6 small bowls and chill until set. Just before serving top each with a spoonful of lumpfish roe or thin slices of cucumber or lemon. Serves 6.

YOGURT CURD CHEESE

4½ c whole milk yogurt, unflavored

Line a large colander with cheesecloth and set over a bowl. Beat the yogurt and pour it into the lined colander. Gather up the corners and tie them together. Hang the bag of curds over a bowl for about 4 hours. Longer draining makes a drier, thicker cheese. Refrigerate the cheese and use within 3-4 days. Makes 1½ cups.

TURTLE SOUP A LA MICROWAVE

4 lbs turtle meat
1 c onions, chopped
¼ c parsley
2 tsp lemon juice
2 tsp salt
2 ozs sherry
1 clove garlic, chopped
½ c celery, chopped
6 hard boiled eggs, sliced
1 c oil
¼ c scallions, chopped
1 c plain flour
½ tsp pepper
1½ gal water
2 tbsp tomato paste
4 slices lemon

On stove top slowly boil turtle garlic and lemon juice in water for about 2 hours. In a 4 quart glass casserole in microwave, cook roux with flour and oil until golden brown (about 13 minutes), stir in onions and celery and sauté 5 minutes. Stir in tomato paste, scallions and parsley. Cook on high 6 minutes. Add stock, meat, lemon and seasonings. Cover. Cook on high 10 minutes, medium speed 30 minutes. Serve with sherry and sliced eggs.

MICROWAVED TURTLE SOUP

1 lb turtle meat, diced
¾ c plain flour
2 med onions, minced
3 bay leaves
½ tsp thyme
1 10¾-can tomato puree
½ c lemon juice
1 tbsp parsley, minced
1 c unsalted butter
4 ribs celery, minced
1½ tsp garlic, minced
1 tsp oregano
½ tsp black pepper, freshly ground
3 10½ oz cans condensed beef broth
5 hard boiled eggs, chopped
1 tsp sherry/serving

Place ½ cup of butter into a 4 quart batter bowl. Microwave on high 40-60 seconds or until melted. Stir in flour well so that no lumps remain and microwave on high for 6 minutes. Then stir and microwave on high 1-2 minutes or until the mixture is a caramel brown color. Set aside. Put the remaining butter in a 4 quart pot and microwave on high 40-60 seconds. Add the turtle meat, celery, onions, garlic, bay leaves, oregano, thyme and pepper. Stir well. Cover with a lid or plastic wrap. Microwave on high 9-10 minutes, stirring after each 3 minutes. Add the tomato puree, recover, and microwave on high 10 minutes. Add the beef broth, recover, and microwave on high 10 minutes. Add the set-aside flour mixture , stir well, and recover and microwave on high 5 minutes or until thickened and bubbly. Add the lemon juice, eggs and parsley. To serve add the sherry to each soup bowl. Serves 8.

TOUT DE SUITE A LA MICROWAVE TURTLE SOUP

3-4 lbs turtle meat
1 tsp + 1 tbsp salt
1 onion, sliced
4 bay leaves
3 slices lemon
⅔ c flour
1 c celery, minced
1 bu scallions, chopped
1 28 oz can tomatoes, chopped
1 qt turtle stock
1 tsp cayenne pepper
2 qts hot water
1 tsp black pepper
3 cloves, whole, stuck into 1 garlic pod
6 hard boiled egg yolks
⅔ c oil
2 c onions, minced
4 cloves garlic, minced
¼ c parsley, chopped
1 6 oz can tomato paste
4 tbsp worcestershire sauce

Place the turtle meat in 1 quart of the water and add 1 tsp of salt and pepper, sliced onion, cloves in garlic pod, bay leaves and 1 slice of lemon. Cover and microwave on high 30 minutes or until the meat is tender. Remove the meat from the bones and mince. Strain the stock and reserve. (This could be done ahead of time and refrigerated.) Mix the oil and flour in a 4 cup measure and cook on high 6-7 minutes. Stir. Add the minced onion, celery, scallions and parsley and cook on high 5 minutes. Stir once while cooking. Now pour into a larger 4-5 quart dish. Add the tomatoes, drained, and paste to this mixture and cook on high 10 minutes. Add the turtle meat, turtle stock, tomato liquid, worcestershire sauce, salt and cayenne pepper and cover with wax paper. Cook on high 30 minutes stirring once. Add the remaining water and lemon and cover and cook on high 30 minutes. Sieve the egg yolks. After serving the soup, place 1 heaping tbsp of yolk in each bowl.

LAC DES ALLEMANDS TURTLE SOUP

Stock

7 lbs turtle meat, bone in
3 lemons, quartered
2 med garlic, peeled
2 gals cold water
1 bu scallions, 2 inch pieces
1 med yellow onion, quartered

Roux

1 c cottonseed oil
3 med yellow onions, chopped
2 bu scallions, sliced
2 qts boiling water
3 16 oz cans diced tomatoes
3 tbsp salt
½ tsp white pepper
4 hard boiled eggs, chopped
2 c all-purpose flour
1 bu parsley, chopped
4 ribs celery, chopped
3 10 oz cans beef consomme
3 16 oz cans tomato sauce
2 tsp cayenne pepper
½ c dry sherry
garlic, onion, and reserved stock

Combine all stock ingredients in a 16 qt pot and cover. Bring to a boil and then turn heat to medium and simmer until turtle becomes tender, approximately 1½ hours. Strain ingredients and set aside reserving the stock. Bone the turtle and chop into coarse pieces. Save the garlic and onion pieces. Discard the remainder. In a heavy skillet heat the oil over medium heat and add the flour and stirring constantly and cook until the roux becomes a dark brown. Add the chopped seasonings except for the parsley and cook until tender, about 15 minutes. Remove the roux to a 16 qt pot and on medium, add the water carefully and stir well and simmer for 20 minutes. Add the reserved stock, consomme, diced tomatoes and tomato sauce and simmer an additional 15 minutes. Add the turtle meat, eggs, salt, cayenne pepper, reserved garlic and onion pieces and stir well. Turn heat to low and simmer 20 minutes. Just before serving add the parsley and sherry and correct for seasoning. Yields 3½-4 gallons.

CLAUDE'S TURTLE ON THE BAYOU

4 lb turtle meat, cubed
1 tbsp Tony's Creole Seasoning*
¼ c water
6 cloves garlic, chopped
1 c onions, chopped
cooked rice

Soak a clay simmer pot in cold tap water for 10 minutes. In the bottom of the pot combine all the ingredients and cover with the soaked lid. Microwave on high for 10 minutes. Stir and again cover and microwave on 50% for 40-45 minutes or until meat is tender. Let stand 5 minutes before serving. Serve meat and gravy over rice.

TURTLE SOUP A LA CHACHERE

3 lbs turtle meat
½ c plain flour
2 c onions, chopped
1 c bell pepper, chopped
3 tbsp whole allspice
4 tbsp worcestershire sauce
parsley, minced
1 tbsp sherry/serving
2 qts water
½ c oil
1 c celery, chopped
6 cloves garlic, chopped
2 lemons, thin sliced, seeded
1½ tbsp Tony's Creole Seasoning*
4 hard boiled eggs, sliced

In a 5 quart casserole combine the turtle meat and water and cover with a plastic lid or plastic wrap. Microwave on high for 20 minutes and then on 50% for 40-45 minutes or until the meat is tender. Set aside. In a 4 cup glass measure combine the oil and flour mixing well. Microwave on high for 6 minutes or until as dark as desired. Stir when there is 2 minutes left and at 1 minute and 30 seconds. Add onions, celery, bell pepper and garlic mixing well and microwave on high for 5-6 minutes or until vegetables are soft. Remove the turtle from the stock and strain the stock and return to the 5 quart casserole. Remove the bones from the heat and dice and return to the stock. Tie the allspice in a cloth bag and add to the stock along with lemon slices, Tony's Creole Seasoning and worcestershire sauce. Cover and microwave on high for 10 minutes and then on 50% for 30-35 minutes. Remove the bag of allspice. Add sherry with each serving and garnish with the eggs and parsley.

MICROWAVE TURTLE STEW

3 lb turtle meat, cubed
3 tbsp oil
2 c onions, chopped
½ c celery, chopped
1 c scallions, chopped
1 6 oz can tomato paste
2 tbsp Tony's Creole Seasoning*
8 whole cloves
1 tbsp sugar
½ c butter
2 qts water
3 tbsp plain flour
1 c bell pepper, chopped
2 cloves garlic, chopped
2 16 oz cans tomatoes, chopped
1 c sherry
4 bay leaves
½ tbsp allspice, ground
6 hard boiled eggs
1 lemon, sliced, seeded

In a 5 quart casserole combine the turtle and water and cover and microwave on high for 20 minutes and then on 50% for 40-45 minutes or until the meat is tender. Set aside. In a 4 cup glass measure combine the flour and oil and mix well and microwave on high for 5 minutes, stirring occasionally, or until as dark as desired. Add the onions, bell pepper, celery and garlic mixing well. Microwave on high for 5-6 minutes or until vegetables are soft but not brown. Remove the turtle meat from the stock and strain and return to casserole. Remove bones, if any, from the turtle and return to the stock. Add the flour mixture, scallions, tomatoes, drained, tomato paste, sherry, Tony's Creole Seasoning, bay leaves, cloves, allspice and sugar. Cover and microwave on high for 10 minutes. Mash egg yolks, chop the whites, and add to the stew. Cover and microwave on 50% for 40-45 minutes. Add lemon slices and butter and cover and let stand for 10 minutes before serving.

McJUNKINS TURTLE SAUCE PIQUANT

1 lb turtle meat, cubed
½ c bell pepper, chopped
2 cloves garlic, chopped
1 5½ oz can mushroom steak sauce
2 tbsp worcestershire sauce
1 cayenne pepper, chopped
¼ c parsley, chopped
1 c onions, chopped
½ c celery, chopped
1 4 oz can mushrooms, chopped
1 10 oz can Ro-Tel Tomatoes
1 tbsp Tony's Creole Seasoning*
½ c scallion tops, chopped
⅓ c water

Soak a clay simmer pot in cold tap water for ten minutes. Combine all ingredients in the pot, cover, and microwave on high for 10 minutes. Stir and again cover and microwave on 50% for 35-40 minutes. Let stand 5 minutes before serving.

DRIED TURTLE SOUP

½ dried turtle
½ tsp sweet basil
juice of ½ lemon
3 qts strong stock
1 wine glass sherry
salt and pepper to taste

Soak the dried turtle in cold water overnight. Put it into a pan and cover with fresh cold water and bring to boil slowly and simmer for 8 hours. Strain and cut into pieces about 1½ inches square. Add these pieces to the stock and simmer until the turtle is quite tender. Five minutes before serving add the basil, sherry and lemon juice.

NANIE'S TURTLE SOUP

6 lbs turtle meat
oil
salt and pepper to taste
3 onions, chopped (optional)
sherry (optional)
2 c roux*
few bay leaves
12 potatoes, diced
3 scallions, chopped
lemon, sliced
hard boiled eggs, sliced
tabasco sauce
12 qts water

Fry turtle meat in a little oil. When fried set it aside. Sauté onion (if using). Add water to roux a little at a time and keep smooth. Add onions (if using), salt, pepper, tabasco sauce and bay leaves. Simmer until meat is tender. Add potatoes and cook until potatoes are done. Just before removing from fire, add the scallions. Serve with sliced eggs and lemons, lightly squeezing lemons in soup. Add sherry if desired.

ERNIE'S TURTLE SOUP

20 lbs turtle meat, boneless
2 c plain flour
10 sprigs parsley, chopped
4 cayenne peppers, whole
5 c bell peppers, chopped
4 ozs worcestershire sauce
2 bu scallions, chopped
5 8 oz cans mushrooms, stems and pieces, drained
10 c celery, chopped
sherry (optional)
2 tsp granulated garlic
17½ qts water
10 c tomato sauce
hard boiled eggs, sliced
hand full of bay leaves
12 lemons, sliced, seeded
2 tbsp sweet basil
20 lge onions, chopped
2 c oil
salt and pepper to taste

Prepare turtle meat. (If with bones boil in measured water until cooked. Bone turtle and reserve stock. Measure. Use stock in place of some water.) Season with salt and pepper and fry in oil until brown. Remove from oil and add the flour and brown slowly until golden. Add onions, celery, bell peppers and mushrooms to flour mixture and sauté until tender. Return the turtle meat to the pot and add the tomato sauce, garlic, and 5 pints water and cook about 4 hours. Add lemon, bay leaves, sweet basil, reserved mushroom liquid, and 15 quarts water and simmer for 5 hours or until soup has reduced to desired thickness. Add worcestershire sauce, parsley, scallions and season to taste and simmer 15 minutes longer. Turn off heat and when oil floats to surface skim off and discard. Serve with slices of hard boiled eggs in soup plates and add ½-1 oz sherry (optional).

SCHEMMEL'S TURTLE SOUP

6 lbs or more turtle meat
2 onions, halved
1 rib celery , chopped
4 carrots, minced
1 11 oz can whole kernel corn
1 16 oz can green beans, diced fine
1 qt tomato juice
6 lemons, rind grated, juiced
1 c plain flour
1 lge potato, diced fine
salt and pepper to taste
3-4 sprigs parsley
1 c pickling spices
½ sm head cabbage, finely diced
1 15 oz can sm sweet peas
1 16 oz can yellow beans, diced fine
1 qt tomatoes, chopped fine
6 hard boiled eggs, diced fine
water

Cover with water the turtle (also shells, if available), onions, celery leaves, parsley, spices (placed in a bag), and seasonings. Boil until the meat is tender and then remove the meat and shell and cool. Reserve the stock. Take the meat off the bones and on the shell and grind or dice it. Strain the stock and add enough water to make 2 gallons. Boil the carrots, cabbage, celery and potato in the stock for 15 minutes. Add the corn, peas, beans, tomatoes, tomato juice and the lemon juice and rind. Simmer and season to taste. Now add the eggs. Brown flour in a skillet over low heat until it is chocolate colored being careful not to burn it. Cool and add enough cold water to make a thin paste. Add to the soup mixture. Add the turtle meat and stir to prevent sticking, adjust seasoning, and simmer for 30 minutes. Makes about 4 gallons.

LADY CURZON SUPPE

2 11 oz cans clear turtle soup
2 tbsp parmesan cheese, grated
1 tbsp brandy
1 egg yolk
2 tbsp heavy cream
¼ tsp curry powder
2 tbsp mushrooms, minced

Heat soup to boiling and add mushrooms to broth and simmer 1 minute. Blend the egg yolk with cream and curry powder. Add a little hot soup to mixture and then combine. Heat, stirring constantly, until soup is consistency of cream. Stir in brandy. Pour into heat proof individual casseroles and sprinkle cheese over the top and then place under broiler until top is lightly browned. Serves 4.

CONSOMME LADY CURZON

1 16 oz can turtle soup
juice of ¼ lemon
2 egg yolks
3 ozs butter
white pepper, freshly ground
½ tsp curry powder
pinch nutmeg
1 oz heavy cream
salt

Open the can of soup and chop the turtle meat into small pieces. Separate the butter into 3 pieces and measure the curry powder into 3. Pour the soup and meat into a pan and simmer a few minutes. Into the top of a double boiler heated over simmering water place the lemon juice, salt and white pepper. Gradually beat in the egg yolks and cubes of butter, stirring constantly. Gradually, whisk in the curry and nutmeg and continue beating. Slowly beat in the cream. The sauce will now thicken fast so lower the heat under the boiler. If the sauce should curdle add an ice cube to it and whip away from heat until it becomes smooth again. Pour hot soup into soup cups and top with the thick cream sauce. Serve at once. Serves 2.

SNAPPER TURTLE SOUP

snapping turtle meat
1 c boiled potatoes, mashed
2 qts water
½ c milk
2 tbsp parsley, chopped
4 tbsp canned pimento
⅛ tsp nutmeg
4 tbsp butter
celery salt
sherry
pepper
turtle eggs (if any)

Place the meat and eggs (if any) of a medium size turtle in a large pot over medium heat. Add water, parsley, nutmeg, celery, and salt and pepper to taste. Boil for 30 minutes or until the meat is tender. Add potatoes, milk, pimentos and butter. Warm and serve at once. Serve with sherry.

SNAPPING TURTLE SOUP

turtle meat, diced
bacon drippings
carrots, diced
1 bay leaf
salt and pepper to taste
potatoes, diced
onions, diced
beef gravy base
water

Brown the meat in the bacon drippings and then place in a dutch oven with the vegetables, beef gravy base and bay leaf and add enough water to cover. Simmer at a low heat for several hours or until meat is tender. Season to taste.

SNAPPING TURTLE SOUP 2

1 10-12 lb snapper
16 + tsp salt
½ c celery, ground
5 c potatoes, ground
2 bay leaves
2 tsp cinnamon, ground
water
2 c cabbage, ground
2 c carrots, ground
1 c onions, ground
1 lb tomatoes, canned
1 tsp cloves, ground
worcestershire sauce

Dress the turtle and wash thoroughly in salted water. Place the meat and a small piece of yellow fat in just enough water with 6 tsp salt to cover and cook until it can be picked from the bones. Pass the meat through a food chopper using the medium blade. Return the meat to the broth and grind the vegetables to make the required quantities after grinding. Place the vegetables in a large soup pot and add the bay leaves, remaining salt, cloves and cinnamon and cover with water. Simmer for 3-4 hours until the vegetables have reduced to a good stock. Now add the turtle and broth with a large shot of worcestershire sauce. Cook for another hour depending on how thick you like it. Makes about 3 gallons.

SNAPPING TURTLE SOUP 3

1 10 lb snapper, dressed
1 c chicken fat or butter
2 ribs celery, chopped
½ tsp thyme
½ tsp marjoram
salt and black pepper to taste
3 qts beef broth
1 c sherry
3 slices lemon, seeded
3¼ lbs veal knuckles*
3 onions, chopped
2 carrots, diced
3 cloves, whole
1 bay leaf
1 c plain flour
2 c whole tomatoes, drained
dash tabasco sauce
1 hard boiled egg, chopped

After dressing the turtle put the meat, skin, bones, shells, veal knuckles which have been cracked into small pieces, into a dutch oven and then add the chicken fat (or butter), onions, celery, carrots, thyme, marjoram, cloves, bay leaf, salt and pepper. Bake in a hot oven (400°F) until it is brown. Remove from the oven and add the flour and mix well. Return to the oven and bake for 30 minutes more. Then place the browned mixture in a large soup kettle and add the beef broth and tomatoes and simmer slowly for 2 hours. Strain it. Remove the bones from the turtle meat and cut into small pieces. Add it back to the strained broth with the sherry, tabasco sauce and lemon slices and simmer for 10 minutes. Correct the seasonings and add the egg. Serve immediately.

PRUDHOMME'S TURTLE SOUP

Seasoning Mix

5 bay leaves
1 tbsp salt
2 tsp white pepper
1¾ tsp garlic powder
1¾ tsp Cayenne pepper, ground
½ tsp cumin, ground
1½ tsp onion powder
1½ tsp dried thyme leaves
1 tsp dry mustard
1 tsp black pepper
1 tsp dried sweet basil leaves

3 lbs turtle meat, boneless
4 tbsp unsalted butter
4 tbsp margarine
½ lb spinach, finely chopped
2 c onions, finely chopped
1 c celery, finely chopped
3½ c tomato sauce
⅔ c plain flour
1 tsp garlic, minced
11 c rich turtle or beef stock
1 c parsley, minced
¼ lemon, seeded, minced
6 hard boiled eggs, coarsely chopped
⅓ c sherry + sherry to add when serving

Combine the seasonings mix in a small bowl and set aside. Finely chop the turtle meat. In a large dutch oven or 5½ qt pot melt the butter and margarine over high heat. Add the turtle and cook until browned, about 6-8 minutes, stirring occasionally. Stir in the seasoning mix, spinach, onions, and celery and cook for about 15 minutes stirring occasionally. Add the tomato sauce and cook an additional 10 minutes, stirring often. Add the flour and garlic and mix well. Cook for 5 minutes stirring almost constantly. Add 2 cups of the stock stirring well. Then stir in 7 cups more stock and bring to a boil and stir occasionally for 5 minutes. Reduce heat to a simmer and cook for about 45 minutes, stirring often and scraping the bottom of the pot well. Mix the parsley, lemon and eggs and add to the soup, stirring well. Add the remaining stock and the sherry. Cook 20 minutes more, stirring often. Remove from the heat and discard the bay leaves. Salt to taste. Serves 8-10.

RICH TURTLE OR BEEF STOCK

1½-2 lbs turtle or beef bones and meat scraps
1 rib celery, in pieces
2 qts cold water
1 med. onion, quartered
1 lge garlic clove, quartered

Add all ingredients in a large saucepan and bring to a boil over high heat, lower the heat and simmer 2-4 hours, adding more water if needed to keep about 1 qt liquid. Strain and discard the solids. To make the stock rich, simmer the liquid until reduced by half.

JUDITH'S TURTLE SOUP

2 c snapping turtle meat, diced
1-4 yellow onions, sliced thin
3 qts chicken or beef stock
1 lemon, sliced thin, seeded
salt and pepper to taste
¼ c butter
1 clove garlic, minced
cayenne pepper flakes
parsley, chopped
1 med onion, diced

Sauté the meat in the butter in a heavy 6-quart pot until browned. Add the sliced onions, garlic, a little cayenne pepper, salt and black pepper to the pot and sauté. Add the stock, cover and simmer for about 1½ hours or until the onions have disappeared and the meat is tender. To serve, sauté the diced onion and add just before serving. Squeeze a little lemon juice and float lemon slices sprinkled with the parsley. (A variation to this soup includes adding flour to the butter and cooking until the flour is slightly brown. Then return the meat and onions to the pot and add 1 quart of whole canned tomatoes and 2 quarts stock. Cover and simmer. Chopped basil, shallots, bell pepper and okra can be added just before serving after vegetables are cooked a little.)

PITTARI'S TURTLE SOUP

2 lbs turtle meat
4 qts water
¼ c oil
2 lge onions, chopped
½ c celery, chopped
1 tsp whole allspice
1 tbsp worcestershire sauce
1 c sherry
hard boiled eggs for garnish
2-3 cloves garlic, minced
1 6 oz can tomato paste
¼ c plain flour
½ c bell pepper, chopped
1 tsp whole cloves
2 tsp turtle herbs*
lemon juice to taste
salt and pepper to taste

Boil the turtle meat in the water for 1 hour. Drain and reserve the stock and chop the meat into small pieces. Make a dark roux with the oil and flour and then add the onions, celery and bell pepper and cook until softened. Add the tomato paste, garlic and the meat and season well. Add the reserved stock. Tie the cloves and allspice in a cheesecloth bag and drop into the soup. Add the turtle herbs* to taste, worcestershire sauce, and the lemon juice and simmer for 1½ hours. Add the sherry and simmer for another ½ hour. Serve with sliced or chopped hard boiled eggs. Serves 6-8.

QUENELLE TURTLE SOUP

10 ozs turtle meat, boneless
1 lge carrot, sliced
1 leek, sliced, white part only
⅓ c madeira
veal quenelles
3½ qts brown stock*
1 rib celery, sliced
1 parsnip, peeled, cut in chunks
salt to taste

Veal Quenelles

10 ozs veal, ground, chilled
4 tbsp creme fraiche
3 egg whites
salt and pepper to taste

Heat the stock in a large saucepan and add the turtle meat and simmer very gently, skimming frequently, for 1 hour. Add the vegetables and simmer for another hour. Strain the broth through several thicknesses of cheesecloth until it is quite clear, discarding the vegetables. Dice the turtle meat. Reserve the turtle meat covered with a little of the broth and store in the refrigerator. Refrigerate the broth overnight. The next day remove any congealed fat from the top of the broth. Gently reheat the broth with the turtle meat. Add the madeira and season with salt if necessary. While the broth is heating, steam the quenelles to warm them. Place 5 quenelles in each soup bowl and ladle broth with the turtle over them. Serves 8.

To make the quenelles place the veal in a food processor and process briefly with the knife blade. With the machine running add the egg whites through the feed tube and process until the mixture is well blended. Add the creme fraiche. Season with salt and pepper. Using 2 tsp dipped in cold water form small oval quenelles and place them so that they do not touch on a platter that has been rinsed in very cold water. Bring a large pot of salted water to a boil and slide the quenelles into the water. Cover the pot and remove from the heat. After 5 minutes remove the quenelles with a slotted spoon and drain on absorbent paper. When the quenelles have cooled, smooth any ragged edges with your fingers and set aside, covered, until ready to be used. Makes 40 small quenelles.

CAJUN TURTLE SOUP

1 lb turtle meat
1 carrot, chopped
1 sm bell pepper, chopped
½ bu shallots, chopped
2 tbsp sherry
1 c plain flour
2 cloves garlic, minced
1 tsp cloves, ground
1 pinch thyme
½ lemon rind, shredded
1 med onion, chopped
½ rib celery, chopped
1 cayenne pepper, chopped
1 hard boiled egg, chopped fine
½ lb butter
1 c tomatoes, crushed
1 tsp allspice, ground
1 pinch bay leaf, ground
4 bay leaves
salt and pepper to taste

Make a stock by boiling together the meat, carrots, celery and half the onion. When the turtle is cooked remove from the stock and refrigerate and reserve the stock. Take the remaining onion, peppers and shallots and sauté in melted butter until vegetables are a golden brown. Add the flour and sauté to a light brown. Add the remaining spices and tomatoes and strain the stock into the mixture, stirring occasionally, and bring to a boil and then lower heat and let simmer. Chop the turtle meat and egg very fine and add to the mixture and bring to a boil again. When boiling again add the sherry and lemon rind and serve. Serves 6.

CAJUN TURTLE SOUP 2

3 lbs turtle meat
3 tbsp peanut oil
2 carrots
4 bay leaves
3 tbsp plain flour
2 ribs celery, chopped
¼ c bell pepper, chopped
1 c tomato puree
3 tbsp lemon juice
¼ tsp tabasco sauce
¼ tsp thyme, dried
¼ tsp white pepper
1 tsp onion powder
salt to taste
2 hard boiled eggs, chopped
1½ tsp seafood seasoning mix
3 qts water
¼ bu parsley
1 rib celery
2 med onions, chopped
2 tbsp carrots, minced
1 tsp garlic, minced
1 c beef stock or broth
1½ tbsp worcestershire sauce
1 tsp basil, dried
½ tsp black pepper
½ tsp garlic powder
½ c sherry
½ c parsley, minced

Remove as much meat from the bones as possible and season the meat with the seafood seasoning Mix. In a heavy saucepan heat the oil over a medium heat and fry the turtle meat until it is nicely browned on all sides. While the meat is browning place the bones in a pot with the water, 2 carrots, ½ bunch parsley, 1 bay leaf and the rib of celery and bring to a boil and then reduce the heat to a rolling simmer. When the meat is browned add the flour and brown to a medium-brown roux*, stirring constantly over a medium heat. Add the onions, celery, carrots, bell pepper and garlic and sauté for 5 minutes. Add the puree and blend in. Add the stock or broth and cook this mixture for 30 minutes over a very low heat. Stir often. Chop the turtle meat into small pieces and replace in the mixture. Remove the bones from the fire and strain the liquid. Add to the mixture the strained turtle stock and add the lemon juice, tabasco sauce, worcestershire sauce and the remaining seasonings to it. Reduce heat and simmer for 45 minutes. Then add the sherry and chopped parsley and simmer for another 5 minutes. Adjust seasonings if necessary. Serve hot in individual serving bowls with chopped eggs. Serves 8.

NEBRASKA TURTLE SOUP

To prepare the meat

2-3 lbs turtle meat
½ tsp salt
2 bay leaves
water
½ lb lean pork
½ tsp coarsely ground pepper
½ tsp cumin

To prepare the soup

3 c turtle and pork, chopped
2 med potatoes, diced
1 tbsp pimento
1 14½ oz can stewed tomatoes
4 carrots, sliced

Place the meat and above seasonings in a pan and add enough water to cover and simmer until the meat is tender. Remove the meat and chop it fine or dice it. Strain off the liquid and set aside. To make the soup mix all the soup ingredients in a large pan and cover with the reserved stock and simmer until the vegetables are done. Adjust the seasoning to taste.

BRANDYWINE SNAPPER SOUP

3 c turtle meat
1 bay leaf
3 beef bouillon cubes
salt and pepper to taste
2 carrots, chopped
2 onions, chopped
1 rib celery, chopped
butter
1 11 oz can cream of tomato soup
1 can consomme
dash ground cloves
dash garlic powder
dash allspice
½ c med dry sherry
2 sprigs parsley, chopped
juice of 1 lemon
1 tsp worcestershire sauce
1 tsp sugar
milk
plain flour

Cook the turtle meat in water to cover with the bay leaf, salt, bouillon cubes, and pepper until the meat is tender. Remove from the broth and cool. Dice the meat fine. Brown vegetables in butter with some salt and pepper. Then place them in a heavy dutch oven with 5 cups of the turtle broth, tomato soup, consomme and the rest of the ingredients except the milk and flour. Simmer until the vegetables are just done. Thicken with milk and flour and add the turtle meat. Serve with additional sherry. Serves 14.

SNAPPING TURTLE AND VEGETABLE SOUP

1 turtle, dressed, cooked
2 c leeks, chopped
3 cloves garlic, minced
24 green beans, ½ inch lengths
1 ham bone
3 qts tomatoes, canned
2 bay leaves
2 sprigs marjoram, chopped
1 tsp basil, dried
4 c potatoes, cubed
8 scallions, chopped
2 carrots, coarsely chopped
salt to taste
1 10 oz pkg cut okra, frozen
1 tsp peppercorns
1 tsp paprika
½ tsp savory, dried
5-6 qts turtle broth

Combine all the ingredients in a large pot and simmer for 1-2 hours. Makes 8-12 quarts.

SOPA DE TORTUGA

1 lb turtle meat, chopped
1 8 oz can tomato sauce
2 tbsp seedless raisins
2 cloves garlic, chopped
1 lge slice ham, diced
1 sprig parsley, chopped
1 tbsp ground almonds
1 qt chicken stock
salt and pepper to taste

Combine tomato sauce, garlic, ham, almonds and raisins and cook for 1 minute. Add chicken stock, turtle meat and parsley. Cook 30 minutes, skim, and serve with heated hard rolls.

Gibbons Snapper Soup

1 med snapper, dressed
6 c water
pinch allspice
juice of 1 lemon
peel of ½ lemon
2 onions, chopped
2 cloves garlic, chopped
1 tbsp plain flour
1 tbsp sherry/serving
hard boiled eggs, sliced
1 bay leaf
¼ tsp thyme, dried
2 cloves, whole
½ tsp salt
little black pepper
pinch cayenne pepper
2 tbsp oil
1½ c tomato juice
1 lemon, sliced thin, seeded

Kill by plunging turtle into boiling water. Scrape off the outer skin and the shields* should come off the bones. The toe nails snap off. Remove the plastron* and remove the intestines and other organs retaining the eggs (if any) and the liver. Remove the gall bladder. Place in the carapace* and place in a large kettle with the water which is sufficient for a medium size turtle. If turtle is larger, a little more water may be added. Season with the bay leaf, allspice, lemon juice, lemon peel, thyme, cloves, salt, black and cayenne peppers. Boil until the meat is tender and falls off the bones. Remove and let cool and then remove all the bones and chop the meat and return to the broth. In a skillet sauté the onions and garlic in the oil until they are just yellow and then add the flour. Cook until the flour is slightly browned and then add the tomato juice. Cook for 10 minutes stirring constantly and then add to the turtle mixture. Bring to a boil and then serve. Stir in the sherry to each bowl and garnish with sliced lemon and eggs.

Turtle Soup with Rum

1½ lbs snapper meat, diced
4 ozs butter
8 ozs onion, diced
4 ozs celery, diced
1 tsp paprika
salt and pepper to taste
1 bay leaf
2 sprigs thyme
3 sprigs parsley, chopped
5 ozs rum
2 ozs port wine
3 ozs sherry
2 tbsp plain flour
1 gal beef stock
1 tbsp worcestershire sauce
4 cloves, whole
6 sprigs parsley

Make a bouquet garni* with the bay leaf, cloves, thyme and 6 sprigs parsley. Sauté the onions and celery in the butter. Add the turtle and fry for 10 minutes and add 3 ounces of the rum. Light the rum and let it burn. Then add sherry, paprika and flour and mix well. Add the beef stock and bouquet garni and season with salt and pepper. Bring to a boil and let simmer for 2 hours. Remove the bouquet garni and add the worcestershire sauce and the remaining rum and port and season to taste. Serve in a large soup tureen. Top with chopped parsley. Serves 8.

SOPA DE TORTUGA, ESTILO CAMPECHE

6 oz can turtle meat, chopped
3 med tomatoes, peeled, minced
1 bell pepper, minced
3 cloves garlic, minced
2 tbsp shortening
¼ tsp cinnamon, powdered
¼ tsp marjoram
⅛ tsp cumin
½ tsp black pepper
3 slices toast
3 med onions, minced
½ tsp salt
10 olives, minced
1 tsp capers, chopped
2 c beef stock
1 tbsp vinegar
2 tbsp white wine
1 tsp flour
3 sprigs parsley, minced
2 tbsp water

Fry the onions, tomatoes, bell pepper and garlic in the shortening. Add the spices, salt, turtle meat and broth, olives, capers, stock, vinegar and wine. Bring to a boil and add, while stirring, the flour which has been blended with the water. Add the parsley. Boil for 5 minutes and serve on small squares of toast. Serves 6.

SOPA DE TORTUGA DEL MAR

1 lb turtle meat or ½ lb cooked
3 cloves garlic, chopped
1 med onion, chopped
½ c tomato puree
¼ c claret or burgundy
⅛ tsp black pepper
⅓ c peas
½ tbsp vinegar
4 sm bell peppers, shredded
½ tbsp marjoram, chopped
½ tsp ginger, chopped
⅛ tsp salt
⅓ c olive oil

Place canned or cooked meat in a saucepan and cover with water and bring to a boil. (If fresh turtle meat is used cut into small pieces and cover with 2 quarts of water and cook until tender. Pour off all but enough water to cover.). Add remaining ingredients and simmer slowly about 15 minutes or until vegetables are tender. Serves 6-8.

TURTLE SOUP

1 lb turtle meat
¼ c onions, chopped
2 tbsp sherry
4 slices bacon
dash cayenne pepper
salt to taste
4 tbsp plain flour
1½ qts water
2 tbsp parsley, minced
1 tsp allspice, whole
2 lemons, sliced, seeded
1 tsp cloves
2 hard boiled eggs, chopped

Put turtle in water and add salt to taste. Tie cloves and allspice in cheese cloth and drop in with turtle. When turtle is tender remove from soup. When cool enough to handle pull meat from bones (if any). Chop in pieces and put back in soup. Fry the bacon crisp and when cold chip up and add to soup. Sauté onions and parsley in bacon drippings and add to stock. Remove the bag of spices. Brown the flour in the skillet. Mix this brown flour with cold water to form a paste and add to stock. To thicken soup make a paste of flour and water and add to stock as desired. Just before serving add eggs and cayenne and at last minute before you serve add the sherry and serve with sliced lemons.

TURTLE SOUP 2

2 lbs turtle meat, cubed
1 wine glass white wine
salt and pepper to taste
a handful of sweet basil
plain flour (optional)
¼ lb fresh mushrooms
¼ lb lean ham, cubed
1 knuckle* of veal
1 tbsp butter (optional)

Put the pieces of turtle, ham and veal into a pot. Cover with cold water and bring to a boil slowly. Season with salt and pepper and allow to simmer gently 4 hours. Add the mushrooms and basil and continue to cook for 30 minutes. Pour in the wine and bring to a boil again carefully. Strain the soup through muslin and serve. This clear soup may be thickened with the butter rubbed with flour, but the clear soup is more delicious. Don't throw away the solids, but eat separately.

TURTLE SOUP 3

4 c turtle meat, diced
dash tabasco sauce
salt and pepper to taste
2 tsp dry mustard
½ tsp sherry
1 qt milk
1 sprig thyme, chopped
2 tbsp butter

Brown and cook turtle meat at low heat in 1 tbsp butter. Stir in the mustard and thyme. Add milk and remaining butter. Season to taste and heat. Pour in sherry just before serving. Serves 4-6.

TURTLE SOUP 4

1 turtle, dressed
1 8 oz can tomato sauce
1 tbsp pearl barley
4 qts water
½ c sweet peas, fresh
½ bu celery, diced
parsley to taste, chopped
1 carrot, diced
1 onion, diced
salt and pepper to taste

Cut up and wash dressed turtle. Cook 3 hours in the water. When tender lift out meat and add the vegetables and cook until tender. Chop meat and put back in soup. Serve.

TURTLE SOUP 5

1 turtle, dressed
water
½ lb butter
2 15 oz cans Veg-All
½-¾ c pearl barley
2 sm onions, chopped
½ bag wide noodles

Cover turtle meat with water and add more water to meat if needed when cooking. Boil, skim, and then add the barley. Cook this until done. Remove meat. Put in Veg-All, noodles and onions and cook until the noodles are tender. Just before serving add the meat which has been browned in the butter. Serve.

TURTLE SOUP 6

turtle meat
calf's feet
1 sm glass madeira
beef
sage
water
pinch of sweet basil
soup vegetables
dash of cornstarch (optional)
hen
pepper
knuckle* of veal
savory
sweet marjoram
salt

Cook the turtle meat 3-4 hours in plenty of salted water. Next a sort of pot-au-feu is made with beef, knuckle of veal, old hen, calf's feet and part of the flesh of the turtle. The liquid of this pot-au-feu comes from the liquid the turtle was cooked in, with the usual soup vegetables added and is cooked until all the meat is done. After the broth has been strained through a cloth, the already cooked pieces of turtle are added, it is cooked another 20-30 minutes and the soup is finished by adding an infusion of aromatic herbs made by dropping a pinch of basil, sage, sweet marjoram and savory into a glass of boiling madeira. Add pepper and serve the soup which is clear or slightly thickened with a dash of cornstarch dissolved in a little turtle broth. This turtle soup is served with pieces of turtle in it. The remaining solids from the pot-au-feu can be served separately.

TURTLE SOUP 7

1 lb turtle meat
1 tsp salt
1 lge onion, chopped
1 bell pepper, chopped
2 tbsp worcestershire sauce
½ c red wine
3 pts water
1 11 oz can tomato soup
1 stalk celery, chopped
½ c barley
2 tsp lemon juice

Add the turtle meat, salt, soup and vegetables to the water and cook for 1 hour. Add the barley and cook 30 minutes more. Add worcestershire sauce, lemon juice and wine. Simmer for 10 minutes and serve hot.

TURTLE SOUP 8

1 lb veal bones, cracked
1 carrot, sliced
¼ c sherry
2 tbsp plain flour
1 onion, sliced
salt and pepper to taste
½ lemon, sliced. seeded
1 c turtle meat, cooked
3 cloves, whole
4 c beef stock, cold
1 hard boiled egg, chopped
1 tbsp butter
5 c tomatoes, chopped

Brown bones, onions and carrots very slowly in butter in a heavy pan. Add flour and stir to blend. Add the beef stock, tomatoes, cloves, salt and pepper and heat slowly to boiling, skim, cover and simmer for 2 hours. Strain stock through a fine sieve and cool quickly. When cold remove the fat and clarify if necessary. Add the turtle meat and sherry and bring to a boil and serve with a slice of lemon and chopped egg. Serves 6.

TURTLE SOUP 9

1 turtle, dressed and diced
1½ tsp poultry seasoning
¼ tsp paprika
1 bell pepper, diced
1 c tomato juice
black pepper
½ tsp celery seed
1 lge onion, diced
1 15 oz can Veg-All
parsley sprigs (optional)
1 tbsp salt
½ bay leaf
2 lge carrots, diced
1½ c fine noodles
1 tbsp whole pickling spices
½ tsp parsley flakes
3 med potatoes, diced
1 c celery, diced
lemon slices (optional)

Combine the turtle and water on a 6-quart kettle and bring to a boil. Simmer for 3 hours or until the meat is tender. Tie the pickling spice in a bag. Add the seasonings and bag to the broth. Add the potatoes, carrots, onion, bell pepper and celery and cook for 20 minutes. Add the noodles and cook for 15 minutes longer. Add drained Veg-All and tomato juice and additional water if necessary. Remove spice bag before serving. Serve in a soup tureen and garnish with lemon slices and parsley sprigs if desired. Serves 8-10.

TURTLE SOUP 10

3-5 lbs turtle meat, diced
1 stick butter, melted
salt to taste
1 c bell pepper, minced
1 tsp dry mustard
dash worcestershire sauce
1 tbsp sherry/serving
water
3 c onions, minced
3-6 hard boiled eggs
peel of 1 lemon, minced
black pepper to taste
½ c scallion tops, minced
½ c parsley, minced
dash tabasco sauce

Wilt the onions in the melted butter and add a sprinkle of salt. When the onions start to soften add the bell pepper, and cook until the onions are clear. Add the turtle meat and fry just until it is no longer raw. Add enough water to cover and simmer an hour, stirring occasionally, and adding more water if necessary. Add mustard, worcestershire and tabasco sauces. Separate the eggs and mash the yolks and chop the whites. Add the yolks to the mixture and then add the lemon peel, salt and pepper to taste. Add a few of the scallion tops and continue cooking until it is fairly thick. Shortly before serving add the egg whites, parsley and remaining scallion tops. Add the sherry to each bowl. Serves 8-10.

MANDY'S TURTLE SOUP

2 lbs turtle meat
1 sm piece ham
2 cloves garlic, minced
1 piece lemon, chopped fine
1 bay leaf
3 sprigs parsley, chopped
water
2 tomatoes, peeled, chopped
sherry or madeira
2 tbsp plain flour
1 tbsp lard
1 onion, chopped
2 pinches cloves, ground
2 pinches thyme, ground
salt and pepper to taste
cayenne pepper to taste
turtle eggs, if any, boiled

Cut the ham into small bits and mash the herbs and seasonings with it and set aside. Boil the meat in enough water to cover for 15 minutes. Remove the meat and chop fine reserving the stock. Brown the onion in the lard and then add the turtle meat and let it brown well. Add the ham mixture, stirring well, and then the flour, again stirring well. Add the stock and an additional 2½-3 quarts of water with additional salt and peppers if needed. Add the lemon and tomatoes and place in a double boiler for an hour or so stirring frequently. A little sherry or madeira can now be added. Chop the eggs fine and add just before serving. If turtle eggs are not available use chicken eggs.

CAJUN SWAMP TURTLE SOUP

Making the stock

3 lbs fresh water swamp turtle
meat, turtle bones (if any)
6 qts hot water
½ tsp cloves, ground
2 lge onions, diced
2 ribs celery with leaves, coarsely chopped
salt and pepper to taste
1 c parsley sprigs
4 cloves garlic, minced
2 bay leaves
½ tsp allspice, ground
½ tsp thyme
½ tsp cayenne pepper

Making the soup

calipash* and calipee*, if any
1 8 oz can tomato sauce
½ lb country ham (or boiled)
4 tbsp plain flour
1 tbsp worcestershire sauce
½ c + sherry
shallot leaves, minced
2 lge tomatoes, peeled
4 scallions with 3 in of tops
1 stick butter
juice of 1 lemon
½ tsp lemon peel, grated
4 hard boiled eggs, chopped
thin lemon slices, seeded

Put the turtle meat and bones (if any) in a heavy soup pot with the hot water and add all the other ingredients in the stock list and bring to a boil. Lower the heat and bring to a slow boil and cook for 2 hours. Stir now and then to prevent it from scorching. If it gets too thick add more water or beef broth. Strain through a triple cheesecloth. Save the meat and discard the remainder. If the country ham is raw cook the slices very slowly in a covered iron skillet browning on both sides for 20 minutes. Put it through the fine blade of a grinder. Grind the scallions and the tomatoes also. Melt the butter and add the flour and cook very slowly until it is a dark brown. Add the ground scallions, tomatoes and ham and cook for 5 minutes. Clean the pot the stock was made in and add the stock and ham mixture to the pot and mix well. Add the diced turtle meat and calipash or calipee, if any. Add the tomato sauce, ½ cup sherry , lemon peel and worcestershire sauce and bring to a boil. Lower the heat and simmer for 1 hour being careful not to let it burn. If it gets too thick add hot water or beef broth. It should be the consistency of a light gravy. At the end stir in the lemon juice and adjust the salt. Put a tbsp of sherry in each heated bowl before ladling in the soup. Sprinkle eggs and minced shallots over the top of each bowl and float a lemon slice in each. Serves 12-14.

(A variation on the above recipe calls for the use of a dark rum and the recipe is called swamp turtle soup au Rhum. After the scallions, tomatoes and ham have cooked in the roux* for 5 minutes make a small "well" in the corner of the skillet and add 2 ounces of dark rum. When it becomes heated, flame it and burn off the alcohol, and then mix it in well with the other ingredients in the skillet.)

SEA BREEZE TURTLE SOUP

5 lbs turtle meat, diced
2 lge onions, minced
½ lb butter
½ gal tomato paste
1 qt dry sherry
salt and pepper to taste
1 rib celery, minced
1 lge bell pepper, minced
½ gal tomato sauce
2 doz hard boiled eggs, chopped
4 gals water

Sauté celery, onions and bell pepper until soft in butter. Add the tomato sauce and paste and simmer for 1 hour. Boil the turtle meat in the water until tender. Add the tomato and vegetable mixture and simmer for 2 hours. Add the eggs and sherry. Season with salt and pepper to taste. Makes 5 gallons.

TURTLE SOUP A LA TECHE

6-10 lbs turtle meat, dressed
3 tbsp oil
6 qts water

Boil the meat in the water and skim as necessary. When cooked about an hour strain through cheesecloth and reserve the stock. When the meat is cool enough to handle cut into small pieces and then fry in the oil until well fried. Set aside.

2 c plain flour
2 sticks butter
2 tbsp oil

Make a roux* with the above and cook until a medium brown.

10 med onions, minced
2 cloves garlic, minced
6 ribs celery, minced
4 10 oz cans beef bouillon
4 bell peppers, minced
2 hot peppers, minced
1 15 oz can tomato sauce
1 bag crab boil*

Add all vegetables to the roux and stir well and cook until all are tender. Add the tomato sauce and blend well. Add to the reserved turtle stock along with the crab boil and beef bouillon. Bring to a boil but not too long or the soup will get too spicy. Add the meat and when boiling again add the following:

salt and pepper to taste
tabasco sauce to taste
worcestershire sauce to taste

Allow to simmer about 1½-2 hours and ½ hour before the end of the cooking time remove the bag of crab boil.

6 hard boiled eggs
1 tbsp sherry/serving
1 c parsley, minced
1 c scallion tops, chopped

Add the egg yolks which have been sieved and the scallions and parsley. When ready to serve add the egg whites which have been chopped and the sherry to each serving bowl. Serves 10-12.

MANALE'S TURTLE SOUP AU SHERRY

1 lb turtle meat, diced large
1½ c oil
1 lge onion, chopped
½ c bell pepper, chopped
6 ozs tomato puree
1 gal water
½ oz tabasco sauce
½ pt sherry
½ c cloves, whole
3 c plain flour
1 rib celery, chopped
3 lemons, minced
1 14.5 oz can tomatoes, diced
1½ ozs worcestershire sauce
salt and pepper to taste

Make a bag and put the cloves in it and secure. Place a 6-quart pot on medium fire and pour oil into the pot and add the flour and stir until brown. Add the onions, celery and bell pepper and cook for 5 minutes. Add the lemons and the cloves. Pour in the tomatoes and puree. Add the water and stir well and add the turtle meat, worcestershire and tabasco sauces, salt and pepper. Mix well and cook for 1 hour and 45 minutes. Add the sherry and heat to serving temperature. Remove bag of cloves. Serve hot. Serves 10.

DELMONICO'S TURTLE SOUP AU SHERRY

1½-2 lbs turtle meat, diced
2 ribs celery
1 onion
4 tbsp plain flour
1½ c leeks, chopped
1 c tomatoes, peeled, chopped
sherry to taste
½ gal water
1 garlic, whole
salt to taste
½ c oil
½ c onions, chopped
2 hard boiled eggs, chopped

Add the water in a large soup pot over moderate heat and then combine the turtle, celery, garlic, onion and salt to the water and cook for about 30 minutes. Skim if necessary. In a medium skillet brown the flour in the oil over moderate heat. Add the chopped onions and tomato and simmer until the onions are brown. Strain the broth of all the meat and vegetables and return the broth to the soup pot. Add the meat back and then add the flour mixture to the broth. Stir in well until the soup thickens. Add more water and salt to suit the taste. Just before serving add a little chopped egg to the dish and sherry to your liking. Serve hot. Serves 6-8.

NEW ORLEANS TURTLE SOUP AU SHERRY

3 lbs turtle meat, diced
⅔ c plain flour
3 c water
8 cans beef consomme
1½ c onions, chopped
5 cloves garlic, chopped
½ c celery leaves, chopped
¾ c ham, cubed, lean
5 tomatoes, peeled, coarsely chopped
8 tbsp dry sherry
½ c parsley, minced
3 hard boiled eggs, sliced thin
2 c butter
1 tsp black pepper
3 pinches cayenne pepper
⅔ tsp thyme, powdered
⅓ tsp cloves, powdered
⅓ tsp allspice
⅓ tsp nutmeg, grated
4 tbsp worcestershire sauce
8 slices lemon, thin, seeded
4 tsp salt
bay leaves

In a large soup pot add the butter and flour and stir constantly until browned on low heat. Add the water, consomme, onions, garlic, celery, ham, turtle meat, tomatoes, salt, peppers, thyme, cloves, allspice, nutmeg and a few bay leaves into the pot and bring to a boil. Lower the heat and cook for 2½ hours. One half hour before being done add the worcestershire sauce, lemon slices and sherry. Ten minutes before finishing add the parsley. Let set for 15 minutes. Just before serving add the eggs.

TURTLE SOUP MELANCON

2½-3 lbs turtle meat, cubed
⅓ c plain flour
½ rib celery, chopped
1-2 cloves garlic, minced
1 6 oz can tomato paste
1-2 tbsp liquid crab boil*
½ tsp cayenne pepper
½ c dry vermouth
⅓ c + 2 tbsp oil
2 onions, chopped
1 bell pepper, chopped
1 16 oz can tomatoes
4 c water
juice of 2 lemons, strained
2 hard boiled eggs, sliced

Heat the ⅓ cup of oil in a heavy pot and fry the turtle until browned. Remove and set aside. Add the remaining oil to the pan drippings and sprinkle in the flour and stir constantly over a medium heat to a dark brown. Sauté the onions, celery, bell pepper and garlic in the browned flour until soft. Stir in the tomatoes and paste and cook for a few minutes. Add the meat, water, crab boil, lemon juice and cayenne pepper and cover and simmer for 2 hours. Just before serving add eggs and vermouth. Serve immediately.

FOLSE'S TURTLE SOUP

10-12 lbs turtle meat, cubed
2 lge bu celery, chopped
2 29 oz cans tomato sauce
2 c oil
2 c plain flour
1 tbsp cloves, ground
1 tbsp allspice, ground
2 gal water
5 lbs onions, chopped
2 12 oz cans tomato paste
6 bay leaves
3 29 oz cans tomatoes, whole
1 doz hard boiled eggs, chopped
1 tbsp thyme, dried
salt and pepper to taste
cayenne pepper to taste

Season the meat with salt and cayenne pepper. Sauté in oil until opaque or most of the water evaporates. Set aside. Make a roux* with the oil and flour until browned. Sauté the onions and celery in the roux until tender. Add the sauce, paste and tomatoes and enough water to prevent from sticking or burning. Stir frequently and bring to a boil and cook for an additional 20 minutes. Add the meat, cloves, thyme, allspice, bay leaves, eggs and the remaining water. Cook on a moderate heat for approximately 2 hours or until the meat is tender. Add salt and pepper to taste.

TURTLE SOUP A LA COMMANDER'S PALACE

1 lb turtle meat, chili ground
¾ c plain flour
4 ribs celery, minced
1 tsp oregano
½ tsp black pepper
3 bay leaves
1 qt turtle stock (or beef stock
  or 8 beef bouillon cubes in 1qt water)
½ c parsley, chopped
6 tsp sherry
1 c butter, melted
½ stick butter
2 med onions, minced (1½ cups)
½ tsp thyme
1½ tsp garlic, granulated
1½ c tomato puree
½ c lemon juice
5 hard boiled eggs, minced
salt to taste

First make a roux* by using the melted butter in a sauté pan and adding the flour. Cook on medium until the mixture reaches a light brown. Remove from the heat and set aside. In a 3-quart saucepan melt the stick of butter and add the turtle meat and cook on high heat until the meat is brown. Add the celery, onions, oregano, thyme, pepper, garlic and bay leaves mixing well. Cook until the vegetables are transparent, about 3-5 minutes. Add the puree and lower the heat and simmer for 10 minutes. Add the stock and simmer for 15 more minutes. Add the roux and cook over low heat until the roux is absorbed. Add the lemon juice, eggs and parsley. Salt to taste and adjust seasonings with pepper and garlic. Cook for 30 minutes more. Serve in 6 individual soup bowls and add 1 tsp sherry to each plate at the table. Serve hot. Serves 6.

TURTLE STOCK

- 2-4 lbs turtle bones
- 2 qts cold water
- 8 cloves, whole
- 3 med carrots, split
- ½ bu parsley
- 3 beef rib bones, broken
- 2 med onions, halved
- 3 ribs celery, cut
- 4 bay leaves
- 4 ozs turtle meat

Preheat the oven to 450°F and brown the turtle and rib bones in the oven. In a 1 gallon sauce pot over high heat add the water and the browned bones. Stick 2 cloves in each onion half. Add all other ingredients and bring to a rapid boil. When it begins to boil lower the heat to a simmer and cook for approximately 4 hours adding water as necessary to keep the level at about 2 quarts. Remove from the heat and strain. Let it cool and skim the fat off. Makes about 2 quarts.

MASSON'S TURTLE SOUP

- 3 veal tails, cut in 2 in pieces
- 1 gal water
- 4 whole cloves
- ½ tsp thyme
- 6 tbsp butter
- 2 c onions, chopped
- 1 c bell peppers, chopped
- ½ tbsp garlic, chopped
- 1 6 oz cans tomato paste
- 1 tsp salt
- 2 c dry sherry
- ¼ c parsley, chopped
- parsley, chopped (for garnish)
- 1½ lbs turtle meat, diced
- 4 bay leaves
- 4 allspice, whole
- peelings and trimmings from soup vegetables used
- 1 c celery leaves, chopped
- 1 c carrots, chopped
- ½ c plain flour
- ¼ tsp cayenne pepper
- juice of 1 lemon
- 2 hard boiled eggs, chopped
- lemon slices (for garnish)

In a large pot simmer the veal bones in the water for 1 hour skimming often. Add the bay leaves, cloves, allspice, thyme and vegetable peelings and trimmings and simmer for 2 more hours. Strain the stock and set aside. In a soup pot brown the turtle meat in the butter. Add the onions, celery, bell pepper, carrots and garlic and sauté until soft, about 20 minutes. Stir in the flour and cook for 10 minutes more. Add the stock, tomato paste, cayenne pepper and salt. Simmer for 1½ hours. Then turn off the heat and stir in the lemon juice, sherry, eggs and parsley. Garnish each serving with a slice of lemon and a sprinkling of chopped parsley. Serves 6-8.

CORINNE DUNBAR'S TURTLE SOUP

2 lbs turtle meat, diced
4 qts + ½ cup water
8 beef bouillon cubes
1 white onion, minced
1 clove garlic, minced
3 ribs celery, minced
3 sprigs thyme, minced
3 bay leaves
½ lemon, sliced
1 c plain flour
dash tabasco sauce
1 tsp Kitchen Bouquet
salt and pepper to taste
2 tsp sherry/serving
lemon slices, seeded
hard boiled eggs, riced
4 cloves, whole

Boil the turtle meat in the 4 quarts of water for approximately 2 hours or until tender. Dissolve the bouillon cubes in the stock and add the onion, garlic, celery, thyme, bay leaves, cloves, lemon slices and tabasco sauce. Make a paste of the flour and remaining water and slowly stir into the soup to thicken. Simmer 2-3 hours. Add Kitchen Bouquet, salt and pepper. Serve in soup bowls and garnish each with sherry, a lemon slice and riced eggs. Serves 6.

ANTOINE'S TURTLE SOUP AU SHERRY

1 lb turtle meat, diced
2 c onions, minced
⅓ c celery, minced
2 c tomato puree
1 lemon
2 tbsp parsley, chopped
white pepper, freshly ground
1 stick butter
1 c sherry
3 c espagnole sauce 2
1 qt beef stock
2 bay leaves
salt

Melt the butter in a large soup pot and sauté the turtle meat until browned. Add onions and sauté until they begin to color. Add celery and cook until soft. Add the puree and cook for 5 minutes. Squeeze the juice of the lemon and add to the pot. Mince the lemon remains discarding the seeds and add to the pot. Add all remaining ingredients to the pot and simmer gently for 1½-2 hours. Serves 6.

ESPAGNOLE SAUCE 2

3 tbsp butter
½ c onions, minced
½ c carrots, minced
3 tbsp plain flour
1 c tomato sauce*
2 cloves garlic, minced
½ tsp thyme, dried
white pepper, freshly ground
1 sm rib celery, minced
2 bay leaves
3 sprigs parsley, minced
1 tsp sugar
3 tbsp vinegar
¼ tsp anchovy paste
salt
1 tsp water

Melt 2 tbsp of the butter in a saucepan and sauté the onions and carrots until they begin to color. Add remaining butter and stir in the flour and cook until browned. Add the stock, tomato sauce, garlic, thyme, celery, bay leaves and parsley. In another small saucepan caramelize sugar with water and remove from the heat immediately and add the vinegar. Mix in the anchovy paste and add to the sauce. Salt and pepper to taste and bring to a boil and simmer for 30 minutes. Strain. Makes 2 cups.

GOWANLOCH'S TURTLE SOUP

3 lbs turtle meat, diced
2 onions, minced
1 c tomatoes, canned
1 clove garlic, minced
2 bay leaves
6 cloves, whole
1 lump sugar
sherry jelly
quenelles of turtle meat*
4 tbsp ham or bacon drippings
4 tbsp plain flour
1 tbsp salt
4 qts stock and water
2 sprigs parsley, chopped
½ tsp mace
2 tbsp lemon juice
hard boiled eggs, sliced
lemon slices

Parboil the turtle meat in enough water to cover for 10 minutes. Strain and reserve the stock. Fry the meat in the fat until browned. Remove and brown the flour in pan juices. Add enough water to the stock to make 4 quarts and add the onions, tomatoes, garlic, bay leaves, cloves, sugar, salt, parsley, mace and lemon juice and bring to a boil. Add the turtle meat and browned flour. Simmer for 3 hours. Strain if desired. Garnish with sliced eggs, quenelles of turtle, lemon slices, and sherry jelly may be added for flavor.

SONIAT'S TURTLE SOUP

2 lbs turtle meat, chopped
8 tbsp plain flour
½ lb lean ham, chopped
1 c onions, chopped
3 med tomatoes, peeled, chopped
¼ c celery tops, chopped
1 bell pepper, chopped
4 cloves garlic, chopped
1½ c water
1 tsp worcestershire sauce
2 thin slices lemon, seeded
⅛ c parsley, minced
2 pinches cayenne pepper
4 bay leaves
½ tsp thyme, powdered
¼ tsp cloves, ground
¼ tsp allspice
¼ tsp nutmeg, grated
4 c beef stock or 3 cans beef consomme
2 tsp salt
⅔ tsp black pepper
4 tbsp sherry
2 hard boiled eggs, sliced

Melt the butter in a heavy soup pot and gradually blend in the flour and stirring constantly cook until a medium brown. Add the ham, onion, tomatoes, celery, bell pepper and garlic. Mix well and cook over a very low fire for about 1½ hours until the vegetables are browned. Add the turtle meat, salt, black pepper, cayenne pepper, bay leaves, thyme, cloves, allspice, nutmeg, stock or consomme and water and bring to a boil. Lower the heat and simmer for 2½ hours. About ½ hour before the soup is done add the worcestershire sauce, lemon slices, and sherry. Ten minutes before add the parsley. If the soup at this point is too thick add a little more water. After removing from fire let it set for 15 minutes before serving. When serving add the eggs. Serves 4-6.

SERPAS' TURTLE-VEGETABLE SOUP

3 lbs turtle meat
2 lge onions
2 tsp salt
2 lge potatoes
2 tsp sugar
2 ribs celery
3 qts water
1 can tomatoes, canned
barley or vermicelli or both
3 carrots, chopped
3 ribs celery, chopped
1 onion, chopped
8 sprigs parsley, minced
black pepper to taste
1 tbsp worcestershire sauce
12 drops tabasco sauce (optional)
1 can lima beans
1 can very tiny white corn

In a 4-6 quart covered soup pot boil the turtle meat in the water with salt, 2 onions, 2 ribs celery, and whole potatoes very slowly until meat is tender. Remove the meat and when cool enough to handle cut to desired size. Take the cooked onions, celery, and potatoes and mash or put through a blender. Return to pot everything except the meat. Add the tomatoes which have been mashed or run through a blender and the other vegetables and cook until the vegetables are done. If barley is used add when the vegetables are returned to the pot. If vermicelli is used add when the turtle meat is returned to the pot. Barley takes about 2 hours to cook. Add parsley, sugar, worcestershire sauce, pepper and tabasco sauce. About 20 minutes before finished return the turtle to the pot. Additional water may be added if needed. Serves 6.

SLIDER TURTLE SOUP

- 1 lge or 2 sm slider turtles
- salt
- 1 lge onion, chopped
- 1 tsp thyme
- black pepper
- 4 qts water

Female turtles are preferable. After the turtle is is killed and bled, dress it by removing its head by plunging into scalding water for a few minutes and scraping off the outer layer of skin. Then remove the plastron* and separate the leg quarters and keeping the liver with the gall bladder removed, shelled eggs (if any) and the ovary strings with immature eggs. After the turtle is dressed bring fresh water to a rolling boil and add the turtle meat and liver along with the seasonings and onion. Reduce heat and simmer for 2 hours or until meat begins to fall off bones. Remove with a slotted spoon and bone the turtles. Chop or grind the meat, coarsely chop the liver, and return to the soup. Correct seasonings and return to a boil and reduce the heat and add the eggs and continue cooking for 20 minutes. Serves 6-8.

SEA TURTLE SOUP

- 1 carapace*
- 1 plastron*
- 4 tbsp roux/quart soup
- turtle meat
- salt and pepper to taste
- soup vegetables
- 1¼ c madeira/quart stock
- stock
- chicken
- 1 bouquet garni*
- peppercorns
- coriander

This soup is made from sea turtle and is prepared with the bony carapace and plastron only. The carapace and plastron is cut into pieces of equal size, blanched for a few minutes in boiling water and cleaned of the outer shields that cover them. Put in a big stew pot with richly flavored stock, (the liquid used to cook the turtle can be a stock made with the flesh of the turtle reinforced with some beef and chicken), soup vegetables and bouquet garni with coriander and peppercorns added, and cook like an ordinary broth for 6-7 hours. After cooking the pieces of turtle are drained, boned, and cut into pieces about 1½ inches square and kept warm in some strained broth. The prepared soup is strained through a cloth, reheated and enriched with 1½ cups madeira to each quart. Just before serving the pieces of turtle are put back in the soup. To thicken (if desired) add a roux of 2 tbsp flour and 2 tbsp butter per quart of soup.

Green Sea Turtle Soup 2

1 40-50 lb green sea turtle
scalding and cold water
2-3 cloves
1 sprig sweet basil
1 sprig marjoram
4 tbsp shallots, chopped
2 tbsp thyme, chopped
2 tbsp marjoram, chopped
2 tbsp savory, chopped
24 whole allspice
5-6 cloves
1 lge ham knuckle*
6 bay leaves
1 sm bu basil
1 lge bu savory
plain flour
cayenne pepper to taste
curry powder
light veal stock
1 onion
½ faggot* parsley
½ faggot* thyme
1 sprig marjoram
1 pt madeira
2 lemons, sliced, seeded
1½ tbsp basil, chopped
4 tbsp parsley, chopped
1 nutmeg
4 blades mace
1 tbsp salt and black pepper
4 lge onions, sliced
1 lge bu thyme
1 lge bu marjoram
salt to taste
lemon juice to taste
1¼ lbs butter

Dress turtle by removing the head and hang by the hind fins. Remove the fore fins and separate the carapace* from the plastron*. Cut off the plastron from the carapace and the lean meat of the plastron. Remove hind fins. Take off the lean meat from the plastron and fins and cut into 2 inch pieces and place into a stew pot. Put the carapace, plastron and fins in scalding water a few minutes which cause the shell to part easily. Remove the shields*. Cut the shells about 6 inches square. Place into a pot with the veal stock. Let boil until meat is tender and then place in cold water and free the meat from the bones and cut into 1 inch pieces. Return the bones to the stock and boil gently for 2 hours. Then strain it. Slice fins into pieces 1 inch wide and boil them in the stock with an onion, 2-3 cloves, a faggot of parsley and thyme and a sprig of basil and marjoram. When tender take out and add this stock to the other. Take the lean meat and put into a stew pan with the madeira, shallots, lemons, 2 tbsp thyme, marjoram and savory, 1½ tbsp basil and parsley. Grind together nutmeg, 12 allspice, 1 blade mace, cloves and 1 tbsp pepper and salt. Mix together with as much curry as will lie on a quarter. Put about ⅔ of this in the lean meat, with ½ pound of butter and 1 quart of stock. Let simmer until the meat is tender. Dice ham knuckle into a stew pot with sliced onions, bay leaves, 3 blades mace, 12 allspice, ½ butter and simmer until the onions are melted. Shred a small bunch of basil and a large one of thyme, savory and marjoram and put these into the onions and keep as green as possible. When simmered sufficiently add flour to thicken the soup. Add by degrees, the stock the bones were boiled in and the seasoning stock from the lean meat. Boil for 1 hour and run through a strainer and add salt, cayenne pepper and lemon juice to taste. Then put in the meat and let it boil gently about ½ hour and if more wine is required it must be boiled before adding. Serve.

GREEN SEA TURTLE SOUP 3

1 20-35 lb green sea turtle
water
sherry
thyme
4 bay leaves
arrowroot
onions
celery
carrots
bell pepper
potatoes

Dress the turtle and set the red meat aside. In a pot of water cook the head and flippers with a court bouillon of vegetables (onions, celery, carrots. bell pepper, potatoes, etc.) for 6 hours. When this is tender remove the meat and cut into small pieces and place in a marinade of sherry. Clarify the stock. Now add the red meat of the turtle in the stock and cook for 1 hour with thyme and bay leaves. Next strain the soup through cheesecloth and thicken it with the arrowroot. Before serving add the cubed green meat and 1 glass sherry for each 2 quarts of soup.

GREEN TURTLE AND PEA SOUP

1 qt commercial green turtle soup
3 lbs green peas
salt and pepper to taste
½ c heavy cream
salt
water
½ c sherry

Shell the peas and cook until tender in a little boiling salted water. Drain and press the peas through a fine sieve. Add the turtle soup to the pea puree and heat the mixture to the boiling point. Add the sherry and season to taste. Ladle the soup into 6 individual casseroles. Whip the heavy cream and put a generous tbsp on each serving. Place the casseroles under the broiler 2-3 minutes or until the cream is lightly browned. Serve immediately. Serves 6.

CREOLE TURTLE SOUP

3 lbs turtle meat, including any available calipash*, calipee*, eggs, and turtle bones
2 qts hot water
2 qts beef broth
1 stick butter
4 tbsp plain flour
2 lge onions, minced
1 bell pepper, minced
2 ribs celery, minced
2 tbsp parsley, chopped, fresh
4 cloves garlic, minced
1 lb can tomatoes, minced
1 tbsp worcestershire sauce
½ tsp cayenne pepper
¼ tsp allspice, ground
½ tsp cloves, ground
¼ tsp marjoram
½ tsp thyme
½ tsp basil
½ c sherry
parsley, chopped
3 hard boiled eggs, chopped
salt and pepper to taste
1 lemon, sliced, seeded

In a large, heavy pot make a roux* with the butter and flour, cooking slowly and stirring constantly until deep brown. Add the onions, bell pepper, celery, 2 tbsp parsley and garlic and cook and stir until the vegetables are tender and translucent. Dice the turtle meat into ½ inch cubes and add the meat. Add the calipash, calipee and bones, if any, to the pot. Add the beef broth, hot water and the bay leaves. Bring to a boil and then lower the heat and simmer for 2 hours. Remove the bay leaves. Add the allspice, cloves, basil, thyme, marjoram, cayenne pepper, lemon slices and worcestershire sauce and mix well. Partially cover the pot and simmer for at least 2-4 hours. Stir now and then scraping the bottom with a metal stirring spoon. If a brown residue shows up on the spoon you are scorching it so lower the fire even more. A ½ hour before it is done add the sherry and the turtle eggs, if any. Remove the lemon slices and the bones and season to taste. Ladle from the bottom into heated soup bowls and sprinkle with the hard boiled eggs and chopped parsley on top of each bowl and float a thin slice of lemon on each. Serves 10-12.

GREEN TURTLE SOUP AU SHERRY

3 lbs green sea turtle meat,
diced calipash* and calipee*,
if available
4-5 lbs turtle shells, chopped (optional)
2 cloves garlic, minced
½ tsp thyme
½ tsp allspice, ground
¼ tsp cayenne pepper
3 egg whites, beaten
salt and pepper to taste
3 med onions, chopped
½ c parsley, chopped
4 scallions, with tops, sliced
1 bell pepper, diced
2 qts beef broth
2 qts hot water
1 1-lb can plum tomatoes, chopped
1 8 oz can tomato sauce
juice of ½ lemon
sherry

Place all of the ingredients except the calipash, calipee, sherry, salt and lemon juice in a large pot with a thick bottom. Bring to a boil and then lower the heat, partially cover, and simmer for 6 hours. Stir now and then to keep from sticking. Add more stock or water if necessary to keep it liquid. At the end strain it twice through quadruple cheesecloth. Place a small amount of the strained stock in a saucepan and boil the calipash or calipee for 1 hour. Set aside to be added to the soup at the end. Save enough of the turtle meat to have 3-4 small pieces for each bowl at the end. To further clarify the stock place it in a clean pot over high heat and add the egg whites and continue stirring vigorously until it boils and then lower the heat and boil slowly for ½ hour. Strain again through quadruple cheesecloth. Add the lemon juice. Add the calipash or calipee, if any, and adjust the salt. Add a cup of sherry for each quart of soup. Heat the soup and place the reserved cubes of meat in the bottom of each preheated bowl and ladle in the soup. Float a thin slice of lemon in each bowl. Serves 10-12.

GREEN TURTLE SOUP 2

1½ lbs green turtle meat, diced
5-6 irish potatoes, diced
water
soda crackers
2 med onions, minced
1 c cream
salt and pepper to taste

Cover turtle meat with water in a good sized pot and bring to a slow boil and cook until tender. Add onions and potatoes and cook slowly until done. Divide the cream evenly into the bottom of each serving bowl. Evaporated milk may be used. Fill bowl with soup, season and serve with soda crackers. Serves 6.

GREEN TURTLE SOUP 3

2 lbs green turtle meat, cubed
3 lge ripe tomatoes, diced
3 med onions, diced
2 bay leaves
thyme
2 qts water
5-6 irish potatoes, diced
salt and pepper to taste
parsley, minced
cayenne pepper (optional)

Add turtle to boiling water and simmer until tender. Add vegetables and seasonings and simmer until the vegetables are cooked. Add cayenne pepper if you like it hot. Serves 6.

FLORIDA GREEN TURTLE SOUP

2 lb green sea turtle meat
2 cloves garlic, minced
6 allspice, whole
2 tbsp plain flour
2 hard boiled eggs, chopped
1 tbsp parsley, minced
1 bay leaf, minced
3 qts water
2 lge onions, minced
6 cloves, whole
1 in square ham
1 key lime, chopped
1 wine glass sherry
1 tbsp thyme, minced
salt and pepper to taste
¼ c butter

Parboil the meat in the water for 10 minutes, remove and save stock. Cube the meat. Chop ham very fine. Brown the onions lightly in the butter and then add the meats, garlic, spices and herbs and sauté stirring constantly to prevent sticking. When browned add the flour which has been blended with a little turtle stock and again season with salt and pepper. Combine with the stock and simmer for 1 hour, stirring often. Add the lime and cook until all is tender. Then add the eggs and sherry and serve at once.

TURTLE SOUP WITH CHABLIS

2 c cooked turtle meat, fresh or canned
1 qt water
2 hard boiled eggs, chopped
1 tsp mint, chopped
1 clove garlic, crushed
1 c chablis
½ bell pepper, chopped
salt and pepper to taste
2 tsp parsley, chopped

Cut meat into small pieces. Put in water with garlic, bell pepper, mint and parsley and bring to a boil and simmer for 15 minutes. Season to taste and add the wine and egg and bring just to a simmer. Serves 6-8.

THICK TURTLE SOUP

4 c canned green turtle soup
3 tbsp plain flour
2 cans condensed beef bouillon
6 tbsp sherry
6 slices lemon, seeded

Cut turtle meat into cubes and add together with the broth to bouillon and simmer 20 minutes. Cool. Mix sherry and flour to a paste and add. Bring to a boil and simmer, stirring constantly, about 3 minutes. Serve with a lemon slice in each portion. Serves 6.

NEW ORLEANS TURTLE SOUP

1 lb green turtle meat, fresh or canned
1 tbsp plain flour
1 tsp parsley, chopped
⅛ tsp thyme
1 sm can tomatoes
2 c bouillon
dash worcestershire sauce
½ med onion, sliced
3 tbsp butter
1 bay leaf
¼ c sherry
salt and pepper to taste
dash angostura bitters

Sauté onions in butter and add flour and brown. Cut turtle meat in 1 inch pieces and sauté until lightly brown. Add tomatoes, bouillon, bay leaf, thyme, parsley, worcestershire sauce, Angostura Bitters and season to taste. Simmer gently 15-20 minutes or until tender. Add sherry just before serving. Serves 4.

TURTLE SOUP A LA NEW ORLEANS

3 lbs turtle meat, chopped
2 bay leaves
1 tbsp whole allspice
3 tbsp oil
1 sprig thyme, chopped
2 qts boiling water
2 cloves garlic, chopped
2 sprigs parsley, chopped
1 lemon, sliced, seeded
1 6 oz can tomato paste
1 tbsp whole cloves
2 tbsp plain flour
salt and pepper to taste
1 c onions, chopped
3 hard boiled eggs, sliced
sherry (optional)

Sauté the onion and garlic in oil and then add the flour and brown lightly. Add the tomato paste and simmer for 3 minutes. Add boiling water, bay leaves, parsley, thyme, salt, pepper and turtle meat. Tie the allspice and cloves in a cloth bag and drop into the mixture. Simmer for 1 hour or until the turtle is tender. Remove the spice bag. Add the eggs and lemon and simmer for 5 minutes. Let stand for 1 hour. sherry to taste may be added when served. Serves 6.

ESCHETE'S TURTLE SOUP

5 lbs turtle meat
1 tbsp salt
1 tsp nutmeg, grated
½ c oil
2 lge onions, chopped
3 ribs celery, chopped
1 15 oz can tomato sauce
salt to taste
3 sprigs parsley, chopped
sherry to taste
5 qts water
1 tsp cloves, ground
1 tsp cinnamon
⅓ c plain flour
2 cloves garlic, minced
1 bell pepper, chopped
5 hard boiled eggs, sliced
tabasco sauce to taste
2 scallions, chopped

Boil the turtle meat in salt water with the cloves, nutmeg and cinnamon until tender. Remove the meat and bone it. Save the stock. Brown the flour in the oil and then add the garlic, onions, celery and bell pepper and cook until tender. Add the turtle meat and the tomato sauce and smother for 15 minutes, stirring often. Add the stock and bring back up to 5 quarts with additional water if necessary, add salt, tabasco sauce, parsley and scallions and a little more of the spices and simmer for 35 minutes. Remove from the fire and serve in soup bowls with sliced eggs and sherry to taste.

TOUPS TURTLE OR ALLIGATOR SOUP

3 lbs turtle or alligator meat
⅓ c plain flour
2 c scallions, chopped
6 cloves garlic, chopped
¼ c parsley, chopped
1 8 oz can mushrooms, sliced
4 qts hot water
⅓ c stuffed olives, sliced
cooking sherry or vermouth
salt and pepper to taste
⅓ c oil
3 c onions, chopped
2 c celery, chopped
½ c scallion tops, chopped
3 tomatoes, peeled, sliced
1 tbsp cloves, whole
8 chicken bouillon cubes
½ lemon, thin sliced, seeded
1 hard boiled eggs, sliced/ serving

Brown the flour in the oil and then add the onions, scallions, celery, garlic and meat (chopped) and cook in a covered pot on low heat until the oil separates. Stir often and add the tomatoes, mushrooms and liquid, cloves, water and bouillon cubes. Simmer on low for 4 hours. Remove the meat if it is tender or when tender. Continue to simmer adding the parsley, scallion tops, olives and lemon and cook for 30 minutes. Return the meat cut into small pieces and season with salt and pepper. To serve place in bowls and garnish each with egg and a jigger of sherry or vermouth to be added as desired.

TURTLE GUMBO

1 smoked ham hock
5 c + 2 c water
2 sprigs parsley
1 sm white onion, peeled
½ tsp dried crushed red pepper
½ tsp thyme, dried
1 tsp salt
1½ c baby okra, fresh or frozen
3 lge ribs celery, in bite size pieces
tabasco sauce to taste
1½ tsp gumbo filé powder*
2 lbs turtle meat, boned
½ c plain flour
3 tbsp oil
4 tbsp butter
6 sm white onions, chopped
1 c long grain white rice
6 sm hard boiled eggs
3 c cherry tomatoes, peeled
1 tbsp worcestershire sauce
¼ c dry sherry

Cover the hock with the water and add the parsley sprigs, peeled onion and crushed red pepper and bring to a boil. Cover and simmer for 2 hours. Strain and discard the remaining ingredients and reserve the stock. In a large non-stick heavy pot over low heat toss and stir the flour until it becomes tan but not scorched. Remove ¼ cup of the flour. To the remaining flour add the oil and butter and blend. Add the chopped onions and simmer a few minutes until softened. Combine the reserved flour with the thyme in a paper bag. Cut the turtle meat into bite size pieces and blot dry. Shake the meat in the bag with the seasoned flour and add to the pot. Over high heat sauté until the meat is brown. Add the reserved stock, cover, and simmer for 2½ hours, stirring occasionally. Cook the rice according to standard procedure. Cut the okra into ½ inch pieces and reserve. Twenty minutes before serving add the worcestershire sauce, salt and okra to the simmering pot. Ten minutes before add the eggs. Five minutes before add the sherry, tomatoes and celery. Just before serving while the broth is boiling remove from the heat and add the gumbo filé. To serve place a ½ c cooked rice in the center of a wide soup bowl. Surround rice with the gumbo and 1 egg in each. Pass the tabasco sauce at the table for individual use. Serves 6.

CAMPER'S TURTLE GUMBO

To prepare the stock

3 lbs turtle meat
2 c onions, chopped
2 carrots, sliced
2 bay leaves
6 cloves, whole
1 c tomato puree
1½ gal water
4 ribs celery, 1 in pieces
pinch thyme
4 peppercorns
2½ c canned tomatoes with juice

Tie the turtle meat in a cheesecloth bag and place all the ingredients in a large pot and bring to a boil. Reduce the heat and simmer for 1½ hours. Remove the turtle meat and bone, if any. Strain the broth and reserve discarding solids.

To prepare the gumbo

¼ c safflower oil
3 ribs celery, diced
½ tsp garlic powder
1 8 oz bottle clam juice
2 c whole plum tomatoes, diced
⅛ tsp cayenne pepper
1½ c okra, cooked
1 tsp gumbo filé powder*
2 c onions, chopped
2 bell peppers, diced
1 gal reserved turtle stock
1 c tomato puree
1 tbsp worcestershire sauce
reserved turtle meat, chopped
1 c cooked long grain rice

Heat the oil in a large pot and sauté the onions, celery and bell pepper until wilted. Add the garlic powder and stir well and then add the stock, clam juice and tomato puree and bring to a boil. Reduce the heat and simmer for 15 minutes and then add the tomatoes, worcestershire sauce, cayenne pepper and turtle meat. Cook uncovered for 15 minutes. Add the okra and rice and salt and pepper to taste. Pour ½ cup of stock into a glass measuring cup and sprinkle in the gumbo filé powder and beat until smooth. Remove the pot from the fire and stir in the filé powder. Mix well and serve.

MINORCA GOPHER TURTLE STEW

1 lge gopher tortoise
1 sm can tomatoes
2 c water/cup meat
1 bell pepper, chopped
2 hard boiled eggs/c meat
salt and pepper to taste
¼ c olive oil/c meat
3 tbsp sherry/c meat
1 lge onion/c meat
½ tsp salt/c meat
dash hot pepper/c meat

Kill turtle in boiling water. Remove the outer skin and claws. The outer skin should peel right off and the claws cut off. Remove meat from the shell. Cut the meat in 2 inch pieces and measure. Simmer until tender in water, salt and pepper. In a deep dutch oven heat the olive oil. Brown in the oil the onions and then add tomatoes and bell pepper. Simmer gently while turtle is cooking. More tomatoes may be added if mixture cooks down too much. When turtle is tender turn the sauce into the turtle pot. There should be enough liquid to make plenty of gravy. Thicken by mashing the yolks of the hard boiled eggs and stirring into the stew. Add more salt and pepper to taste if needed. Stir in sherry and add chopped egg whites. Serve at once.

STEWED TORTOISE (SOUTH AFRICAN FIELD STYLE)

1 tortoise*
salt and pepper
wine glass of wine
miscellaneous available herbs
little fat (or lard)
1 tbsp plain flour
bread crumbs

Remove flesh carefully from a tortoise being careful not to break the gall bladder. Separate the meat from the bones and gristle and cut up into neat pieces. Place in a pot with the fat, flour, salt, pepper and whatever herbs may be available and allow to simmer gently until tender. Thicken the gravy with bread crumbs and add the wine and serve.

STEWED TORTOISE (SOUTH AFRICAN HOME STYLE)

1 tortoise*, dressed, boned
4-5 tbsp beef stock or other kind
1 tbsp lemon juice
1 tbsp plain flour
salt and pepper
1 pinch chili
2 pinches ginger, powdered
1 c cream, boiling
lard

Place the meat in a stew pot with a little lard, the flour, salt, pepper, chili, ginger and stock. Mix well and allow to stew. When the meat is almost done add the lemon juice and simmer 15 minutes longer. Remove the meat and arrange on a dish. Add the cream to the gravy and pour over the meat. It is generally served with cooked rice and some sweet dish, either stewed fruit or sweet potatoes.

Paprika Turtle Goulash

- 1 lb turtle meat, cubed
- garlic, chopped
- salt and pepper to taste
- butter
- white wine
- onions, chopped
- paprika
- plain flour
- water

Add the meat, onions, salt, pepper, paprika and a little garlic in a roasting pan with some butter and roast it for several hours at a low temperature. Add water to keep the meat moist. Just before removing add a little wine to taste.

Creole Turtle Stew

- 2 lbs turtle meat
- ½ lb lean boiled ham
- 1½ tsp garlic, minced
- ¾ tsp black pepper
- 3 tbsp brandy
- ½ tsp lemon juice
- 2 tsp salt
- 1 c dry red wine
- ⅛ tsp cayenne pepper
- 1 tbsp plain flour
- 2 bay leaves
- 1½ c onions, chopped
- 2 c water
- 2 tbsp butter
- 1 lge tomato, cut up
- ¼ tsp mace

Melt butter and add the turtle meat, 1½ tsp salt and ½ tsp pepper. Cook over medium high heat, stirring frequently, about 8-12 minutes. Remove from heat and drain off liquid. Return to high heat, add ½ tsp salt and ¼ tsp pepper and cook, stirring constantly until meat is browned, about 7-9 minutes. Add the onions and cook until the onions are browned. Stir in the flour and mix thoroughly. Cook 2 minutes more, then add the ham and tomato. Continue cooking until a thick gravy begins to form. Add the wine and brandy and mix well. When the liquid comes to a boil add the water, bay leaves, garlic, mace, cayenne and lemon juice. Bring to a boil again, stir to mix, then reduce heat to a simmer. Cover the pan and cook about 45 minutes or until turtle is tender. Add additional seasonings if needed. Serve over rice. Serves 4-6.

TURTLE STEW 1

4 lbs turtle meat
pinch cayenne pepper
1 c dry spanish sherry
parsley, chopped
1 tbsp paprika
salt to taste
bacon crumbs
½ c butter
1½ oz plain flour
1 grated lemon rind
¼ c olive oil
1 stem celery, chopped
4 tomatoes, peeled, diced
2 sm onions, chopped
1 bouquet garni
hard boiled eggs
1 clove garlic, chopped
2 pts beef stock
juice of 1 lemon

Cut turtle meat into 1½ inch cubes and brown in oil and butter. Add garlic, onions, paprika, cayenne pepper and celery. Sprinkle with flour and cook slowly for 5 minutes before adding the stock. Insert the bouquet garni (consisting of 6 peppercorns, thyme, 2 bay leaves, 3 cloves, 6 corns of allspice) and remove in 1½ hours whenever the meat is tender. Correct seasonings and add sherry, tomatoes, lemon peel and juice. Garnish with hard boiled eggs, bacon crumbs and chopped parsley. Serve over rice.

TURTLE STEW 2

3-4 lbs turtle meat
1 8 oz can sliced mushrooms
2 c consomme
3 stems celery, chopped
3 onions, chopped
2 tbsp lemon juice
⅓ c plain flour
⅓ c oil
salt
3 cloves garlic, minced
1 bell pepper, chopped
cayenne pepper
1 tbsp worcestershire sauce
3 hard boiled eggs, chopped
1 c cream (optional)
½ tsp tabasco sauce
1 c sherry (optional)

Brown the flour in hot oil until very brown. Add onions, celery, bell pepper and garlic while stirring constantly. Add the consomme and turtle meat, which has been seasoned with salt and cayenne pepper. Stir constantly and add seasonings. If necessary add more consomme until meat is just covered. Simmer very slowly until meat is tender. Add eggs and sliced mushrooms. When all has blended well add the sherry or if you prefer the cream instead.

TURTLE STEW 3

- 3 lbs turtle meat
- 3 med onions, chopped
- 2 bell peppers, chopped
- ½ tsp powdered allspice
- 4 cloves garlic, minced
- 1 c sherry
- 1 tbsp sugar
- 3 tbsp oil
- 2 no. 2 cans tomatoes
- 6 hard boiled eggs
- 4 ozs butter
- 1 6 oz can tomato paste
- salt and pepper to taste
- 1 lemon, sliced, seeded
- 3 tbsp plain flour
- 1 rib celery, chopped
- 4 bay leaves
- boiling water
- 1 bu scallions, chopped
- 8 whole cloves

Parboil the turtle meat with just enough water to cover. Make a brown roux* of oil and flour. Add onions, garlic, bell pepper, tomato paste and tomatoes and slowly cook for 20-30 minutes. Add mixture to the turtle meat along with additional boiling water to cover meat. Boil down. Add celery, scallions, salt, pepper, sherry, bay leaves, cloves, allspice and sugar. Cook covered over high heat for 30 minutes. Mash egg yolks and chop the whites and add to stew. If stew gets too thick add a little more water. Cook slowly for about 3 hours. One half hour before serving add sliced lemon and butter. Serves 6.

TURTLE STEW 4

- 2 lbs turtle meat, diced
- 4 tbsp butter
- plain flour
- 1 c potatoes, diced
- 2 qts water
- 1 clove garlic, diced
- 1 lge onion, chopped
- 1 11 oz can zesty tomato soup

Add the turtle to the water and boil for 20 minutes. Remove the meat reserving the stock. Melt the butter in a dutch oven and add the garlic and cook slowly until lightly browned. Lightly coat the turtle with flour and add the meat and onion to the dutch oven and carefully turn until brown. Pour part of the reserved stock over the meat mixture and simmer for several hours until the meat is tender. Add the potatoes and the rest of the stock and blend in the soup and cook 30 minutes longer.

TURTLE STEW 5

3 lbs turtle meat
3 tbsp plain flour
4 cloves garlic, minced
1 6 oz cam tomato paste
1 rib celery, minced
2 bell peppers, minced
6 hard boiled eggs
salt to taste
½ tsp allspice
¼ lb butter
1 10 oz can Ro-Tel Tomatoes
½ tsp chili powder
3 tbsp salt
4 lge onions, chopped
1 lb canned tomatoes
boiling water
1 bu scallions, minced
1 c sherry
cayenne pepper to taste
4 bay leaves
1 tbsp sugar
1 lemon, sliced, seeded
½ tsp black pepper
1 tsp basil

Parboil the turtle meat. Brown the flour in the oil and then add onions, garlic, tomato paste and tomatoes and cook slowly for 20-30 minutes. Add the turtle and add enough boiling water to cover the meat and boil again. Add celery, scallions, bell peppers, sherry and seasonings and cook over high heat for 30 minutes. Mash the egg yolks and chop the whites and add to thicken the stew. If the stew gets too thick add a little more water. Cook slowly for about 3 hours. A half hour before serving add the sliced lemon and butter. To increase amount add ½ lb turtle meat per person.

TURTLE STEW 6

24 lbs turtle meat, boned
1 c plain flour
10-12 cloves garlic, minced
2 10 oz cans Ro-Tel Tomatoes
8-10 bay leaves
2 lbs mushrooms, fresh
cayenne pepper to taste
cooked rice
water
oil
14-16 lge onions, chopped
2 8 oz cans tomato sauce
1 bu celery, minced
1 pt sherry
5 lge bell peppers, minced
salt to taste
worcestershire sauce to taste

Fry meat in enough oil to cover the bottom of a skillet and brown well. Sauté onions, garlic, bell peppers and celery in oil in a separate skillet. Put the turtle in a big iron pot and add enough water to cover. Add the sautéed vegetables and the tomatoes and tomato sauce. Cook slowly over a low heat at a slow boil. Brown the flour in a little oil in another skillet until browned and add to the stew. Throw in the bay leaves and cook for 3-4 hours or until the meat is tender. Add the sherry about 30-40 minutes before done. About 20 minutes before done add the mushrooms, salt, pepper and worcestershire sauce to taste. Throw in the scallions and parsley at the last minute. Serve over rice. Serves about 30.

TURTLE STEW 7

- 2½ c turtle meat, diced
- 3 med onions, diced
- salt and pepper to taste
- 1 med onion, sliced
- 1 c lima beans, fresh
- water
- 2 c celery, diced
- 1 c tomatoes, diced
- 4 tbsp butter
- 3 med potatoes, diced
- ½ c parsley, chopped

Place onion, lima beans and celery in a dutch oven and cover with water. Bring to a boil and simmer 30 minutes. Sauté the turtle meat in the butter until brown on all sides. Add the meat, butter, potatoes, carrots, tomatoes, parsley, salt and pepper to the mixture in the dutch oven. Simmer for 45 minutes or until all the vegetables are tender. Serves 6.

TURTLE STEW 8

- 6 lbs soft shell turtle meat
- 1 lb onions, diced
- 1 med onion, chopped
- 2 c oil or shortening
- ½ bu parsley, chopped
- ¼ pt sherry
- 1 4 oz can mushrooms, drained
- 1½ c tomato puree
- 2 sm spice bag
- 1 rib celery, diced
- 1 c brandy
- ¼ pt white wine
- ½ lb bacon, diced and fried

Braise the turtle meat in the oven until about half-cooked. Add the celery, onions and garlic and cook. Make a gravy with the oil or shortening and flour until hazel brown. Add a little water or stock to the tomato puree and add to your turtle meat. Make a small spice bag with various spices and add along with sherry and wine to the turtle mixture. Simmer until thoroughly cooked. Shred the soft part of the shell of the turtle and boil until tender and add to the stew. Add bacon, drained mushrooms and brandy. Serve with dry toast or rice.

GORDON'S TURTLE STEW

2 lbs turtle meat, cubed
2 carrots, chopped
2 sweet potatoes, chopped
2 pkg mixed vegetables
2 10¾ oz cans cream of mushroom soup
oregano
soy sauce
worcestershire sauce
1½ c catsup
1 med onion, chopped
rice, cooked
thyme
seasoned salt
garlic powder
italian seasonings
MSG
1 8 oz can mushrooms, stems and pieces, drained
1 stem celery, chopped
½ bell pepper, chopped
little barbecue sauce

Mix well some thyme, seasoned salt, garlic powder, italian seasonings, MSG, oregano, soy and worcestershire sauces and sprinkle over the meat. Pressure cook the meat for several hours at a low temperature. Cook together the carrots and sweet potatoes until tender. Cook together the mixed vegetables, soup, mushrooms, celery, bell pepper and onions until vegetables are tender. Mix everything together with the catsup and barbecue sauce and simmer briefly. Serve over cooked rice. (This can also be placed in a pie crust and baked until crust is browned.)

SNAPPING TURTLE STEW

30 lbs turtle meat
salt and pepper to taste
2 gal whole tomatoes
corn-on-the-cob (optional)
3 gals white cream style corn
2 hens
10 lbs onions
15 lbs boiling potatoes
½ c sherry
3 lemons

Boil turtle meat until reasonably tender, drain, and remove all bones. Do the same with the hens but don't use the livers or gizzards. Save both stocks and use in some other recipe. Peel and boil both the potatoes and onions until tender enough to drain without coming apart. Grind the corn, tomatoes, potatoes, onions, turtle and hens. Place all this in an adequate pot. Squeeze lemons and drop rind halves into pot along with the juice and sherry. Cook slowly over medium heat while constantly stirring until done in approximately 1½ hours. Salt and pepper to taste about ½ way through this cooking process. Under no conditions allow the stew to stick. As an added embellishment corn on the cob may be added the last 30 minutes of cooking time. Serve with saltines or bread. Yields about 9 gallons.

CAJUN TURTLE STEW

- turtle meat
- ¼ c bell pepper, minced
- 4 bay leaves
- 1 c scallions, minced
- few cloves
- black pepper
- ½ c celery, minced
- ¼ c cayenne peppers, chopped
- salt
- pinch thyme
- lime juice
- 6 tomatoes, cut up
- orange juice
- few whole allspice

Marinate hunks of turtle meat in salt, pepper, orange juice and lime juice for several hours. Drain and cook with water to cover with scallions, bell pepper, cayenne pepper, celery, allspice, tomatoes, cloves, bay leaves and thyme. Cook until meat is tender.

MALTESE TURTLE STEW

- 2½ lbs sea turtle meat, diced
- 2 onions, chopped
- 2 ozs olive oil
- 1 tbsp tomato puree
- 2 bay leaves
- fresh mint leaves, chopped
- salt and pepper to taste
- ½ c water
- 2 ozs red wine
- ½ c nuts, chopped
- 4 ozs seedless raisins
- 8 ozs olives, chopped
- 2 apples, peeled, sliced fine
- 2 chestnuts, peeled, chopped
- 1 tbsp capers
- croutons

Heat the olive oil and fry the onions in it lightly until they turn translucent. Add the puree diluted with the water, bay leaves, mint and seasonings. Simmer for 5 minutes. Then add the remaining ingredients except wine and continue cooking for 30 minutes, adding a little more water to cover if needed. Add the turtle and wine, cover tightly and cook at least another 30 minutes. Serve with croutons. Serves 6.

TORTUGA ESTOFADA (STEWED TURTLE)

1 lb canned or cooked turtle meat
2 anchovies, chopped
1 tbsp white raisins
2 tbsp olive oil
1 pimento, chopped
pinch of saffron
pinch of allspice
½ tsp cloves, ground
pinch of epazote* (if not available, parsley)
6 capers
10 lge green olives, pitted
¼ c almonds, blanched, chopped
1 lge tomato, peeled, chopped
1 bell pepper, chopped
pinch of ginger
1 tsp cinnamon, ground
salt and black pepper to taste
½ c rum

If fresh turtle meat is used wash well and cook thoroughly in salted water until almost done before going on with recipe. Cook turtle meat for 10 minutes in liquid from can or a bit of its own liquid if fresh turtle is used. Add the capers, anchovies, olives, raisins and almonds and set aside. In the oil sauté the tomato, pimento and bell pepper and add the spices except for the salt, pepper and epazote and sauté the vegetables until soft. Add the turtle meat mixture and season with salt, pepper and epazote and heat to the boiling point. Add the rum and cook for another few minutes, stirring well. Serve immediately.

TORTUGA ESTOFADA 2 (STEWED TURTLE)

¾ lb turtle meat, cooked
12 capers
8 lge green olives, pitted
1 tbsp raisins
⅛ c almonds, chopped
2 tbsp olive oil
1 med tomato, peeled, chopped
salt and pepper to taste
2 bell peppers, chopped
1 pimento, chopped
⅛ tsp saffron
½ tsp allspice
1 tsp cinnamon, ground
½ tsp cloves, ground
½ c sherry

Place the turtle in a saucepan and cook about 10 minutes. Add the capers, olives, raisins and almonds and set aside. Put in 1 tbsp of the oil in a saucepan and fry briefly the tomatoes, peppers, pimento, saffron, allspice, cinnamon and olives. Add to the turtle meat mixture and heat to boiling and add remaining olive oil, sherry, salt and pepper to taste. Cook for 5 minutes and serve very hot. Serves 6.

STEWED TURTLE WITH CREAM

1 pt turtle meat, cooked
2 tbsp butter
1 tbsp rice flour
¼ tbsp nutmeg, grated
4 egg yolks, well beaten
1 pt light cream
1 tbsp salt
½ tbsp white pepper
pinch cayenne pepper
1 tbsp lemon juice

Place in a saucepan the butter and flour and stir over fire until it bubbles and then stir in the cream, salt, peppers, nutmeg and turtle meat (drained) and stir over a moderate heat until scalding hot. Keep hot but not boiling and stir in the egg yolks. Pour at once into a tureen containing the lemon juice. Serve.

MOISE'S SNAPPING TURTLE STEW

- snapping turtle meat
- salt and pepper
- ½ c + fat
- 1-3 cloves garlic, minced
- 1 c plain flour
- 1 6 oz can tomato paste
- 3 tbsp worcestershire sauce
- 2 8 oz cans tomato sauce
- 6-8 onions, chopped
- 1½ c celery, chopped
- 2 bell peppers, chopped
- juice from 2-3 lemons or limes
- 2 dashes tabasco sauce
- water as needed

Salt and pepper the meat and sauté in fat until just browned and then pour into a colander. Wipe pot out and cover the bottom with a half cup of melted fat and garlic and bring up the heat. Then stir in the flour until the flour is nice and brown. Then add the tomato paste, stirring. When this is good and thick add tomato sauce, onions, celery, bell peppers, juice, worcestershire and tabasco sauces. Keep stirring over a medium-low heat for half an hour and then add the turtle meat with a little water and adding water as needed. Cover and simmer for another 2 hours, stirring occasionally. Add salt and pepper to taste. Serve over rice.

TURTLE STEW WITH CORN

- snapping turtle meat
- fresh corn cut off cob
- onions, chopped
- salt and pepper
- potatoes, cubed
- water

Boil down the turtle in just enough seasoned water to cover and when cooked pull the meat off the bones. Return the meat to the stock and add corn, potatoes and onions and simmer until done.

FOXFIRE TURTLE STEW

- turtle meat
- cayenne pepper
- sweet milk
- black pepper
- salt
- water
- butter

Parboil the turtle with salted water and cayenne pepper. When done, bone and stew in the milk with butter, pepper and salt until warmed thoroughly. Serve like oyster stew.

WEST INDIAN TURTLE STEW

2 c cooked turtle meat with stock or 1 can turtle meat
½ c celery, sliced thin
½ c mushrooms, chopped
1 med tomato, peeled, dice
1 clove garlic, crushed
¼ tsp marjoram, crushed
1 tbsp flour
½ c onion, minced
1 c spanish style tomato sauce
¼ c olive oil
1 c white table wine
1 tsp parsley, chopped
1 tsp salt
¼ tsp basil, crushed
¼ tsp thyme, crushed
¼ c brandy

Heat the olive oil in a kettle and add the onion, celery, mushrooms, tomato which has most of the seeds removed and garlic. Cook over a low fire until the vegetables are soft. Then stir in the flour and add the tomato sauce, wine, parsley and salt. Stir until the sauce comes to a boil and then turn the heat down to a simmer. Add the crushed herbs to the sauce and simmer for 30-40 minutes. Cut the turtle meat into small cubes and add meat to sauce along with about ½ cup stock and continue to simmer another 30 minutes. Just before serving pour the brandy over the stew and set afire. Stir into stew. Serve with steamed rice. Serves 6.

TURTLE STEW WITH DUMPLINGS SEMINOLE INDIAN STYLE

leatherback turtle meat
water
salt
plain flour

Using the fleshy parts of the legs and back, boil in salt water to cover until meat is tender. When the meat is tender, mix together flour and water to make a soft dough. Pinch off marble size dough balls and pull out thin and drop into the hot turtle liquid. Cover and let simmer about 15 minutes. Flour from the dumplings thickens the gravy. Serve hot with fried bread.

PATRICK'S LANDING SNAPPER STEW

1 lge snapper turtle
onions, quartered
cayenne pepper
potatoes, quartered
salt
water

Dress and bone a snapper and cut up the meat. Pressure cook with a litttle water until tender. Add the onions, potatoes, and seasonings and cook until the potatoes are done and the gravy thickened.

SONIAT'S TURTLE STEW

- 4 lbs turtle meat
- 4 tbsp oil
- 4 tbsp plain flour
- 4 lge onions, chopped
- 3 cloves garlic, minced
- 1 24 oz can tomatoes
- 1 6 oz can tomato paste
- 3 ribs celery, minced
- 1 bell pepper, chopped
- ½ c dry sherry
- ¼ tsp cayenne pepper
- ½ tsp black pepper
- 4 bay leaves
- ½ tsp cloves, ground
- ½ tsp allspice
- ½ tsp chili powder
- 1 tsp basil, dried
- ½ tsp thyme, powdered
- 1 tsp sugar
- 4 hard boiled eggs
- ¼ lb butter
- 1 lemon, sliced, seeded
- 1 bu scallions, minced
- salt to taste

Boil the turtle in enough water to cover the meat for 5 minutes. In a heavy pot heat the oil and add the flour and brown. When browned add the onions, garlic, tomatoes and tomato paste. Let this cook slowly for about 25 minutes and then add the turtle meat which has been cut into fairly large pieces. Add the stock and any additional water to cover contents and bring to a simmer for ½ hour. While it is cooking add the celery, bell peppers sherry and all other seasonings. After 30 minutes mash the egg yolks and chop the whites and add to the stew and let cook slowly for 3 hours. If the stew gets too thick just add a little more water. About ½ hour before serving add the butter, lemon and scallions and correct for salt and pepper. Serves 6-8.

LOGGERHEAD SEA TURTLE STEW

- 2 lbs loggerhead sea turtle meat
- ½ c salt pork, diced
- 4 tbsp butter
- 1 can condensed tomato soup or tomato puree
- 2 qts water
- 1 onion, chopped
- 1 clove garlic, chopped
- 1 c potatoes, diced
- plain flour

Cut the meat into bite size pieces and boil in the water for 20 minutes. Remove the meat from the water and save the stock. In a dutch oven fry the salt pork and add butter and garlic and cook slowly until light brown. Then add the onion and lightly floured turtle meat turning gently until browned. Pour part of the stock over this and simmer slowly for several hours until the meat is tender. Add the potatoes and the rest of the stock and cook for 30 minutes more. Add additional water for desired consistency and thicken with flour. Serves 6.

CAYMANIAN TURTLE STEW

- 3 lbs turtle stewing meat
- ½ hot country pepper
- 1 onion, chopped
- salt and pepper

Cut the meat into ½ inch cubes and put into a pot with the peppers and onion. Cook for about an hour and a half. If the meat is tough add about 1 cup of water. If you are using a crock pot cook about 6 hours. (A variation suggested by the chef at the Lighthouse Club is to use 2 pounds of stewing meat and 1 pound of lean fin meat. The fin meat makes the stew thicker.) Old style Caymanians would add the liver and tripes.

SAVOURY TURTLE AND ONION STEW

½ lb turtle steak, chopped
1 oz plain flour
1 pt meat stock
5 cloves, whole
salt and pepper to taste
2 lge onions, sliced
Pickappeppa Sauce
3 bay leaves
1 tbsp vinegar
1½ ozs shortening

Brown the onions and flour in the shortening and gradually blend in the stock. Add the bay leaves and cloves and simmer for 7 minutes covered. Add the chopped meat and vinegar and simmer for a further 30 minutes. Thicken if desired and garnish with green and red bell peppers.

TERRAPIN STEW

1 prepared terrapin*
1 lb butter
4 hard boiled egg yolks
1 pt white wine
salt and pepper to taste

Bring prepared terrapin meat to boil in enough water to cover and cook until tender and water reduced to almost nothing. Rub the butter and egg yolks to a cream and add cold to the terrapin. Season to taste and stew for a few minutes. When ready to serve add wine and keep very hot.

SOUTHERN TERRAPIN STEW

3 prepared terrapins*
1 tbsp worcestershire sauce
2 tbsp onion, grated
1 pt heavy cream
3 c turtle stock
3 tbsp plain flour
6 hard boiled eggs
¼ tsp nutmeg
grated rind of 1 orange
salt and pepper to taste
terrapin eggs (if any)
½ lb butter
juice of 1 lemon
1 c madeira

Rub the yolks of the hard boiled eggs through a sieve and combine them with flour and butter to form a paste. Season with nutmeg, onion, worcestershire sauce, the orange rind and the lemon juice. Heat turtle stock until boiling and then stir in the seasoned paste and reduce the heat and continue stirring until well blended. Add madeira and heavy cream and continue stirring until the mixture is thickened and smooth. Be careful not to burn or curdle it. Add the heated terrapin eggs (if any), meat and the chopped whites of the hard boiled eggs. Heat thoroughly and taste for seasoning. Serve with hot buttered toast.

MUSHROOM TERRAPIN STEW

2 c prepared terrapin*
1 c milk
1 can cream of mushroom soup
2 tbsp butter
1 c mushrooms, sliced
salt and pepper to taste

Sauté mushrooms in butter for 5 minutes, stirring frequently. Add the soup and milk. Stir well and season to taste with salt and pepper. Add the turtle meat. Serve hot on biscuits or toast. Serves 4-6.

KATHERINE'S TURTLE STEW

3 lge terrapins
salt to taste
1 onion, chopped
6 hard boiled eggs
1 c butter
juice and rind of 1 lemon
1 tbsp worcestershire sauce
1 tsp nutmeg, ground
cayenne pepper to taste
2 c half and half cream
3 tbsp plain flour
3 pts water

Remove the heads of the terrapins and allow to bleed by hanging head end down for a while. Place the terrapins in boiling water and remove the outer skin and shields* from the plastron* and carapace*. Remove the plastron and leave the legs attached to the carapace but carefully remove entrails and gall bladder retaining the liver for other recipes. Place the shelled eggs (if any) and the strings of eggs in cold water. Wash thoroughly. Place the turtle carapace with attached parts in the water with salt and onion and simmer for about 45 minutes. Do not boil. When tender remove the meat from the carapace and discard bones. Cut up the meat across the grain. Chill the stock until ready to prepare the stew. To make the stew combine the egg yolks, butter and flour and blend well. Set aside. Chop the egg whites and set aside. Bring the reserved stock to a boil and add the egg yolk mixture, lemon juice and rind, nutmeg, worcestershire sauce, cayenne pepper, salt and reserved terrapin meat and egg. Stir in half and half and the egg whites. Heat thoroughly. Remove the lemon rind before serving. Serves 6.

EUELL GIBBON'S TERRAPIN STEW

2 terrapins
salt
1 onion, chopped
1 handful celery leaves
2 cloves garlic
1 c cream, scalded
½ c sherry
water
1 bay leaf
1 sprig parsley
1 carrot, diced
5 hard boiled eggs
2 tbsp butter
cayenne pepper

Kill the terrapins in boiling water and then remove all the outer skin, claws and shields*. After cleaning place the terrapins in a kettle with just enough salted water to cover. Add bay leaf, onion, parsley, celery leaves, carrot and garlic and boil for 1 hour. Then remove the terrapins and allow them to cool enough to handle. Turn on their backs and remove the plastron*. Strain the inside juices into a container and save. Discard the head, tail, entrails and heart. Save any eggs and the liver but remove the gall bladder carefully. Remove all shells and bones. Chop the meat, liver and eggs and mix with the inside juices. Press the egg yolks through a ricer and work them into a paste in the butter. Chop the whites. Now combine the terrapin meat and juices with 1 cup of the stock the terrapins were cooked in and simmer for 5 minutes. Stir in the yolk mixture and stir and simmer for five minutes and then add the egg whites. Add the cream and sherry, salt to taste and dust with cayenne pepper. serve hot but do not boil after adding the cream. Serves 4-5.

TURTLE ON THE SHELL

1 turtle, 1-2 lbs per portion
bacon, sliced
biscuits
sage dressing
shoestring potatoes
honey

Dress a turtle but do not remove quarters, neck or tail from the carapace*. Stuff with sage dressing and replace the plastron*, wrap limbs and neck with bacon slices and hold in place with small skewers or toothpicks. Bake well in a moderate oven. Remove plastron and serve in its own shell with shoestring potatoes, biscuits and honey.

SAGE DRESSING

¼ c onion, minced
¼ c butter
2 tbsp celery leaves, minced
3 c 3 day old bread, cubed
1 egg, beaten
½ tsp sage
⅛ tsp black pepper
¼ tsp salt
½ c milk

Add onions to melted butter and sauté until transparent and yellow. Add the celery, bread and seasonings. Toss to lightly coat with butter. Remove from heat. Combine milk and egg and drizzle over the bread mixture. Blend. Pack loosely into the turtle cavity. The quantity you make depends upon the size of the cavity to be filled.

STUFFED TURTLE

1 turtle
2 c water
1 tbsp salt

Stuffing

½ c stewed tomatoes
1 c milk
1 c turtle broth
2 tbsp margarine or butter
1 tsp bourbon
1 tbsp sherry
cracker crumbs
2 hard boiled eggs, chopped
2 c croutons, crushed
1 tbsp worcestershire sauce
1 tsp celery salt
½ tsp mace
½ Cayenne pepper
½ tsp black pepper

Drop live turtle into boiling water and cook for 45 minutes at 475°F. Remove and let cool. Remove plastron*. Remove the meat, fat and liver. Remove the toes. Place the meat in a pot and add the 2 cups of water and salt. Simmer for 1 hour or until the meat is tender. Reserve the stock. Remove the meat from the bones and cut up. In a large bowl add the meat and the stuffing ingredients except the cracker crumbs and mix well. Place the stuffing into the carapace* which has been thoroughly cleaned. Cover with the cracker crumbs and dot with margarine. Bake at 375°F for 1 hour.

TORTOISE IN ITS SHELL SOUTH AFRICAN STYLE

1 tortoise, dressed
1 clove garlic, minced
salt
milk
1 c orange juice
fat or lard
1 sm onion, chopped
1 chili pepper, crushed
bread crumbs

Dress a tortoise and cut the meat into bite size pieces and braise in fat with the onion, garlic, chili pepper and salt. The shell is carefully cleaned and the inside rubbed with fat and filled with the braised meat mixed with bread crumbs which has been steeped in milk and the orange juice. It is put in the oven and baked until the meat is tender and served with rice and chutney.

GREEN TURTLE PIE

1 10-20 lb green sea turtle
4 med potatoes, peeled
4 lge onions
3 c onions, sliced
2 bay leaves
thyme
basil
worcestershire sauce
oil
4 onions, minced
cloves
salt and pepper to taste
pie pastry
1 lb carrots
2 stalks celery
parsley
5 lge tomatoes, diced
few peppercorns
rosemary
allspice
water
2 c sweet peas, cooked
bread crumbs, loaf dry bread
sage
1 egg yolk, beaten
butter

Dress turtle by removing all meat from the carapace*. Save carapace to bake the pie in. Plunge it in boiling water to remove the shields*. Remove all the fat and cut into ½ inch cubes. Make a broth by boiling carrots, potatoes, celery large onions and a little parsley in a lot of water. Remove vegetables when done and add the turtle fat. Simmer until the fat is tender. Make a sauce by grinding 1½ pounds of turtle meat and sautéing in butter with the sliced onions. Add tomatoes and simmer and then add a little thyme, bay leaves, peppercorns, a little rosemary, basil, allspice and worcestershire sauce to taste. Add a quart of the stock that the fat was cooked in. Simmer for about 10 minutes. Grind meat from the flippers and mix well with the minced onions and bread crumbs along with cloves, sage, allspice, salt, pepper and any other seasonings you desire. Roll into small balls and fry in oil until browned. Drain and add to the sauce. Simmer over low fire and add the cooked potatoes and carrots. Fill the carapace with the remaining turtle meat cut in bite size pieces and add the peas (canned will do). Save a bowl of the sauce without meat balls but pour the remaining sauce, meat balls and vegetables over the turtle meat in the shell. Cover with a rich pastry of your choice rolled thin and crimp along the edges of the shell. Make several slits in the pastry to let out steam. Brush with the egg yolk and bake slowly in a moderate oven until a golden brown. Serve from the shell adding a little of the sauce to each serving. A 20 pound turtle will serve 15 people.

TURTLE PIE I

2¼ lbs turtle meat, ground
½ tsp dried marjoram
1 c hot water
½ c onion, chopped
⅛ tsp ground cloves
1 c celery, chopped
⅛ tsp mace, ground
pastry for a 2-crust pie, unbaked
1 clove garlic, crushed
½ tsp black pepper
¼ c parsley, chopped
2 tbsp plain flour
3 tbsp butter
1 tsp salt
2 beef bouillon cubes

Sauté onion, celery, garlic and turtle in butter in a large skillet until turtle is brown and vegetables are tender. Stir in parsley, salt, marjoram, cloves, mace and pepper. Cover and simmer over low fire for 30 minutes. Drain excess butter from skillet. Blend flour into the meat mixture. Add bouillon cubes that have been dissolved in hot water. Return to heat and bring mixture to a boil and simmer for 1 minute, stirring constantly. Remove from heat and set aside to cool. Pile meat mixture in pie shell, seal and flute edges. Make small slits in upper pie pastry. (Brush with egg if desired). Bake at 400°F for 45 minutes or until golden brown.

TURTLE PIE 2

- 1½ c turtle meat, cubed
- 1 onion, diced
- 3 tbsp butter
- 3 tbsp plain flour
- 1½ c water
- salt and pepper to taste

Brown the turtle meat in butter, add onions and season to taste. Add the water and let simmer for 1 hour and then remove from the water and put in a greased casserole. Make a flour paste and add it to the turtle meat. Make a baking powder biscuit dough and cover the casserole with it. Bake in hot oven until the biscuits are brown.

BAKING POWDER BISCUITS

- 2 c self rising flour
- ⅔ c milk

Add milk to flour all at once and stir with a fork into a soft dough. Beat dough vigorously 20 strokes until stiff and slightly sticky. Roll dough around on cloth-covered board lightly dusted with flour to prevent sticking. Knead gently 8-10 times to smooth up dough. Roll ½ inch thick. Dip cutter in flour and cut biscuits.

TURTLE PIE 3

- 3 lbs turtle meat
- 1½ tsp salt
- 2 eggs
- ¾ tsp thyme, dried
- 2 c hot water
- 1 onion, chopped
- ½ c tomato sauce
- 2 egg yolks
- ½ c dry bread crumbs
- 1 c sherry
- 2 c plain flour
- ½ c shortening
- 2 tsp cold water
- ½ tsp marjoram, dried
- 4 tbsp butter
- 1 tbsp cornstarch
- 1 tbsp butter, melted
- ¼ tsp pepper
- 4 tbsp fat

Sift the flour and salt into a bowl and cut in the shortening with a pastry blender or two knives until the consistency of coarse sand. Beat 1 egg and the cold water together and add it to the flour mixture, tossing lightly until a ball of dough is formed. Wrap in wax paper and place in the refrigerator. Divide the meat in half and cube half and grind half. Place the cubed meat in a saucepan and cover with water. Bring to a boil, drain, and rinse with cold water. Combine in the saucepan with the thyme, marjoram, and hot water and cook over a low heat for 45 minutes. Drain, reserving the stock. Set the cubed meat aside. Melt the butter in a saucepan and add the 2 onions and sauté for 10 minutes. Add the cornstarch and stir well. Add the tomato sauce and the reserved stock and mix well. Cook over a low heat for 30 minutes. Strain and combine with the cubed meat and correct the seasoning. Combine the ground meat with the remaining onion. Add the melted butter, egg yolks, pepper, and remaining salt. Mix well and shape into 2 inch balls. Beat the remaining egg and dip the balls in it and then into the bread crumbs. Beat the remaining egg and dip the balls in it and then into the bread crumbs. Heat the fat in a skillet and fry the balls in it until brown on all sides. Preheat the oven to 350°F. Pour the cubed meat mixture into a 2-quart casserole. Place the balls on top and pour the sherry over it. Roll out the dough on a lightly floured surface to fit the top of the casserole. Place on top sealing the edges well. Brush the tip with a little beaten egg or milk and make several incisions in it. Bake in a 350°F oven for 35 minutes. Serves 8-10.

COOTER PIE

1 med cooter*, dressed
1 c milk
1 tbsp whiskey
cayenne pepper to taste
2 slices toast, crumbled
2 tbsp butter
mace
salt
cracker crumbs
½ c stewed tomatoes
1 c liquor from stew pot
celery salt
2 hard boiled eggs, chopped
worcestershire sauce
1 tbsp sherry
black pepper to taste
water

Drop a live cooter into a pot of boiling water and cook for 45 minutes. Open the shell with a saw and remove the meat, fat, liver and eggs being careful not to break the gall bladder. Remove the meat from the feet and legs. Put the meat in a pot with a little water and a little salt. Then simmer until the meat is tender, usually about 1 hour. Remove the meat and dice the meat, liver and eggs. Add the tomatoes, eggs, milk, liquor from the stew pot, butter, whiskey, sherry, eggs, bread crumbs and seasonings. Put into the shell which has been thoroughly cleaned. Cover with cracker crumbs, dot with butter, and bake in a 375°F oven about 45 minutes.

PASTEL DE HICOTEA

1 lge hicotea*
½ glass dry sherry
2 tbsp raisins
1 bell pepper, sliced
4 tomatoes, chopped
1 jalapeno pepper, dried
cumin
saffron
lemon juice
water
2 tbsp bread crumbs
rice
olives
capers
8 almonds, chopped
1 lge onion, chopped
4 cloves garlic, chopped
cinnamon
oregano
salt
6 tbsp oil
vinegar
5½ ozs flour

Dress and cut the turtle in chunks keeping the legs and thighs intact. Wash it with lemon juice. Grind the jalapeno, cinnamon, cumin, oregano and saffron together and marinate the turtle pieces in it for 15 minutes adding salt and capers. In a saucepan, add 4 tbsp of oil and garlic. Sauté the marinated turtle until brown and transfer to a pot and cover with water. Cover and cook on a low fire. Sauté separately the onion, tomatoes, bell pepper, capers, almonds, olives and raisins. Then add to the turtle in the pot along with the sherry, salt and a touch of vinegar. Cook together and add the bread crumbs to thicken the gravy. Cook for 1 hour until tender. Wash and clean the turtle shell thoroughly inside and out with lemon juice. Make a dough of the flour and 2 tbsp of oil. Roll out and cover the entire shell bottom using salt water to attach the dough to the borders. Put the cooked turtle meat into the shell and then cover with criss crossed strips of dough and cook in a 300°F oven until the dough is brown. Serve with saffron rice.

TURTLE SAUSAGE

1 lb turtle meat
1 lb pork with some fat
sausage casings
sage
granulated garlic
salt and pepper to taste
cayenne pepper

Grind and mix together thoroughly the meats and season to taste and add a small amount of sage. Stuff into sausage casings. Cook as for other fresh sausage or smoke it with hickory until done. Then it can be used like smoked pork sausage.

BULK TURTLE SAUSAGE OR PATTIES

2 lbs turtle meat, ground
salt and pepper to taste
cayenne pepper to taste
parsley flakes
oil
2 eggs, beaten
thyme, dried
marjoram, dried
granulated garlic
sausage casings (optional)

Boil turtle meat in a little water until tender. Drain thoroughly and combine with the eggs and seasonings to taste. Make patties and drop into hot oil or fry off a spoon in a skillet until brown in very little oil. Can also be put into casings and used as a sausage.

TURTLE BURGERS

turtle steaks
butter
salt and pepper to taste

Chop turtle steaks until flattened and about the size of the hamburger bun you are using. Sauté the steak in butter, salt and pepper to taste and place on bun. Garnish with usual hamburger dressings.

TURTLE GOULASH

1½ lbs turtle meat, cubed
1½ tsp salt
8 carrots, sliced
1 c water
1 onion, diced
3 tbsp butter
3 tbsp plain flour
2 sprigs parsley, chopped
4 potatoes, peeled, sliced
dash of pepper

Brown meat and onion in butter. Salt and pepper. Cover with water and let simmer for 1 hour or until meat is tender. Add the vegetables about ½ hour before serving. After the vegetables are cooked add flour paste to thicken.

TURTLE CROQUETTES

1 lb turtle meat, ground
oil + 2 tbsp
¾ c milk or water
1 tsp salt
½ c bread crumbs, softened
1 tsp water
3 tbsp prepared biscuit mix
¼ tsp tabasco sauce
1 tsp parsley flakes
1 egg, beaten
bread crumbs

Sauté turtle gently in a little oil until tender. Set aside in a bowl. In the same skillet heat the 2 tbsp of oil and add the biscuit mix and stir. Gradually pour in the milk (or water), stirring constantly. When thickened remove from the heat. Add to the mixture salt, tabasco sauce and parsley and mix well. Add the turtle and the softened bread crumbs and mix well. Divide into 4-6 portions and shape into patties or croquettes and refrigerate 30 minutes before dipping into the egg with the tsp water added. Roll in bread crumbs and refrigerate again for 15 minutes. Fry in deep oil until browned on all sides. Drain and serve with tomato sauce 2 or horseradish sauce

TOMATO SAUCE 2

3 tbsp butter
1 sm carrot, diced
1 sm onion, diced
2 cloves garlic, crushed
1 tsp sugar
a little black pepper
4 sprigs parsley
¼ c plain flour
2½ c tomatoes, chopped, fresh
1½ c brown stock* or water
½ tsp salt
pinch of thyme
½ bay leaf
1 rib celery with leaves

Melt the butter in a heavy saucepan and add the carrot and onion and sauté until onion is soft but not brown. Stir in the flour and cook until golden. Add the tomatoes, stock (or water), garlic, salt, sugar, thyme, pepper and a bouquet garni* of parsley, celery and bay leaf tied together. Bring the sauce to a boil stirring constantly until it thickens. Continue cooking stirring occasionally and skimming the surface when necessary for 1-1½ hours or until the sauce is reduced to about 2 cups. Discard the bouquet garni and rub the sauce through a fine sieve. Bring again to a boil and cook for 5 minutes while stirring.

HORSERADISH SAUCE

½ c sour cream
½ c mayonnaise
1 tsp sugar
2 tbsp prepared horseradish
2 tbsp capers, drained
1 tsp caper liquid

Combine and blend all ingredients well. Cover and chill. Serve.

FRIED TURTLE

turtle meat
glass of wine
plain flour
catsup
a few mixed spices
butter

Cut turtle meat into small pieces and dredge in flour. Fry in the butter until browned. When browned add some catsup, spices and wine and cover and steam about ½ hour until tender.

FRIED TURTLE MISSISSIPPI STYLE

2 lbs turtle meat, cubed
plain flour
salt and pepper to taste
¾ c water
1 c oil
1 tbsp onion, minced
¾ c red wine

Salt and pepper the turtle and roll in flour. Brown in really hot oil. When browned reduce the heat and add the wine, water and onions and simmer until tender.

FRIED TURTLE MISSISSIPPI STYLE 2

1 med turtle, dressed
1 tsp salt
½ tsp pepper
water
2 bay leaves
seasoned flour*
butter

Dress and disjoint the turtle into medium-sized pieces and place in a stewing pot with the salt, pepper and bay leaves. Add enough water to cover the meat and boil for 15 minutes. Drain well. Reserve the stock for future use. Dust the meat with seasoned flour and fry in a heavy skillet in the butter until it is tender and golden brown. Make a cream gravy from the residue in the skillet.

MCCOY'S FRIED SNAPPING TURTLE

snapping turtle meat
1 cayenne pepper
1-2 eggs
butter
1 bay leaf
1 clove garlic
cracker crumbs

Cut up the turtles leaving bones in joints and parboil until tender but not falling off bones with bay leaf, cayenne and garlic. Drain and allow to cool a little. Flatten the pectoral girdle pieces with a cleaver to make them flat. Dip the pieces in a little beaten egg and then roll in cracker crumbs and fry in butter until brown and crisp on the outside.

BASHLINE'S FRIED TURTLE

2 lbs turtle meat
1 bay leaf
1 beef bouillon cube
salt and pepper
½ c plain flour
lemon juice
5 tbsp milk
2 eggs, beaten
¼ c + 2 tsp oil
catsup
horseradish
water

Cook the turtle meat in water to cover with the bay leaf, bouillon cube and a little salt and pepper. When tender cut the meat into 2 to 4 inch pieces. Make a batter with the flour, a little salt, milk, eggs and 2 tsp oil. Dip the turtle pieces in the batter and fry in remaining oil until brown, about 5 minutes. Serve with hot chili sauce made of catsup, horseradish and lemon juice. Serves 4.

FRIED SOFT SHELL TURTLE

2 lbs turtle meat, cubed
1 tsp + ⅛ tsp salt
¼ c + 1 tbsp milk
2 tsp olive oil
water
½ c vinegar
½ c plain flour
2 eggs, separated
vegetable oil

Combine the turtle, vinegar and salt with enough water to cover and simmer for 1 hour or until the meat is tender. Drain and set aside. Combine the flour, milk, egg yolks, olive oil and ⅛ tsp salt and mix well. Beat egg whites until stiff and fold into the batter. Dip the turtle pieces into the batter and fry until golden brown in deep oil heated to 375°F. Drain well on paper towels. Serves 4-6.

TORTUE DE MER A LA SAUCE TOMATE

2½ lbs sea turtle meat
3 tbsp vinegar
2 tbsp tomato paste
1 tsp coriander
2 bay leaves
salt and black pepper to taste
1½ pts salted water
½ c oil
6 cloves garlic, chopped
1 tbsp flour
2 cloves
1 cinnamon stick
5-6 peppercorns
½ qt water

Bring the salted water, vinegar, peppercorns, 1 bay leaf, cloves, cinnamon stick, and half the garlic to a boil, then simmer on a low fire for 45–60 minutes. Clean the turtle meat of all fat and cut into bite size pieces and place in the boiling water, let cook for 1 hour. Take off the fire and let cool. Heat the oil and brown the remaining garlic, then add the tomato paste, flour, coriander, remaining bay leaf, and season with salt and pepper to taste. Add the water and bring to a boil, then simmer for 15 minutes. Drain the turtle and put into the sauce and finish cooking for 15–20 minutes. Serve warm.

TURTLE VINDALOO

Marinade

8 sm whole cloves
2 tsp tumeric
1½ c white wine vinegar
1½ tsp cayenne pepper
3 tbsp olive oil
¾ tsp ginger
1 tbsp cumin
2 tbsp coriander
2 med white onions, chopped

Prepare the marinade by putting all of the above ingredients in an electric blender. Spin at high speed a few seconds and repeat until the marinade appears even in color and rather thin.

2 lbs turtle meat
4 tbsp butter
½ tsp cardamon
3 tbsp honey
½ tsp pepper
¼ tsp cayenne pepper
1 tsp salt
8 bay leaves
4 tsp coriander
½ tsp cloves
½ tsp ginger
1½ in stick cinnamon, crushed
½ tsp mace
½ tsp nutmeg

Cut up the meat into 1½ to 2 inch chunks and rinse. Place the meat in a porcelain or stainless steel bowl and then pour the marinade over. Mix well and pack the turtle meat so that it is covered with the marinade. Cover and refrigerate for 24 hours. To cook the vindaloo, melt the butter over medium heat in a large saucepan. Add the turtle meat and marinade and keep the heat at medium and add the remaining ingredients one at a time. Stir well to mix after each addition. Bring mixture to a boil and then lower heat to a simmer. Partially cover the pan and cook, stirring occasionally, for 1 hour or until turtle is tender. Serve over rice. Serves 4.

TURTLE CASSEROLE MEXICAN STYLE

3 lbs turtle meat, cubed
oil
water
1 c milk
1 10¾ oz can cream of mushroom soup
1 7 oz can green chilies, diced
salt
black pepper, freshly ground
1 10 oz can cream of chicken soup
1 onion, chopped
12 corn tortillas, cut in 1 inch strips
1 c hard yellow cheese, grated

Brown turtle meat in a little oil and sprinkle with salt and pepper. Add some water and cover and simmer until tender. Mix the soups, milk, onion and green chilies in a bowl. In a greased casserole layer tortilla strips, meat, soup mixture and cheese. Cover with foil and refrigerate for 24 hours. Bake in a 325°F oven until heated through. Serve.

LIZARDS

Preparation—Only the larger species of lizards are of any value as food. By the time you skin and gut small forms, you have nothing much left to eat and large quantities would have to be used to do any good. Not worth it.

Skin and gut the animal. In removing the body contents be sure not to break the gall bladder. Retain the liver and strings of yellow eggs and shelled ones, if any. Incubate the shelled eggs. After washing thoroughly, cut up the meat according to the described method of preparation.

In Mexico, I have seen vendors on the side of the road and in marketplaces selling Green Iguanas (*Iguana iguana*) and Rock Iguanas (*Ctenosaura pectinata*). They have the lizards' feet trussed upwards over the backs by one of two methods: some use a heavy string, and others cut the toes in such a way that the tendons can be pulled and then used to truss up the lizard. A bar or pole is inserted between the body and legs for transport.

IGUANA STEW

1 iguana*, dressed, cut up
2 carrots, diced
1 bay leaf
8 peppercorns, crushed
water
plain flour
2 med onions, sliced
2 white potatoes, cubed
3 fresh tomatoes, quartered
salt
oil

Place dressed iguana in a glass bowl, cover with salted water, and keep in refrigerator overnight. Remove from water and place in a stew pot containing rapidly boiling and salted water; parboil for 30 minutes. Remove and drain. When dry, roll in flour and brown in hot oil. When browned, remove to a serving platter. Pour off all but about 2 tbsp of the oil where meat was browned. Place the pan back on a slow heat, add the vegetables, and cook 6 minutes, stirring constantly. Lay cooked meat on top and pour enough hot water to almost cover. Cover and simmer until all is tender.

CAYENNE IGUANA STEW

1 fat female iguana*
coconut oil
garlic
cayenne peppers
black pepper to taste
salt

Catch a fat female iguana, preferably in March or April. Kill, skin, and remove the entrails. Save the eggs, liver, and heart. Cut up the body down the backbone and divide the sections in three parts and legs in two parts. Place in a pot containing a little heated coconut oil and brown lightly over a low fire. Pour in enough water to cover the meat and drop in a cayenne pepper and garlic to taste. Meanwhile, in another pot of salty water, boil the shelled eggs with a cayenne pepper for 30 minutes. Drain and add to the meat along with the diced liver, heart, and yellow eggs. Cook everything until the liquid has almost disappeared. Pour the remaining broth over red beans and rice, and heap the stewed meat on top.

IGUANA MOLÉ

3 lbs iguana* meat
boiling water
juice of 1 lge lime
2 tbsp oil
1½ lbs tomatoes
¼ c raisins, seedless
½ tsp salt
¼ tsp cloves, ground
1 oz chocolate, unsweetened
4 tbsp sesame seeds
1 lb bell peppers
1 med onion, quartered
2 cloves garlic
½ tsp dried red peppers, crushed
1 lge tortilla, toasted, crumbled
¾ c almonds, ground
¼ tsp anise seed
¼ tsp cinnamon, ground
16 lge tortillas

Cover iguana meat with boiling water and lime juice. Simmer for 30 minutes or until tender. Tear the meat into long slender strips and reserve the stock. Cut bell peppers into quarters, discarding seeds and stems, and place quarters skin side up on a tray under broiler until skins blister. Peel off skins. Add onion, bell peppers, and garlic in a blender and puree. In a flameproof earthen pot over low heat, combine oil, red peppers, and blender mixture. Simmer, stirring occasionally with wooden spoons while preparing the molé. Peel tomatoes, cut in half, and seed. Place tomatoes, toasted tortilla, raisins, almonds, salt, anise, cloves, and cinnamon and puree. Add to the earthenware pot. Cut the chocolate into slivers over the pot and add 1 cup of the stock. Simmer for 1 hour thinning with additional stock if necessary. Tightly wrap the tortillas in foil and heat in preheated 325°F oven for 20 minutes. Heat the meat in a tightly covered dish. Serve meat and molé on hot tortillas, sprinkled with sesame seeds. Serve with cactus salad and refried beans. Serves 8.

CHICKEN FRIED IGUANA

1 iguana*, cut into bite size
1 egg, well beaten
oil
plain flour
cracker crumbs

Dredge iguana pieces in flour, then dip in the egg and roll in cracker crumbs. Skillet fry in oil until golden brown all over.

FRICASSEE OF IGUANA (GUATEMALA)

1 large iguana*
Achiote Paste*
onions, chopped
lime juice or vinegar
salt and pepper to taste
2 lbs chopped tomatoes
butter
1 clove garlic, minced
bay leaf
fresh coriander or parsley, chopped
plain flour
thyme

Skin and dress a large iguana, saving the heart, liver and eggs**. Divide the back, sides and upper portions of the tail into pieces. Make a marinade of the chopped onions, salt, pepper, thyme, garlic, coriander or parsley leaves and a good pinch of Achiote Paste; mix with a sufficient quantity of lime juice or vinegar to moisten the meat evenly. Rub the pieces of the iguana, heart, and liver with the marinade and let stand 2-4 hours. Drain, dry, and coat lightly with flour. Drain the marinade and simmer it in a little butter for 5 minutes. When the onions are transparent and the butter very hot, add the iguana and cook until browned on all sides. Add a bit more flour to make a light roux*, and correct the flavoring with more coriander, thyme, bay leaf, Achiote Paste, salt, and pepper, as needed. Stir in tomatoes and simmer until the meat is tender and the sauce is reduced to a thick gravy. Serve with rice, fried plantains and hot tortillas. ** See dried iguana eggs for use.

DRIED IGUANA EGGS

Iguana* eggs
salt and pepper to taste

Remove strings of iguana eggs. Use the remainder of the iguana for another dish. Season the yellow egg strings to taste and hang up to dry in the sun for 2-3 days. These are considered a delicacy and are eaten dried.

AJIACO (SOUP) OF IGUANA

3 lbs iguana*, cut into pieces
1 tsp salt
1 pinch thyme
3 ears corn cut in 1½ inch rounds
1 avocado, peeled and sliced
1 lb beef bone cut in 3 pieces
2 tbsp capers, rinsed
1 bay leaf
¼ tsp pepper
1 c heavy cream
2 onions, cut in half
1 pinch cumin, ground
4 med potatoes, peeled and cut in half

In a heavy 5 quart casserole combine the iguana, beef bones, and water. Bring to a boil over light heat, skimming the scum that rises to the surface. Add the onion, bay leaf, cumin, thyme, salt, and pepper. Reduce the heat to low, cover, and cook for about 30 minutes or until the iguana is tender. Transfer the iguana to a platter and remove the beef bones and onion. Strain the stock and cut the iguana meat into strips. Return the strained stock to the casserole and bring to a boil over moderate heat. Add the potatoes. Cover and boil 30 minutes or until potatoes are soft. Mash them against the side of the pan until soup is thick and fairly smooth. Add the corn and the iguana and simmer uncovered for 5-20 minutes, depending on tenderness of corn. To serve, pour 3 spoons of cream and about ¾ teaspoon of capers into each of 5 deep soup bowls. Ladle the soup into the bowls and float the sliced avocado on top. Serves 6.

FRICCASSEE OF MONITOR (AFRICA)

1 fat female monitor
thyme
1-2 bay leaves
plain flour
butter
few small onions
parsley, chopped
water

Dress and cut up monitor into pieces, reserving the eggs. Brown the pieces in hot butter in a casserole. When browned, sprinkle a little flour over them. When this is browned, add a little water along with the parsley, bay leaf, thyme, and the onions. Simmer about 45 minutes. Add the eggs and simmer a few minutes longer. Serve.

SATEB VARANUS SALVATOR WITH A PIQUANT PEANUT SAUCE

Varanus* meat, cubed
2 cloves garlic, chopped
sweet soy sauce
3 onions, chopped
juice of ½ lemon

Combine onions, garlic, lemon juice, and enough soy sauce to completely cover the meat and marinate for 3 hours. Place 6 pieces of meat on skewers and roast over hot coals on the barbecue pit. Serve with peanut sauce. Side dishes include thin, sliced cucumbers in olive oil, vinegar, chopped onions, pepper, and salt; baked bananas; and whole small chili peppers in vinegar. Beer also goes well with this dish.

PEANUT SAUCE

peanut butter
2 onions, minced
4 cloves garlic, minced
1 tbsp oil
1 tsp sugar
2 + tbsp chili pepper paste*

In a skillet add onions and oil, and saute until onions are transparent. Add the peanut butter according to the quantity of sauce you want, and stir rigorously. If sauce is too thick add a little water. Lower heat and add the rest of the ingredients; simmer 15 minutes. Keep sauce warm.

SNAKES

Preparation—If using venomous snakes, decapitate about three inches behind the head. If nonvenomous forms are being used, decapitate immediately behind the head. Cut off the tail about one inch before the vent (anus). Then peel off the skin and strip the body cavity of all organs. Wash thoroughly and process according to the recipe. Scent glands in the anal area are removed by discarding the tail. If you retain the tail, be sure to cut these glands out before cooking.

SNAKE SOUP AND ASTRALAGUS HENRYL

½ lb snake
Chinetta*
1 tbsp cooking wine
*Astralagus henryl**
5 c water
½ tsp salt

Dress snake and cut into 2 inch pieces. Add snake and seasonings in water and steam for 30-60 minutes. Serve.

RATTLESNAKE SALAD

1 med rattlesnake, cut up
2 cloves garlic
1 tsp poultry seasoning
½ stalk celery, minced
½ onion, minced
1 c mayonnaise
2 bay leaves
1 tbsp salt
2 tbsp sherry
4 hard boiled eggs, diced
½ c sweet pickles, chopped
water

Combine the rattlesnake, bay leaves, garlic, salt and poultry seasoning; boil until the snake is tender. Drain and remove meat from bones, chop finely and mix well with remaining ingredients. Serve as sandwiches or over quartered tomatoes and lettuce.

351

OPHIDIA A LA MINTON

1 4-5½ foot snake, dressed
3 med onions, chopped
pinch of thyme
butter
¼ tsp cloves, ground
kosher salt*
parsley, chopped
1 clove garlic, smashed
lemons, sliced thin, seeded
msg
½ tsp marjoram
freshly ground black pepper
8 ozs dry white wine

In dressing remove the tail with the scent glands. Wash thoroughly and rub inside and outside with mixture of equal parts of kosher salt, pepper, and msg. Liberally butter casserole; layer bottom with onions and the garlic. Coil the snake in the casserole, then dot with butter, lemon rounds and parsley. Add wine. Sprinkle with the marjoram, ground cloves, and thyme. Cover casserole or seal with aluminum foil and bake 30 minutes in a 360°F oven, or until snake flakes with a fork. Serve with buttered new potatoes and artichokes.

CREOLE RATTLESNAKE

1 med rattler, dressed
¼ c butter
1 c white wine
½ tsp basil, ground
1 onion, diced
½ c water
salt and pepper to taste
¾ c plain flour
1 4 oz can mushrooms, sliced
1 8 oz can tomato sauce
4 chicken bouillon cubes
2 bell peppers, diced
¼ c cornstarch
paprika to taste

Clean meat from bones and cut into bite size pieces. Season flour with salt, pepper, and paprika. Dredge meat in flour and brown in butter. In a saucepan add mushrooms liquid, wine, tomato sauce, basil and bouillon; simmer 15 minutes. Stir in mushrooms, onion, pepper, and meat. Cover and cook over low heat for 30-45 minutes or until meat is tender. Combine cornstarch and water, and stir into creole until thickened. Serve over rice.

BROILED RATTLESNAKE

1 rattlesnake, dressed
crab boil*
lemon juice
garlic, minced
salt
oil
parsley, minced
water

Remove vertebrae and ribs and cut meat into 1½ inch pieces. Boil in water with a little salt and crab boil for a few minutes. Drain and coat meat with oil. Place on rack in broiler pan and broil each side for about 10 minutes or until tan. Then sprinkle with lemon juice, parsley and garlic. Serve.

ORANGE FRIED RATTLESNAKE

1 rattlesnake, dressed, in 1 inch chunks
salt and pepper to taste
juice of 1 lemon
¼ tsp black pepper
bread or cracker crumbs
¼ c milk
2 eggs
juice of 4 oranges
dash nutmeg
oil

Make a marinade of pepper, nutmeg, orange and lemon juice. Soak meat in marinade for 3 hours. Drain and dry. Season with salt and pepper, dip in beaten egg and milk, and then roll in crumbs. Fry in oil until brown.

FRIED RATTLESNAKE

rattlesnake meat
eggs
plain flour
milk
salt and pepper to taste
oil

Make a batter of the eggs, flour, and milk (or a beer batter*). Salt and pepper the rattlesnake meat (which has been cut into 2 to 3 inch chunks) and pass through the batter until coated. Have oil heated and place the battered meat in the oil; fry until nicely browned. Serve.

FRENCH FRIED RATTLESNAKE

1 rattlesnake, dressed, in 1 inch chunks
1 c cracker crumbs
1 tsp salt
oil
2 eggs, beaten
1 clove garlic, minced
¼ c plain flour
⅛ tsp pepper

Dip pieces of rattlesnake in eggs in which the garlic has been added. Combine crumbs, flour, salt and pepper. Roll the dipped pieces in the crumb mixture. Fry until golden brown. Serves 4-6.

RATTLESNAKE STEAK

2-4 ft rattlesnake, dressed and cut in chunks
1 c plain flour
black pepper
salt
1 c oil, very hot

Salt and pepper the meat and then roll in the flour. Drop in the oil and fry until a golden brown. Serve hot. Serves 4.

PRAIRIE SHRIMP (RATTLESNAKE)

1 rattlesnake, dressed
pancake flour
oil
seasoning salt

Fillet meat from the bones and cut into 2 inch long strips; wash thoroughly and dry. Roll strips in pancake flour to which seasoning salt has been added. Deep fry until golden brown.

DEEP FAT FRIED RATTLESNAKE

1 rattlesnake, cut in 3 inch pieces
1 tsp salt
½ c milk
oil
2 tbsp lemon juice
¼ c vegetable oil
1 egg
½ c + 1-2 tbsp self rising flour

Marinate rattlesnake meat in refrigerator overnight in lemon juice, vegetable oil and salt, basting occasionally. To make a batter, beat together egg and milk and stir in the flour. Let rest about 20 minutes. When used it should be runny, because only a thin batter will fry crisp. Dry meat and dip in batter. Deep fat fry in hot oil in a heavy skillet.

HOT FRIED RATTLESNAKE

5 lbs snake meat, dressed
salt and pepper to taste
tabasco sauce
plain flour
oil
vinegar

Cut meat in 1 inch steaks; soak in vinegar for 10 minutes. Remove and sprinkle with tabasco sauce, salt, and pepper. Roll in the flour and fry. Serve immediately.

ITALIAN FRIED RATTLESNAKE

1 lb snake, dressed, in pieces
½ tsp black pepper
¼ c lemon juice
plain flour
1 c sherry
½ tsp Season-All
½ c italian salad dressing

Marinate snake pieces in a mixture of sherry, pepper, Season-All, lemon juice, and salad dressing for 2 hours. Drain and dredge in flour. Fry for about 15 minutes, turning, until browned. Drain and serve hot.

MARINATED FRIED RATTLESNAKE

rattler meat, in 2 in cubes
salt to taste
oil
buttermilk
pancake mix

Soak meat in salted buttermilk for 1 hour. Drain and dredge in the pancake mix and fry 7 minutes in oil or until browned. Drain well and serve.

ROUNDUP RATTLESNAKE

1 rattler, dressed, filleted
1 egg, slightly beaten
salt and pepper to taste
1 c + pancake mix
1 sm bottle 7-Up
oil

Combine the cup of pancake mix, egg and 7-Up. Mix well. The batter will be thin; set aside. Dip prepared meat in dry pancake mix and allow to stand and dry for 20 minutes. Sprinkle the fillets with salt and pepper to taste. Dip in batter and pan-fry in deep oil until golden brown.

SAUTEED SNAKE

1 rattlesnake, dressed
water
plain flour
salt and pepper to taste
oil

Steam snake cut into chunks in a pressure cooker with a little water and salt at about 15 pounds pressure for 15 minutes or until tender. Remove and pepper the snake to taste and dredge in flour. Saute in a little oil until lightly brown.

ITALIAN BARBECUED RATTLESNAKE

rattlesnake meat
barbecue sauce
italian salad dressing

Marinate the cleaned meat which has been cut into 4 inch chunks in the italian salad dressing for several hours. Place meat over coals, and mop the pieces with the marinade while cooking. When near done, add your favorite barbecue sauce and finish cooking. Serve.

BARBECUED RATTLESNAKE

rattlesnake fillets, 2-3 in pieces
water
barbecue sauce

Boil fillets for about 45 minutes in a small amount of water. Add barbecue sauce (your favorite) to meat and let come to a boil. Serve with toothpicks.

CHINESE BARBECUED RATTLESNAKE

1 snake
1 tbsp soy sauce
1 tsp ginger powder
1 tbsp barbecue powder
1 tbsp sugar
1 scallion, chopped

Dress snake and cut into 3 inch pieces. Soak with all seasonings for 1 hour. Baste with sauce while barbecuing. Serve.

MARINATED RATTLESNAKE

1 rattlesnake, dressed
dash nutmeg
juice of 1 lemon
2 eggs, beaten
oil
¼ tsp freshly ground pepper
juice of 4 oranges
salt
bread crumbs

Cut snake into 1 inch pieces and marinate in a sauce of nutmeg, lemon juice, pepper, and orange juice for 3 hours. Remove the meat, drain, and wipe dry. Salt and dip in eggs and roll in bread crumbs and saute until brown.

RATTLESNAKE CHILI

2½ lbs roast beef, cut in 1 inch cubes
¼ c cornmeal
1 tsp salt
1 4 oz jar chili powder
3 onions, chopped
2 tbsp cumin
4 15½ oz cans beef broth
½ c masa flour
2 c rattlesnake meat, cubed
1 tsp black pepper
2 c beef suet, ground
4 cloves garlic, crushed
1 tsp oregano

Combine meats and let stand at room temperature for 2 hours. Combine flour, cornmeal, salt, pepper, chili powder, and blend well. Dredge meat in flour mixture. Render suet at high heat. Drop several pieces of meat into hot fat, browning well. Continue until all the meat is browned. Remove meat; add onions and garlic to drippings. Saute on low heat until onions are tender. Return meat and add beef broth, cumin, and oregano. Add remaining flour to mixture, stir to blend, and simmer for 2½-3 hours, stirring frequently. Serves 6-8.

SNAKE GUAM STYLE (COCONUT MILK)

1 lb snake, dressed, in pieces
coconut milk from 2 grated coconuts
1 c water
3 cloves tumeric, grated
salt to taste
1 onion, sliced thin

Cook snake pieces in water about 30 minutes or until tender. Add coconut milk, salt, tumeric and onion and simmer 5 minutes. Do not boil. Serve hot.

SNAKE A LA VIETNAMESE

- snake meat
- 2 tbsp oil
- 1 tsp black pepper
- 1 tsp msg
- onion (optional)
- 1 tbsp chili powder
- 1 tbsp garlic powder
- 2 tsp sugar
- ½ coconut, grated

Prepare the snake by skinning, removing the scent glands at the tail, intestines and wash thoroughly. Cut the meat into pieces, removing the bones. In a medium bowl full of meat add sugar, msg, pepper, garlic and chili powder and mix thoroughly. Heat the oil in a sauce pan large enough for the meat and add the meat, turning often until all the oil is absorbed. Grate the coconut after removing the brown outer layer. Pour all the coconut juice from the grated coconut (if no juice, add water to the grated coconut and squeeze) into the sauce pan with the meat and mix thoroughly until juice absorbed. The coconut has to be ready before the cooking begins. After a while the meat is ready. Remove from heat and add onion (optional) and serve.

SNAKE ADOBO

- 1 lb snake, dressed, in pieces
- 1 tbsp vinegar or lemon juice
- black pepper to taste
- water
- ½ tsp sugar
- ½ c soy sauce
- 2 cloves garlic

Boil snake pieces in a little water for 30 minutes. Drain and brown in pan. Add remaining ingredients and cook for 30 minutes.

BAKED SNAKE WITH BERRIES

- 1 snake
- berries

Dress the snake by removing its head. Do not slit the skin but peel and keep it intact and attached below the vent and when peeled back remove the intestines and the scent glands at the vent and wash thoroughly. Pack the body cavity with as many berries as you can. Berries that are not quite ripe will keep their shape better when cooking and will give flavor to the meat. Pull the skin back over the snake and stuff with more berries if possible. Build a fire in a hole as for ordinary underground roasting. Scoop out a place and coil the stuffed snake into it, cover with coals and ashes and cook for 30-60 minutes. Let it cool in the ground until the fire is cold. Strip off the skin and shake out the berries if they were too green to cook well. They can be eaten although they may taste bitter. The berries were used mainly for improving the flavor of the snake. The meat should be quite tender and will taste something like frog legs.

BAKED RATTLESNAKE

1 rattlesnake, dressed
1 tsp rosemary, chopped
1 tsp sweet basil, chopped
2 limes, sliced thin, seeded
1 4 oz can mushrooms, sliced
1 tbsp parsley, chopped
1 tsp white pepper
cream of mushroom sauce

Wash well in cold water. Cut into 3 inch sections and place in a large baking dish. Spread sliced mushrooms, limes, basil, parsley, rosemary, and pepper over top; then cover with cream of mushroom sauce. Cover tightly. Bake in 300°F oven for 1 hour or until done.

CREAM OF MUSHROOM SAUCE

2 10¾ oz cans cream of mushroom soup
3 scallions, chopped
1 12 oz can evaporated milk
1 8 oz can mushrooms, drained, chopped fine

In a saucepan add the cream of mushroom soup, evaporated milk, and mushrooms and heat. Simmer about 5 minutes. Remove from heat and add scallions.

HERB BAKED RATTLESNAKE

1 med to lge rattlesnake, dressed
1 tbsp plain flour
1 c half and half cream or whole milk
1 tsp white pepper
1 tbsp lemon juice
¼ c capers
1 tbsp butter
½ tsp salt
¼ lb mushrooms, sliced
1 tsp basil
1 tsp rosemary

Cut the snake into 3 inch pieces and place in a large baking dish. To make a sauce melt the butter and stir in flour and salt. Gradually stir in cream or milk until thickened and smooth. Pour sauce over meat and add fresh mushrooms, basil, pepper, rosemary, and lemon juice. Cover and bake at 300°F for 1 hour or until done. Garnish with capers.

CORN BAKED RATTLESNAKE

1 rattlesnake, dressed
milk
garlic to taste
fresh corn husks
salt and pepper to taste

Cut into 3 to 4 inch lengths, and soak in cold milk for 1 hour. Drain and add salt, pepper, and garlic to taste. Place in corn husks and seal to keep the juices from seeping out. Bake about 1 hour in a preheated 350°F oven. If corn is out of season aluminum foil can be substituted.

MADISON BAKED RATTLESNAKE

1 rattlesnake, dressed
¼ lb fresh mushrooms, sliced
1 tsp basil
1 tsp rosemary
1 recipe thin cream sauce
2 limes, thinly sliced, seeded
1 tsp white pepper

Cut dressed rattlesnake into 3 inch sections. Place in a large baking dish and cover with the thin cream sauce; add mushrooms, limes, basil, pepper, and rosemary. Cover tightly and bake in 300°F oven for 1 hour or until done.

THIN CREAM SAUCE

1 tbsp butter
1 c milk, scalded
1 tbsp plain flour
seasonings to taste

Blend the butter and flour and stir over low heat for 5 minutes. Do not let it take on color. Slowly add milk and continue stirring until the sauce is smooth and thick. Season to taste.

ROSS ALLEN'S DIAMONDBACK RATTLESNAKE MEAT

The following six recipes presented are given for their historical value. For a number of years Diamondback Rattlesnake Meat with Supreme Sauce was canned by Ross Allen at Silver Springs, Florida. Since this item is no longer available these recipes became moot. Searching through various cookbooks a Supreme sauce was found. It is not certain if this is the recipe Ross Allen used nor his method of preparing his canned product. It was stated that this sauce was piquant and has cayenne pepper in it. Possibly gourmet experimentalists can duplicate the recipes with reasonable success. The recipes are as follows with the addition of a Supreme sauce, velouté sauce and white stock.

RATTLESNAKE MEAT IN PATTY SHELLS

Remove contents into double boiler. Heat slowly to boiling point. Just before serving add one tbsp dry white wine. Serve in individual patty shells and top with finely chopped chives or parsley.

RATTLESNAKE MEAT WITH HARD BOILED EGGS

Empty contents into saucepan, heat to boiling point. Remove from fire and add one tbsp dry sherry wine, four quartered hot hard boiled eggs. Mix carefully so as not to break up eggs. Serve on waffles or sour cream pancakes.

RATTLESNAKE MEAT CROQUETTES

Mix contents of can with two tablespoons cracker meal. Let stand 15 minutes to thicken. Chill, then form into small croquettes not over one inch in diameter. Dip in beaten egg, roll in cracker meal and deep fry quickly in piping hot fat until brown. Garnish with watercress or parsley. Serve immediately.

RATTLESNAKE MEAT HORS D'OEUVRE

Chill can in refrigerator for several hours before serving. Empty contents of can into small chilled bowl, spread on crackers or toast, garnish with paprika and parsley. Serve at once.

RATTLESNAKE MEAT SAUCE FOR FISH

Add cream to contents of can until of medium sauce consistency. Heat thoroughly, then add one tsp capers and one tbsp dry white wine. Serve over boiled filets of fish or place boiled or baked fish on platter and pour sauce over it. Garnish with strips of pimento, a few more capers and watercress.

SUPREME SAUCE

2 c chicken stock
3 fresh mushrooms, sliced
salt
1 cup velouté sauce 3*
1 c sweet cream
cayenne pepper

Cook the chicken stock with the mushrooms until it is reduced by ⅔. Combine with the velouté sauce and bring to a boil and reduce it to about 1 cup. Gradually add the cream, stirring constantly. Correct seasoning with salt and a little cayenne pepper and strain through a fine sieve. If not used immediately fleck the surface with butter.

VELOUTÉ SAUCE 3

¼ c butter
¼ c plain flour
1 cup fresh mushrooms, chopped
1 sprig parsley
6 c boiling white stock
3 white peppercorns
salt

Melt the butter into a saucepan and add the flour and mix well. Cook for a few minutes but do not let it color. Add white stock, 2 cups at a time, stirring vigorously with a wire whip. Add mushrooms, peppercorns, a little salt as necessary and parsley. Cook stirring frequently for about 1 hour, skimming it from time to time, until it is reduced by ⅓ and has the consistency of heavy cream. Strain through a fine sieve. If it is to be stored, stir it occasionally with the whip as it cools and dot it with bits of butter. If it is too thick, add a little more stock.

WHITE STOCK

2 lbs veal bones
2 tsp salt
1 leek
1 carrot, peeled
6 peppercorns
chicken bones (if any)
5 qts cold water
5 celery stalks with tops
1 onion, peeled
5 sprigs parsley
veal trimmings (if any)
water

Parboil bones, cut in small pieces until the scum rises and then drain them. Put the bones in a kettle with the 5 quarts of water, salt, celery, leek, onion, carrot, parsley and peppercorns. Add any available veal trimmings or chicken bones and bring to a boil removing the scum as it accumulates. Simmer for 3 hours, cool, and remove the fat from the surface. Strain through several thicknesses of cheesecloth. Store in refrigerator. Will keep 4-5 days.

RECIPES OF INDIGENOUS PEOPLE

Introduction—In the old days, *Podocnemus* turtles were a staple, all species, from the small *sextuberculata* to the giant *expansa*. They are now protected. So the following recipes are given more in a historical than a culinary spirit.

SARAPATEL

Decapitate the turtle. Let the blood flow into a bowl and coagulate. Save it. Mince together the meat (with skin), the liver and the blood clot. Leave some fat (a bit chopped, a bit in blobs). Season with salt, black pepper, cayenne, garlic, chopped onions, a little vinegar (or lemon juice), chopped green peppers. Then heat in a pot enough oil to fry the mixture. When it is frying evenly, taste, correct the seasoning and serve. The gravy should be thick and rich.

SARAPATEL 2

Collect the blood from decapitated turtle in a bowl containing a little vinegar and allow to coagulate. Take the heart, lungs and liver (remove gallbladder) and carefully wash them in vinegar and a little lemon juice. Slice them into very small pieces and combine with the coagulated blood together with chopped onions and coriander. Add salt and vinegar to taste. Boil for more than 1 hour. Serve hot with white rum.

STEAKS

Cut thin steaks and leave them overnight (or at least for some hours) in a sauce of vinegar (or lemon juice or, better dry white wine) liberally seasoned with salt, black pepper, cayenne, chopped parsley (plenty), chopped onions, and some oil. Remove from the sauce, let run for a bit, and fry in very hot oil, turning the steaks to fry evenly on both sides. When ready, dry on a piece of wrapping paper and serve.

BROWN SAUCE

Decapitate the turtle. Let the blood run into a bowl with some vinegar, stir to avoid clotting. Save it. Chop the meat in 1 inch long pieces, with the skin. Mince the liver. Fry in a pot, in hot oil; salt, black pepper, garlic, chopped onions, cayenne, chopped parsley, 1 bay leaf. Add the minced liver. When this seasoning is frying well, add the meat and let fry for some 10 minutes. Add the blood and let simmer until the meat is soft.

Adjust the sauce to an almost velvety consistence and serve with rice.

PAXICA

Take a cleaned carapace* and put it to roast over a low fire, so it will not split. Mince the skin and cook it as in steaks above. When it is frying nicely, add water. Pour the mixture in the carapace and let it boil for some minutes. Thicken the broth with strained white flour until it barely runs. Add cayenne to individual taste and serve.

PACHICA

Cut the heart, liver and kidney into ½ inch cubes and fry in the turtle's own fat until crisp.

"MOSQUECA" OR TURTLE STEW

5 lbs turtle meat
green vegetables
2 c coconut milk
lemon juice
palm oil
black pepper
salt

Chop the meat into cubes and place in a pot with the green vegetables. Add the coconut milk, pepper and lemon juice. Simmer for 30 minutes. Add a generous amount of palm oil and boil for 5 minutes. In Brazil it is served with dumplings of manioc meal which has been dunked in the liquid of the stew and then crisply fried.

ROASTED HAUNCHES

Skin the haunches. Prepare a sauce as in brown sauce above; place the haunches in it and punch plenty of holes with the point of a sharp knife. Fry on the side some condiments (as, for example, in brown sauce that is universal here, you can add or avoid according to your taste), add the haunches, fry them for a few minutes, then add water to barely cover. Let cook but remove before the meat is too soft. Remove the haunches and let them run. When moist, rub well with butter seasoned with garlic, onion, black pepper and chopped parsley (essential); tomato paste if you like. Roast in a hot oven. When the meat is golden, serve.

Geochelone is also protected. Many think it a pity. The male is indifferent fare, but the female is well liked.

IN NUT MILK

Prepare some Brazil nut milk by grinding the nuts very fine, wrapping them in a white napkin (preferably linen) and squeezing. Save. Alternatively do the same to a couple of coconuts or obtain coconut milk. Chop your tortoise, with skin and claws, in pieces about 1 inch long. Fry the meat and the chopped liver as in brown sauce above. When the seasoning has caught, put everything in a pressure cooker and cook for 30 minutes. Serve with cassava flour or rice.

WHOLE

Throw tortoise from the second floor, to make the liver swell. Open a square hole in the plastron*. Through it gut the tortoise, leaving the fat and saving the liver (it can be sliced thin and french fried for a nice canape, that is why it is made to swell). Put through the opening all the seasonings you have in the house. Roast in a very hot oven until it smells delicious.

FROGS (TOADS)

Catch toads, twist off their heads, pull off the skin while all the time holding the animals under running water lest the meat becomes very bitter. Parboil, then cook as any other meat. The above from "Cherokee Cook-lore". 1951. Mary Ulmer and Samuel E. Beck, eds. The Stephens Press, Cherokee, N. C.

ROASTED POISON DART FROG

Epipedobates petersi frogs
manioc
salt

These frogs are collected by the Campa Indians in Peru who eat these as well as other genera of frogs. They are prepared by salting whatever quantity they have and are wrapped in a leaf and well roasted in campfire ashes to remove the poison and then eaten with manioc. After roasting they may be chewed. Taken from an article by Rodriguez, Lily and Charles W. Myers. 1993. A New Poison Frog from Manu National Park, Southeastern Peru (Dendrobatidae, *Epipedobates*). American Museum Novitates, (3068):1-15.

BETUTE

1 frog
garlic, ground
salt
soy sauce to taste
onions, chopped
cayenne pepper, ground
msg
oil

Dress and grind the leg meat of the frog and mix with a little chopped onions. Grind some cayenne pepper and garlic and add to the mixture. Add salt, msg and soy sauce. Combine well. Put the mixture into the frogs body cavity and let it dry under the sun so that when it is fried it is a little bit crispy.

ADOBO

1 monitor*, dressed
cayenne pepper, ground
salt
msg
water
garlic, ground
vinegar
soy sauce
onions, diced

Parboil a dressed cut up monitor in a little water until tender. When parboiled saute with garlic and onions. Serve with the cayenne pepper, garlic, vinegar, salt, soy sauce and msg made thick enough for a sauce. Pour over meat and serve.

BROILED RATTLESNAKE

If you are camping out, you might try an old Indian way of cooking rattlesnake. After you have caught your snake cut off the head and tail, and gut it. Wash. Throw it on the campfire coals without skinning. Let it roast until it is tender, then cut it in pieces and let each diner remove the roasted skin and eat the meat with his fingers.

BLACKFEET INDIAN JELLIED SNAKE

1 med snake
2 c indian vinegar*
1 handful mint
2 fingers* coltsfoot salt*

Cut off the head, skin, and take out the intestines. Remove the glands in the tail region. Cut into 1 inch pieces. Wash in cold water. Put the vinegar, mint and coltsfoot salt in some kind of container; put the pieces of snake on top and cover with cold water. Let stand overnight. Put the container over the hot coals in the morning and simmer slowly for about 35 minutes. Remove from the fire and cool. The dish is ready to eat when the jelly has set. This recipe taken from Walker, Herb. 1977. Indian Cookin'. Baxter Lane C., Amarillo, Texas.

BAKED SNAKE PAPUA NEW GUINEA STYLE

A dressed snake is slowly smoked over a fire.

BAKED SNAKE AUSTRALIAN BUSH STYLE

Heat the whole snake slowly over a fire, stretching it continuously. This ensures that the juices are retained without the snake contorting. Then make shallow incisions along both sides close to the backbone to cut the sinews. The snake may then be rolled up and tied like a rolled roast and baked in a low oven or on an open fire. The intestines should be removed before serving. The size of the serving, of course, depends on the snake.

TURTLE AND TORTOISE PAPUA NEW GUINEA NATIVE STYLE

Sometimes tortoises and turtles are cooked whole; the shell is broken open and the meat and juices eaten from the shell. Otherwise they are cut up and distributed and baked or boiled in pieces. Turtle eggs are considered a delicacy.

CROCODILE PAPUA NEW GUINEA STYLE

The flesh is usually baked or broiled. Crocodile eggs are regarded as something of a delicacy. The eggs boiled are said to have a creamy consistency. This Papua New Guinea section is taken from May, R. J. Kaikai Aniani A Guide to Bush Foods Markets and Culinary Arts.

GOANNAS*

There is an art in cooking goannas. They must always be felt tenderly by an expert to see that the fat is inside, because on the quantity of fat depend its value as a food and delicacy. If it has plenty of fat it must not be held up too long by the neck, or the thin membrane holding the fat will break. The secret is to keep the fat in the membrane bag inside the belly.

Before it is cooked it is placed on a fire and turned over until all the scales become crisp. A native will tell you that this is intended to drive the juices back into the flesh. The vent is held to the fire to make it tender, then sprinkled with sand, and the entrails are carefully drawn out with an expert twist. When it is clean, another incision is made under the forearm. A little hooked stick is put in and turned round to get the liver and gall bladder out, very gently. The gall bladder is then broken off, and the liver is pushed back.

The goanna is then cooked lengthways in the ground oven. The best parts are the ribs and the arms, though, with typical politeness, the natives always give the tail to the white men, who imagine it the best part to eat. The fat is always taken out and everyone eats some. You can eat up to a tea cup full and it will not give you indigestion, and the eggs are excellent. Like most other good, rich food, goanna eggs are taboo to the young, so are crocodile eggs. These are the prerogative of the old men.

BLUE-TONGUES (TILIGUA)*

Blue tongues are cooked exactly like goannas.

MALAYAN DRY CURRY

Turtle meat, after it has been cooked in the oven, can be cut into little cubes, put into the frying pan and cooked as a curry, but it must always be baked in the ground oven first. The cook melts turtle fat in the pan, adds chopped onions and curry powder, and cooks the meat in the pan until it becomes crisp.

Then he thickens the curry with a little flour and adds some water. This is a fine malayan dry curry, made with the delicious turtle meat and cooked in fresh turtle fat.

TURTLE EGGS

I have already described how unlaid turtle eggs are made into a sausage. Now, should large quantities of laid eggs be discovered, these can be broken into a real paperbark dish, which is then wrapped up, tied with bush string, and baked on the coals of the fire. The result is like a large cake of hardboiled eggs which will last for several days and is a good stand-by in an emergency. Any eggs can be baked in the ashes, but care must be taken that they are only cooked on the soft white ashes, and that a hole is driven into the upper end of the egg. The eggs are stood up around the fire so that they will cook in the ordinary way, but if there is no hole pierced in the top, you may have an explosion.

TURTLE EGG OMELETTE

Take 12 turtle eggs, (there are usually 60 in a nest), wash them well to remove the sand, then break into a dish. Add half a pint of milk, pepper, salt, and any other flavorings you require, such as chopped bacon, cheese or herbs. Beat again for two or three minutes. Put a little butter in the frying pan, and when it is smoking hot, pour in the egg mixture, stirring it continuously with a fork or an egg slice. At first it will be very watery, like custard, but with continuous stirring it becomes crisp. Turn it over and over for about 15 minutes. This is really a beautiful dish.

TURTLE EGG GRIDDLE CAKES

Take up to 12 eggs and add four spoonfuls of flour, and a little cream of tartar and baking soda. Beat up well, add a little water, and make it a fairly thick batter.

Pour some batter into a hot greased frying pan, to a thickness of about ½ inch. It will immediately double or treble in thickness. When it is perfectly brown on the underside, add a little more batter and turn it over to brown on the other side. Turn the griddle cakes on to a hot plate and sprinkle with sugar.

TURTLE "SOUP"

The natives generally watch the beaches, and capture the turtles as they come up to lay their eggs. The captured creature is first turned over on its back and killed. Next, they make a hole at the base of the neck and through the hole remove all the entrails, including the liver and the heart. The gullet is cleaned and turned inside out, and carefully set aside, as also the long, cleaned gut.

A big fire is blazing away and as it burns down, the cooks grill their special "perks", the liver and smaller pieces of the edible organs. Then the main cooking begins. The eggs that are still left inside the turtle, that is, the soft eggs, are taken out and stuffed into the prepared gullet, which swells to the size of a pineapple. If there are any eggs left over, they are put into the gut to form a large sausage that may be anywhere up to 2 ft long.

When the fire has burned down, it is time to start cooking the turtle, which has been thoroughly washed and cleansed with salt water. Hot stones are now put into the cavity from which the entrails have been removed, then the turtle is laid on its back in the coals and more hot stones and coals are laid upon it. The opening at the neck is made steam proof with leaves well rammed home. The whole is covered in the usual way and left to cook. Meanwhile, the gullet and the sausage are put into hot sand or hot ashes, and inside an hour or two, when the eggs are hard cooked, they are taken out and laid aside to cool. These egg sausages, which last for several days, are eaten in slices. And very good they are, too.

After three or four hours, the ashes are scraped off the turtle; a knife is used to cut round the carapace of the belly, which is torn off and put aside; the stones are removed, and the cooked flesh is exposed. All the stones that were inside are carefully removed, and the shell now looks like a great inverted bowl, only the legs and arms remain to adhere to the sides. In the bottom of the shell are all the juices that have run out during the cooking. Now salt and pepper are added to the steaming broth, and everyone present dips in his pannikin for a taste of the first course.

After the soup comes the meat. The green turtle fat is solid, like beef fat, and in the eating a piece of flesh is wrapped round a smaller bit of green fat. If you are still hungry, you can have a slice of turtle egg sausage later on, as a savoury.

In other places, the sea turtle is cut up before it is cooked, of course, it is sealed over the hot fire to drive the juices into the flesh, according to native custom, before it is cut up. Then it is cooked in the stones, the carapace being used to cover the meat instead of paper bark. For this method of cooking, turtle beef is flavoured with gum leaves or grevillea, and potatoes and onions are cooked alongside.

TURTLE SOUP

Take the fore-flippers and, after washing them thoroughly, boil them, skin and all, in half a kerosene tin of water, keeping the water constantly up to this level. After a time the skin becomes separated, and can be pulled off the flesh. When this occurs, keep the dish simmering until the whole thing becomes like a jelly. Add some salt. This makes a perfect stock, a basis for turtle soup. All you do to complete the cooking is to add vegetables and flavourings as desired, and finish off.

Freshwater turtles are cooked the same as sea turtles.

The above on goannas through turtles is excerpted from Harney, Bill. 1960. Bill Harney's Cook Book. Landsdowne Press, Melbourne. Collaboration with Patricia Thompson.

TURTLE OIL

The following is excerpted from Cutright, Paul Russel. 1940. The Great Naturalists Explore South America. The Macmillan C., N. Y. xii-339, map. Reprinted 1968 by Books for Libraries Press, Freeport, N. Y.

"As soon as all the eggs have been unearthed, those not intended for food are picked up and thrown into a canoe. They are usually broken with wooden implements manufactured for the purpose. But occasionally the Indians, particularly the children, unable to resist the opportunity of a lifetime, jump into the canoe and continue the maceration with their feet. After the eggs are well squashed and everyone is besmeared with yolk and albumin, water is poured on the mass, which is then left in the sun until the heat induces the oil to come to the top. As it reaches the surface, it is skimmed off and boiled. If properly and carefully prepared, the oil is clear and odorless and very like olive oil. It is used in lamps, and also for cooking, being an excellent substitute for lard and butter."

TURTLE (VARIOUS)

The following is excerpted from Bates, Henry Walter. 1863. The Naturalist on the River Amazons. John Murray, London. 2 vols. reprinted, abridged in 1864, 466 pp., maps, by John Murray, London. It was again reprinted by the University of California Press in 1962.

"I became so sick of turtle in the course of two years that I could not bear the smell of it, although at the same time nothing else was to be had, and I was suffering actual hunger. The native women cook it in various ways. The entrails are chopped up and made into a delicious soup called sarapatel, which is generally boiled in the concave upper shell of the animal used as a kettle. The tender flesh of the breast is partially minced with farinha, and the breast shell then roasted over the fire, making a very pleasant dish. Steaks cut from the breast and cooked with the fat form another palatable dish. Large sausages are made of the thick-coated stomach, which is filled with minced meat and boiled. The quarters cooked in a kettle of Tucupi sauce form another variety of food. When surfeited with turtle in all other shapes, pieces of the lean part roasted on a spit and moistened only with vinegar make an agreeable change."

GLOSSARY

Achiote Paste—See Fricassee of Iguana (Guatemala) for recipe.

Achoques—The larval form of the Lake Patzcuaro Salamander (*Ambystoma dumerilii*).

Allemande Sauce—See Frog Legs Mariniere for recipe.

Amphiuma—A large eel-like salamander found in the southeastern United States. Also the genus and common name of this salamander.

Andouille—A smoked, highly seasoned pork sausage made with meat chunks instead of ground meat.

Astragalus henryl—A species of milk vetch, a legume, used as an herb in oriental cooking. Available in some oriental food stores.

Axolotl—The larval form of mole salamanders.

Bechamel Sauce—See Frog Legs a la Bechamel for recipe.

Beer Batter—See Fried Alligator in Beer Batter for recipe.

Bouquet Garni—See Grenouilles a l'Indienne for recipe.

Brown Veal Demi Glaze Sauce—See Gold in Veal recipe.

Brown Stock—See Variation 1: Frog Legs Osborn for recipe.

Cajun or Creole Seasoning—A combination of seasonings packaged for commercial sale. It varies slightly among brands.

Calipash—A fatty, gelatinous, dull greenish substance next to the carapace of a turtle.

Calipee—The fatty, gelatinous, light yellow substance attached to the plastron of a turtle.

Capipee—Green turtle fat.

Carapace—The dorsal shell of a turtle.

Chili Pepper Paste—Called Sambal. Best is Trassi or Badjak. Available in oriental stores.

Chinese Stock—Combine 8 oz each of chicken and lean pork in 1 pint of water and cook until liquid is reduced by half. Remove fat and meat.

Chiretta— A plant used in India and Nepal as an herb and a medicinal. Overdose has side effects. *Sertia chirata* or Indian Genetian or Indian Balmony.

Clarified Butter—Melted butter (not margarine) which has been poured off the milky sediment at the bottom of the pan. A clear liquid. Also called Drawn Butter.

Coltsfoot Salt—Coltsfoot is a plant often found along streams and swamps. Flowers bloom before leaves appear. The underside of the leaves are covered with a dense fuzz. The Indians formed the green leaves into balls and laid them out in the sun to dry. Then they put them on flat stones and burned them to ashes. The ashes are very salty and is a good substitute for salt.

Congo Eel—A common name for *Amphiuma*—See *Amphiuma*.

Cooled Alligator—See Poached Alligator Tail and Boiled Alligator for recipes.

Cooter—Common name for fresh water turtles of the genera *Pseudemys* and *Trachemys*.

Crab Boil Liquid—A highly seasoned liquid used in boiling seafood such as crabs, shrimp and crawfish. Also available in bags of dried mixed seasonings instead of liquid.

Creole Tartar Sauce—See Broiled Frogs for recipe.

Crepes—See Alligator Crepes for recipe.

Drawn Butter—See Clarified Butter in glossary.

Dried or Black Mushrooms—Available in chinese stores.

Epazote—A plant (*Chenopodium ambrosiodes*) called "goosefoot, chichiquelite, epazote" used in mexican dishes as a seasoning. Parsley is a substitute.

Faggot—A small bundle.

Fermented Salted Black Beans—Available in chinese stores.

Fingers—A unit of measurement used by the Indians and used to measure dry materials. Using the number of fingers called for hold the fingers together and dip into the material called for and lift out with the fingers and thumb without turning the hand.

Fish Stock—Place 2 lbs chopped fish bones and trimmings in a buttered pan with 1 minced onion, a few sprigs of parsley and 10 peppercorns. Add to this 1 qt each of white wine and water and season with a pinch of salt. Bring to a boil and then simmer for 25 minutes. Strain through cheesecloth before using.

Five Spices Powder—Available in chinese stores.

Fried Vegetable Steak—A vegetarian food that looks like a small beefsteak but made from wheat gluten. Available canned in chinese stores.

Goannas—See monitors. An Australian name for monitors.

Gumbo Filé Powder—Leaves of the sassafras tree which is ground to a powder. Used as a seasoning in gumbos. Added at the last minute when dish is taken off the fire or served at the table as cooking it makes it slimy.

Half Glaze—See Frog Legs Italian Style for recipe.

Hicotea—Spanish common name for fresh water turtles of the genera *Pseudemys* and *Trachemys*. Also spelled Jicotea.

Hoi Sin Sauce—This is a sweet, brownish red sauce made from soy beans, flour, sugar, water, spices, garlic and chili. It is available in chinese stores.

Iguana—A large lizard found in Mexico and Central and South America. Also the genus and common name of this large lizard.

Indian Vinegar—This is made from the sap of the sugar maple or birch tree. Also buds, twigs and sap were allowed to ferment in the sun. After fermenting it is strained through a cloth.

Kneaded Butter—See Frog Legs Poulette for this recipe.

Knuckle—A joint such as a knee or elbow.

Kosher Salt—A salt prepared according to Orthodox Jewish rules.

Marinade—See Marinaded Frog Legs with Creole Sauce for recipe.

Maître d'Hôtel Butter—See Frog Legs a L'Anglaise for recipe.

Monitor—A small to large lizard found in Africa, Asia and Australia.

Oyster Sauce—A thick brown sauce with a rich flavor made from oysters, soy sauce and brine. Available in chinese stores.

Panne—The fat which covers the pig kidney and fillets and when rendered makes a superior quality lard.

Pavere Margarine—A margarine prepared in the kosher tradition.

Plastron—The bottom shell of a turtle.

Prepared Terrapin—See Prepared Terrapin for recipe.

Roux—Roux is a combination of oil and flour. A White Roux is made by combining equal amounts of oil and flour mixed well and heated over a very low fire for a few minutes, stirring constantly. Do not brown. A Brown Roux is made as above but browned to the desired color. Roux a la Microwave uses 2½ c flour to 2 c oil and mixed together thoroughly in a glass or microwave dish. Set on high for 3 minutes. Stir and repeat each 3 minutes until you have cooked it about 15-17 minutes or when the desired color has been reached. Different quantities will require different times. An Oven Roux mixes flour and oil equally and placed in a 375°F oven and baked until desired browness is achieved stirring occasionally. A variation of this is to leave out the oil using flour only. Some even brown flour on the stovetop at a high heat stirring constantly. Roux can be stored in bottles and used at a later date. In all rouxs, butter or margarine can be substituted for the oil. Sometimes the butter or margarine is preferred.

Sauce Mornay—See Frog Legs a la Mornay for recipe.

Scallions—Also called green onions or shallots. Both the green stems and white bases are used; sometimes together or separately.

Seasoned Flour—Flour in which at least salt and pepper has been added. Other seasonings can also be added. Also cracker or bread crumbs are handled this way.

Shields—Chitinous plates on the carapace and plastron of turtles covering the bony elements.

Tasso—A very highly seasoned smoked piece of meat, usually pork, that is almost dried and used as a seasoning in various creole or cajun dishes. See also turkey tasso.

Terrapin Butter—See Terrapin Butter recipe.

Tiligua—A genus of Australian Blue-Tongued Skinks.

Tomato Sauce—See Frog Legs Italian Style and Turtle Croquettes for recipes for 2 versions.

Tony Chachere's Creole Seasoning—A commercial brand of cajun or creole seasoning.

Tortoise—Usually refers to a land turtle but also sometimes to an aquatic one.

Turkey Tasso—A highly seasoned smoked piece of meat that is almost dried. Also a pork tasso is made and used in creole cooking. Used as a seasoning. See also tasso.

Turtle Herbs—Equal parts of dried basil, thyme, bay leaves and marjoram.

Turtle Quenelles—See Quenelle Turtle Soup for recipe. Replace with turtle meat for veal as in the recipe.

Turtle Soup au Sherry—See Manale's, Delmonico's, New Orleans, Green and Antoine's Turtle Soup au Sherry for a recipe.

Varanus—See monitor. A generic name for monitors.

Velouté—A sauce that has the consistency of heavy cream. See Rattlesnake Meat Sauce for Fish, Brochetons "Tatan Nano" and Terrapin Philadelphia Style for 3 versions.

White Sauce—See Creamed Frog Legs for recipe.

Winter Melon—A round green melon having a translucent and white pulp. Its flavor resembles zucchini or other soft skinned squash. Zucchini or cucumber can be substituted.

INDEX

B

C

D

E

F

G

H

I

J

K

L

M

N

O

P

Q

R

S

T

V

W

Y

Z